DO THE MATH!

For Belén

DO THE MATH!

ON GROWTH, GREED, AND STRATEGIC THINKING

JOHN K. WHITE

Los Angeles | London | New Delhi
Singapore | Washington DC

Los Angeles | London | New Delhi
Singapore | Washington DC

FOR INFORMATION:

SAGE Publications, Inc.
2455 Teller Road
Thousand Oaks, California 91320
E-mail: order@sagepub.com

SAGE Publications Ltd.
1 Oliver's Yard
55 City Road
London EC1Y 1SP
United Kingdom

SAGE Publications India Pvt. Ltd.
B 1/I 1 Mohan Cooperative Industrial Area
Mathura Road, New Delhi 110 044
India

SAGE Publications Asia-Pacific Pte. Ltd.
3 Church Street
#10-04 Samsung Hub
Singapore 049483

Printed in the United States of America

Library of Congress Cataloging-in-Publication Data

White, John K.

Do the math! : on growth, greed, and strategic thinking /
John K. White.

p. cm.
Includes bibliographical references.

ISBN 978-1-4129-9959-5 (pbk.)

1. Economics—Moral and ethical aspects. 2. Statistics.
3. Cooperation. I. Title.

HB72.W477 2013
330.01′51—dc23   2011050634

Publisher: Vicki Knight
Associate Editor: Lauren Habib
Editorial Assistant: Kalie Koscielak
Production Editor: Brittany Bauhaus
Copy Editor: Megan Granger
Typesetter: C&M Digitals (P) Ltd.
Proofreader: Victoria Reed-Castro
Indexer: Kathy Paparchontis
Cover Designer: Gail Buschman
Marketing Manager: Nicole Elliott
Permissions Editor: Adele Hutchinson

MIX
Paper from
responsible sources
FSC® C014174
FSC
www.fsc.org

This book is printed on acid-free paper.

12 13 14 15 16 10 9 8 7 6 5 4 3 2 1

In Do the Math!

- How do pyramid scams work?
- When did the real downturn in the economy occur?
- Do inflation measures work in fast-changing times?
- Are reality television shows fair?
- Can one win game shows using simple probabilities?
- Which state was really responsible for George W. Bush's 2000 win?
- Why do disagreements persist in Northern Ireland?
- How can you avoid paying interest?
- Is the economy expanding or contracting at any given moment?
- Can we calculate the cause of reduced crime?
- Why are North American sports leagues better than those in Europe?
- How do stock pickers win at the market?
- Why is the average person not the most likely person?
- Do birth dates indicate success in life?
- Where did the concept of lucky numbers originate?
- How does Arnold Schwarzenegger send secret messages?
- Is debt destroying the world?
- Who is profiting from your debt?
- Can No. 1 chart toppers explain increasing world poverty?
- How much do advertisers add to the cost of goods?
- Why are sports so uncompetitive?
- Who really broke the bank in 2009?

Brief Contents

Detailed Contents

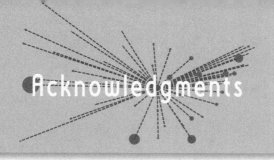

Acknowledgments

I am indebted to many people who helped and advised me during the writing of *Do The Math!*, especially to those who provided essential background, read parts or all of the manuscript, and prepared the manuscript for publication. Writing *Do The Math!* has been an immensely rewarding journey, one where I have looked under many mathematical stones to see the teaming world of numbers at our fingers. Happily, I was ably helped throughout.

In particular, I would like to thank Doug Alderman for his creative comments and for keeping me on my toes, Vicki Knight for her constant guidance and careful shepherding from draft to book, Tom and Pat Lawson for inspiring me to think creatively about issues, Tom McCormack for our many discussions about the intersection of physics and economics, Owen Priestly for reminding me that economics is about common sense, and David Thomas who helped make the theoretical real.

I would especially like to thank the academic reviewers for the thoroughness of their remarks and many excellent comments, in particular, John Bohte of University of Wisconsin Milwaukee, James I. Bowie of Northern Arizona University, Steve B. Lem of Kutztown University, Stephen N. Kitzis of Fort Hays State University, Liza L. Kuecker of Western New Mexico University, Terry F. Pettijohn II of Coastal Carolina University, Nancy Sonleitner of University of Tennessee at Martin, and Guillermo Wated of Barry University. The final result is all the more rigorous for their attention. I would also like to thank Danny Heffernan, Kirk Junker, Belén Roza, and Gerry O'Sullivan for their invaluable comments, as well as Maurice Coyle, David Lockey, Rosa Monleon, Emma Sherry, Laurie Snell, Avelino Valle, and Peter Whitney for their kind assistance along the way.

I am also indebted to my many colleagues at the School of Physics, University College Dublin, for their help and support. Many kernels of ideas began at UCD and were first sounded there. In particular, I would like to thank Ann Breslin, Peter Duffy, Padraig Dunne, David Fegan, Andrew Fitzpatrick, Bairbre Fox, Alison Hackett, Lorraine Hanlon, Colm Harte, Brian McBreen, Tom McCormack, Alex Montwell, Fergal O'Reilly, and Gerry O'Sullivan. I would also like to thank everyone at Sage, in particular Vicki Knight. Thanks to Vicki I was able to enjoy the whole process as we shared many ideas about writing and learning. I would also like to thank Kalie Koscielak and Brittany Bauhaus for their generous help and careful attention, and Megan Granger for her expert editing. I am indebted to Megan for her thorough attention to detail as well as her many helpful suggestions. I would also like to thank Gail Buschman, Nicole Elliott, Lauren Habib, Adele Hutchinson, Kelley McAllister, and Dory Schrader for their help.

This book would not be possible without Belén Roza, who kick started the process and offered many thoughts and inspirations along the way. She is my partner and my constant guide. Thank you for sharing your beautiful spirit. Daisy White has been a constant source of inspiration to me, from her indomitable *joie de vivre* to her always positive approach to life, and I would like to thank her for her unwavering support.

It has been a joy to be involved in the writing of *Do The Math!*, and I thank everyone involved for their encouragement and invaluable contributions. I hope I have advanced the discussion about how to live in a fairer world, and I thank you all for helping.

About the Author

John K. White received his B.Sc. in Applied Physics from the University of Waterloo, Canada, and a Ph.D. from University College Dublin, Ireland. He has worked around the world as a physicist, lecturer, project manager, and computational analyst over a 25-year career. He has worked as a project manager and technical writer for Sun Microsystems, The Netherlands Organization, and Berminghammer Foundation Equipment, a consultant for Interactive Image Technologies, ScotiaBank, and the Ontario Government, and as a lecturer and research fellow at University College Dublin. He is also active in promoting physics and numeracy in schools, and has published widely in academic journals, contributed chapters to edited volumes, and authored numerous technical publications. A series of *Do The Math!*-related downloads, exercises, and links are available on his website at www.johnkwhite.ie/DoTheMath.html.

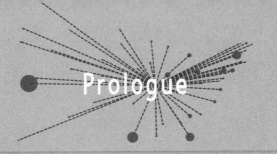

Prologue

When I was growing up in the '60s and '70s, my parents often reminded me and my siblings of leaner times after World War II, when supplies were scarce and food was rationed. Although we wanted for little, our lifestyle was nonetheless economical, as we were taught to measure our wants and not to expect too much. We could always ask for more, even for an occasional luxury item such as a new bike or the latest toy, but we were taught that the primary goal of acquiring and using was need and not want; in so doing, we learned the value of our material world.

The ongoing economic troubles in the world's financial centers, however, suggest that something is not quite right, from the questionable ethics of trusted institutions to a poor understanding of numbers and systems—although the mathematics of collateralized debt obligations, credit default swaps, and derivatives would baffle many a PhD mathematician. Complicated financial instruments and creative accounting have now become the norm, contrary to the well-taught ethics of the past, such as hard work, fairness, and public good, as most of us understand them. In the process, our world, which we call egalitarian, meritocratic, and religious, has been usurped by those in the financial know—aided by government inadequacy—creating a culture defined more and more by money and less and less by the common good.

One wonders if the age in which we live—one of scientific mastery over resources that produces newer technological gadgets by the day—has created such a world, where our wants are manufactured as needs. We do seem to be in a hurry to have it all as we marvel at the lifestyles of the materially well-endowed. *Growth* is the buzzword, apparently at any cost, and greed is its motive. But to what end? Constant economic uncertainty? A permanently stressed-out workforce? Everyday stories of government and corporate corruption and abuse? What's more, are we falling behind in the race to get ahead as we deplete the earth's finite and dwindling resources?

Limping from one crisis to another upsets and destroys the fabric of our lives and that of our communities, as witnessed by dramatic increases in unemployment and mortgage defaults during the now-regular economic downturns. There are better ways to proceed, but we need stronger knowledge of basic systems to understand our world and to analyze the more complicated problems of our times.

One place to start is with counting and mathematics, perhaps our oldest mental constructs, on which numerous systems have been created— from ancient calendars, weights and measures, and games of chance to today's world of finance, banking, and statistics. Our modern way of living, from household bills to the details of modern economics, would be scarcely possible without the age-old language of counting. And yet many of us recoil from even the simplest math, seemingly afraid to grasp the basics, believing the sums too hard or better left to others—no matter that it is impossible to communicate in today's world without a proper mathematical grounding.

Perhaps the first battle is to overcome our fear and cast away the notion that some can and some can't. Mathematics is easier than we think. As with any skill, one must practice. We train our bodies and souls in regular sessions, so why not train our minds to acquire better understanding? Problem solving need not be left to uninformed trial and error, where old-world ideas and new-age logic trump informed thought.

Without an understanding of mathematics, we simply can't make the right decisions. Fuzzy thinking keeps us in the dark about consumer costs, enslaves us to lending institutions, leads to poor choices, and confuses wisdom with jargon. But, more important, by shying away from the numbers, we can't make important decisions about essential issues such as government bailouts, oil supplies, or global warming. Our inability to analyze numerical information keeps us from understanding issues that are incumbent on our being and have exceedingly important ethical implications for our future.

To be sure, lustful gurus abound, telling us how to interpret our fast-changing technological landscape, and many of us are unsure to whom we should listen or which data to trust. Life is full of "asymmetrical information," we are told, where those in the know pass on insider info via hidden networks, leading many to believe that real knowledge is unattainable. But that belief is false. Any person on the street can communicate about today's number-filled world without an advanced degree in business or statistics. All it takes is the right data, the right tools, and some straightforward techniques to analyze the numbers.

Do the Math! is not a mathematics refresher, nor does it shy away from the numbers. Instead, many creative examples are given to explain our number-filled world, such as pyramid scams and economic growth (Chapter 2); the stock market and auto racing (Chapter 3); cost-benefit

xvi DO THE MATH!

analysis and reality television shows (Chapter 4); the Electoral College and inflation (Chapter 5); Shakespeare's plays, the death penalty, and sports competitiveness (Chapter 6); Monopoly, blackjack, and roulette (Chapter 7); astrology, codes, and lottery odds (Chapter 8); looking under the hood of modern financial practices (Chapter 9); consumer rip-offs (Chapter 10); and cooperative thinking (Chapter 11). Along the way, we look to mathematics to explain geometric series, small-world phenomena, power laws, continuous variables, derivatives, error, statistical methods, correlation, computing, probability, game theory, and more. Just as Lynne Truss stresses the importance of grammar for better literacy in her book *Eats, Shoots & Leaves*, numbers and analytical methods are stressed here for better numeracy. Indeed, being innumerate is no more acceptable than being illiterate.

Aesop's well-known fable "The Tortoise and the Hare" can be used to explain sophisticated theories such as inflation and derivatives. Other fables show how exponential growth left unchecked leads to disastrous results, and help us better understand the pyramid scams of Enron, Bernie Madoff, and Allen Stanford, or how Bear Stearns, Lehman Brothers, and Ameriquest played their parts in a near global economic meltdown.

Other stories help us question past methods, such as the tale of the woman who cut off the sides of the family roast before putting it in the oven as her mother had done, not knowing that her mother did so only because her roasting pan was too small. Or the story of the woman who put all her money—a *farthing*—in the collection plate, thus giving less and yet more than all the others.

Here, a simple calculation shows how one-size-fits-all justice contradicts the notion of a shared society—which we have been taught to believe exists—where the *relative* difference can be more than the *absolute* difference and a flat rate is not the same for everyone. Indeed, a $500 fine is nothing to a millionaire, and were a fine charged in like proportion to someone earning $25,000 a year, it would be only $12.50—hardly a credible deterrent. For a fine to have the same weight to the millionaire as the $500 fine does to the person earning $25,000 (2%), it would need to be $20,000—an amount perhaps not as easily scoffed at by today's typically undeterred rich. Similarly, are thousand-dollar fines for cutting corners any better? For example, is such a fine sufficient for an oil spill, where cleanup costs can amount to tens of billions of dollars, or for a bank disaster that requires public money to ensure private liquidity, where costs are socialized but benefits are privatized? What about uneven tax codes, accounting loopholes, or preferential trading practices, which subsequently disadvantage others? Mathematics can quantify the disparities.

Obviously, we don't have everything right if half the world lives on so little (Europe, the United States, China, and Japan account for 60% of the world's estimated $73 trillion GDP[1]), a small percentage controls most of the world's wealth (in the United States, one estimate states that 1% owns one third of the country's wealth, whereas in the United Kingdom 1% owns more than two thirds), and the debt load of the world's poorest countries is so great they can never repay it. Nor can it be right that the divide is getting worse (relative GDP of the bottom countries has decreased sixteenfold), that average consumer debt is greater than average income (in 2011, American personal debt was more than $16 trillion, or $50,000 per person), or that failed financial institutions have destroyed people's savings and cost jobs by the millions. Perhaps the so-called trickle-down wealth system said to benefit everyone actually polarizes society. But for whom does the world exist— for you and me, or for the financial elite?

Numbers always have a moral side, hidden behind a society that increasingly rips people off. What is a 50% sale when the markup is 100%? Why do we pay twice as much for goods before Christmas as after Christmas? Airlines tack on so many extras that the final price bears no resemblance to the advertised price (one such $0 out and $0 return was more than $100). Our culture has become one of hard-to-read numbers, hidden logic, and dubious claims, where in the absence of proper regulation, one must doubt every claim and question every transaction, vigilantly acting as one's own regulator. In such an unfair world, it seems that more than a cursory understanding of numbers is needed to identify the traps, the red queens, the wicked witches, and even our own gluttony.

By looking at the numbers, we expose the myths, the half-truths, and the out-and-out lies routinely trotted out by suspect sources wielding made-up facts and figures. Banks affect our well-being. How does a bank calculate compound interest? How much are you actually spending on those "no money down, no interest" loans? Can we understand a banking system that loaned 100 times its capital to speculate on mortgages, which dramatically impacted the lives of millions when it failed and resulted in a trillion-dollar banking bailout?

How about the lottery, seemingly simple by comparison? Should we teach a work ethic to our children, yet devalue that same ethic by encouraging get-rich-quick, instant-freedom, one-in-a-million chances at amounts in excess of $100 million? No matter how altruistic the cause, lotteries are regressive taxes that disadvantage the less well-off, who use them most. Or what about a now-ubiquitous gambling culture spawned by our fascination with numbers and chance that exists solely to profit from other people's ignorance and failure? Should those of us in the know permit such practices, or do we keep a competitive advantage by allowing the less-informed to be so abused?

Behind a number, a calculation, a story lie useful methods, and not only mathematicians and economists should understand them. Should our economies rumble on, stumbling from one spike to the next, hell-bent on producing unsustainable growth? What is inflation and how is it measured? To whom do we owe such large national debts, given that economics is a closed system? We must become better informed, working with known facts and proven methods, so it becomes easier to challenge bad practice.

Socrates said, "The unexamined life is not worth living"—wise words that I hope still apply in our fast-changing world of globalization and the Internet. Or are we stuck with an out-dated, superstitious, fad-ridden society that continually falls victim to the latest swindle and unsubstantiated claim? Can we expose garbled facts, wonky myths, and screwed-up tales that serve only to separate our money from our wallets? One hopes so, and that we can tackle the harder questions and put into practice better thinking and better strategies.

In *Do the Math!*, we learn about such strategies, some as stern and autocratic as the tides, some that apply in certain conditions, and others as downright whimsical as the weather. Of course, many competing ideas and conflicting methods suggest the best road forward, but how we should progress to achieve the best ends and to whom the world belongs are questions we must ask and for which we must have the right information and tools to proceed. Using less by reevaluating our wants is one solution. Better cooperation using organized strategies is another and is essential in any working system, whether to reduce, reuse, and repair, or to expand continually at increasing peril. Indeed, the goal of committed citizens and their political institutions must be to apply strategies that work to create a better living for all as set out in the great parchments of our times.

Our world has improved enormously since the uncertain and impecunious postwar times of my parents, but does it serve the needs of everyone or even a majority of its citizens? If we say we live in a fair, democratic, and enlightened world, where everyone has equal opportunity, then we must learn the systems and count the costs. We must work the numbers, lest we be sold a bill of goods.

Do the Math! puts together various thoughts about numbers and noise, statistics and probability, and money and economics that are making the rounds as we begin to understand the limitations of our world and its many varied systems. The presentation is purposefully eclectic, allowing the mind to synthesize ideas from different areas. As such, I have borrowed liberally from the anecdotal and the academic, from literature and the newspaper, and from the stock market and the casino—all in the hopes of improving numeracy.

What qualifies me for this task? Well, I am good with numbers and have spent a career as a computational physicist, analyzing many diverse systems, from nuclear reactors to rocket propulsion, using anything from a Commodore 64 (with its fantastically slow input cassette drive) to the latest multicore parallel computing architectures. I have also delved into the world of sports scores, gaming, and financial trading, creating numerical models and statistical analyses, all to see how the numbers work.

Perhaps more important, I have worked in different areas, including academia, industry, and business, where I have seen firsthand how numbers are used and abused. I am also a generalist, constantly looking for the common in the seemingly disparate, what overlaps and can be agreed on—and not just the mathematics of our daily world, but how the numbers jibe with what we are told.

I also want to stand and be counted, literally, as someone who is trying to assess the increasing discrepancy between fact and fiction at play in today's seemingly more complicated world. The intention is to look more closely at the perceived wisdom and to see the real mechanisms at work. I hope you enjoy the journey.

Note

1 **Gross national product and gross domestic product**: Gross national product (GNP) is the value of all goods and services produced annually by citizens of a country regardless of where they live. Gross domestic product (GDP) is the value of all goods and services produced annually within a country regardless of whether by citizens or foreign nationals. GNP numbers throughout are from the Central Intelligence Agency's *World Factbook* (2011). Barber (2007) cited a report that such measures overestimate the health of an economy, whereas the GPI (genuine progress indicator) measures the quality as well as quantity of an economy (p. 148).

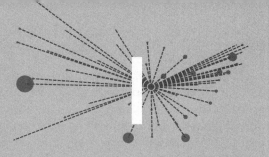

THE DOUBLING GAME

From Thomas Malthus to Bernie Madoff

To understand how our world has become stuck in a seemingly endless series of economic crises, we first look at a trend that is increasingly causing distress to the planet, is most certainly man-made, and is no longer sustainable given the nonfrontier global world in which we live. In fact, the early seeds of capitalism in the British Industrial Revolution followed by the frontier thinking of the expansionist United States (as witnessed in the opportunist battle cry "go west, young man" and in the extraordinary increase of industrial output during the two world wars) has allowed this trend to embed in our modern economic thinking and to become the leitmotif of an entire civilization, the implication of which is serious for all and has at its core our own greed.

I am referring to the doubling game, which was first told as a tale in ancient India and has since fueled the ambitions of many a fortune seeker, from tulip traders to property speculators to stock-market dealers. Here, we look at its origins in ancient India and how it played a role in a few modern examples, such as the 1929 stock market crash, the 2009 credit crisis, and the billion-dollar pyramid scams of New Yorker Bernie Madoff and Texan Allen Stanford. All hold the belief that continuous growth can last forever, endlessly unchecked.

Many examples of doubling appear in our lives. Backgammon has a doubling cube to a maximum of 64 ($2 \times 2 \times 2 \times 2 \times 2 \times 2$ or 2^6 for a six-sided die). James Bond's favorite game, baccarat, allows a doubling of the bet ad infinitum. European paper sizes[1] are made in exact doubling ratios, where the standard-size writing paper, A4, equals one half the next size, A3, and so on. The television quiz show *Who Wants to Be a*

Millionaire has bits of geometric-like doubling in its cash prizes ($100, $200, $500, $1,000, . . .), as do population sizes over time (3.3 billion in 1965, 6.5 billion in 2006), cells doubling by mitosis, and even infectious diseases. Even the tribbles of *Star Trek* fame grew exponentially, with just one multiplying to 1,771,561 in 3 days via a tenfold growth every 12 hours, according to Spock's calculation.

The semiconductor industry has generally followed "Moore's Law"— first proposed in 1964 by Gordon Moore (1965), cofounder of Intel— where the number of transistors on a chip (or chip density) doubles every 2 years. The number of transistors on a standard computer processor now stands at more than 4 billion, or 32 doublings ($2^{32} = 4,294,967,296$). Correspondingly, the transistor size has shrunk and will eventually reach atomic dimensions.

But can such doubling be sustained? How long before there are no more sides to a die, the bank says no, or atomic dimensions reach their limits? How long before a piece of paper can no longer be folded? Is there a limit to the doubling game?

A series of coin tosses or roulette spins (or any two-outcome or binomial test) is the mathematical basis for doubling logic, where one can easily count the sample space (number of total outcomes) versus all possible outcomes. For example, we'll see later how to calculate the probability of getting any mixture of heads and tails on a series of coin tosses, the chances of landing on red or black in a number of roulette spins, or the possibility of correctly predicting an up/down stock market index over a number of weeks. For now, each flip, spin, or pick is simply a doubling of outcomes (e.g., either heads or tails, red or black, win or lose).

In the same way, a doubling or geometric progression can inform us about a pyramid scam, where she told two friends, who told two friends, who told two friends, and so on, and before you can say "Bob's your uncle" or "Bernie Madoff is a crook," there are no more friends, thanks to the exponential growth of doubling: $2 \times 2 \times 2 \times 2$ . . . . Some also see the doubling game in government pensions, which continually pull in young workers to pay older retirees,[2] or in continuous bond issues to pay off debt in a seemingly perpetual payout machine that postpones a future reckoning.

Thomas Malthus (1798), the English mathematician and clergyman, wrote about such progressions in his *Essay on the Principle of Population*, stating that the world's population would double (geometric) while the food supply would increase only linearly (arithmetic), spelling future disaster and eventual world famine. "A slight acquaintance with numbers will show the immensity of the first power in comparison of the second," wrote Malthus (p. 4), basing his theory of the inability to maintain subsistence for a constantly growing population in part on the rapid growth of the Irish population due to the success of the potato plant, prior to its subsequent failure in the 1840s.[3]

The effect of the two functions is easy to see when plotted (see Figure 1.1). The *geometric* function is written as $y = 2^x$ (also referred to as an exponential function because of the raised exponent x). The *arithmetic* function is a straight line and is written as $y = Ax$, where the constant A is the slope, indicating how fast the line rises for every x (in the figure, A = 1). In Malthus's description, $y = 2^x$ is the world population and $y = x$ is the food supply, where a simple comparison of the two ($2^x/x$) shows how quickly the geometric "doubling" (2, 4, 8, 16, 32, . . .) swamps the arithmetic "stepping" (1, 2, 3, 4, 5, . . .).[4]

As can be seen, after three steps (x = 3) there is almost no difference (8/3), but soon the geometric progression outpaces the arithmetic

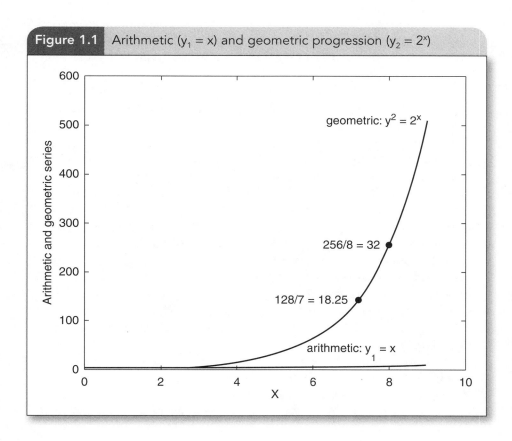

Figure 1.1 Arithmetic ($y_1 = x$) and geometric progression ($y_2 = 2^x$)

geometric: $y^2 = 2^x$

$256/8 = 32$

$128/7 = 18.25$

arithmetic: $y_1 = x$

X

progression, and all the more as they both increase. After seven steps ($x = 7$), the ratio of the geometric to the arithmetic progression is almost 18 times greater (128/7), while after eight steps the ratio is already 32 times greater (256/8).

Here, a fable helps highlight the mathematics. As the story goes, a clever courtier gave an Indian king a beautiful chess set, and the grateful king asked what the courtier would like in return. According to the tale, the courtier asked for a grain of rice doubled every day for each square on the chessboard. Thinking the request simple, the king agreed—laughing, it is said—and on the first day, the courtier was given a single grain of rice. The next day, he was given two grains of rice, and the next four. After 1 week, the courtier received 64 grains of rice ($1 \times 2 \times 2 \times 2 \times 2 \times 2 \times 2 = 2^6 = 64$) for a total of 255 ($2^7 - 1$). The king, it is said, was still laughing.

Continuing with their agreement, after 2 weeks the courtier received 2^{13} or 8,192 grains, although by then the king didn't seem to be laughing as much. By the end of week 3, the courtier received 2^{20} or 1,048,576 grains, and the king wasn't laughing at all. In fact, the king was downright annoyed, as his rice supply was noticeably diminishing and they were still only at square 21. Soon, after most of the rice had been delivered to the smiling courtier, the king was beside himself.

Alas, by the end of week 4, the jig was up, although it is said that the king got the last laugh—he had the all-too-clever courtier executed. The story has also been told with pieces of gold instead of rice, for which the courtier was executed far sooner.

To help clarify the impossibility of sustaining such geometric progressions, the payout for each square on the king's chessboard is shown in Figure 1.2, where it can be seen that had there been enough rice to continue until the last square and had the king continued to pay, the courtier would have received 2^{63} or 9,223,372,036,854,775,808 grains of rice for a total of $2^{64} - 1$ or 18,446,744,073,709,551,615 (more than 18 trillion million). To illustrate how large 18 trillion million is, a 500-gram box of Uncle Ben's rice (about four servings) has about 32,000 grains, which would require every person in the world to eat two boxes every day for more than 1,000 years to equal 18 trillion million—numbers that boggle the mind, and readily show the impossibility of endless doubling.

Of course, the rate at which doubling occurs is also a concern. It is one thing to hand over a bag or two of rice but quite another to start delivering loads in ever-increasing sizes—first wheelbarrows, then trucks, then container ships. At the start, the means may be simple, but

Figure 1.2	Chessboard doubling ($2^{64} = 18,446,744,073,709,551,616$)						
Square 1 Day 1 1	Square 2 Day 2 2	4	8	16	32	Square 7 End week 1 64	128
256	512	1024	2048	4096	Square 14 End week 2 8,192	16,384	32,768
65,536	131,072	262,144	524,288	Square 21 End week 3 1,048,576	2,097,152	4,194,304	8,388,608
16,777,216	33,554,432	67,108,864	Square 28 End courtier 134,217,728	268,435,456	536,870,912	1,073,741,824	2,147,483,648
4,294,967,296	8,589,934,592	17,179,869,184	34,359,738,368	68,719,476,736	137,438,953,472	274,877,906,944	549,755,813,888
1,099,511,627,776	2,199,023,255,552	4,398,046,511,104	8,796,093,022,208	17,592,186,044,416	35,184,372,088,832	70,368,744,177,664	140,737,488,355,328
281,474,976,710,656	562,949,953,421,312	1,125,899,906,842,624	2,251,799,813,685,248	4,503,599,627,370,496	9,007,199,254,740,992	18,014,398,509,481,984	36,028,797,018,963,968
72,057,594,037,927,936	144,115,188,075,855,872	288,230,376,151,711,744	576,460,752,303,423,488	1,152,921,504,606,846,976	2,305,843,009,213,693,952	4,611,686,018,427,387,904	9,223,372,036,854,775,808

eventually the infrastructure is unable to keep up. Simply put, such growth cannot continue unabated.

Figure 1.3 shows the doubling of the world population in time (United Nations, 2011) and clearly shows a trend of exponential growth (circles mark the doubling periods). What is most revealing here is that the doubling period has been shortening from one double to the next—300 years from 500 million to 1 billion (1500 to 1800), 125 years from 1 billion to 2 billion (1800 to 1925), and 50 years from 2 billion to 4 billion (1925 to 1975), until recently leveling off (54 years from 4 billion to a projected 8 billion, 1975 to 2029). In the process, the population has been doubling faster and, as such, consuming resources faster, with significant consequences to the environment.

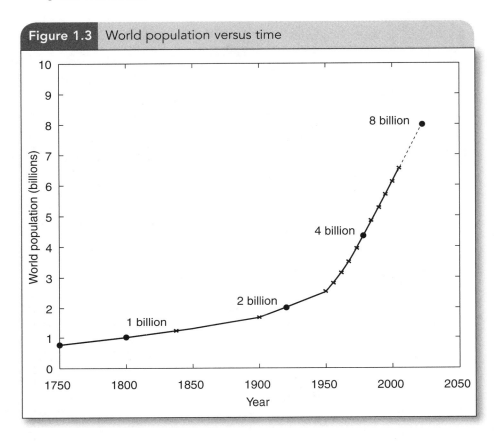

Figure 1.3 World population versus time

Indeed, economic productivity and national production have grown enormously since the beginning of the Industrial Revolution in the early 19th century, which has prompted new questions with regard to the earth's resources: Are there limits to continued growth and, if so, when will it end? Will we run out of coal, oil, gas, and other natural resources? What's more, will we see the end coming?

Oil output is a good example of a doubling game, showing huge increases throughout the 20th century, doubling in the United States every 8.7 years from 1880 to 1930 (Hubbert, 1956, p. 7), and prompting questions about its continued sustainability. The concept of peak oil (defined as half the available oil in the ground to be extracted) originated with the American geophysicist M. King Hubbert (1956), who predicted that the United States would reach maximum production (*peak* oil) in the 1970s, which it did (p. 24). Figure 1.4 shows his further prediction for world peak oil in 2000.

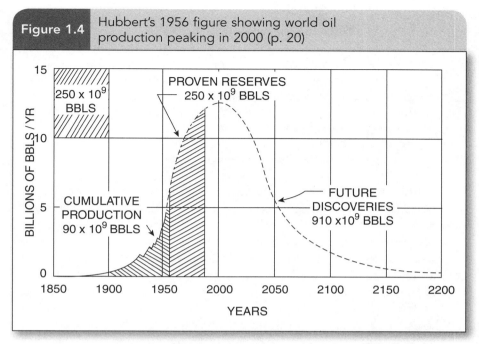

Figure 1.4 Hubbert's 1956 figure showing world oil production peaking in 2000 (p. 20)

SOURCE: Marion King Hubbert Collection, Box 85, Folder 4, American Heritage Center, University of Wyoming.

Globally, peak oil is still debated, although annual discoveries of new oil have declined since the 1960s (Hobbs, 2010, p. 33) and may be declining by as much as 9% per year (Jackson, 2009, p. 32). Some estimate that oil supplies have already peaked (Hobbs, 2010, p. 36; Hubbert, 1956, p. 20), others that they soon will (Fleming, 2011a). The International Energy Agency predicts "a significant potential gap between supply and demand by 2015" (Moyo, 2011, p. 165). Note, however, that what has already been extracted is the easier half—the upside—and that the downward half will be much more difficult, not to mention scarcity- or security-based price increases as we approach "end" oil.[5]

In his 1973 book, *Small Is Beautiful: A Study of Economics as if People Mattered*, the British economist E. F. Schumacher continued in the tradition of Malthus, warning us that we are using up fossil fuels at an alarming rate. He suggested that fossil fuels, what he called "natural capital," should be treated as capital and not as income items, thus necessitating our concern for their conservation. Natural capital is not limitless, doled out forever like collecting $200 every time one passes "Go" in Monopoly. Fossil fuels cannot be replenished, and, thus, the question is not *will* we run out of resources—which of course we will—but *when* and what damage will be done in the process? Furthermore, will we ultimately destroy the world? As Schumacher (1993) noted,

> We are not in the least concerned with conservation; we are maximising, instead of minimising, the current rates of use; and, far from being interested in studying the possibilities of alternative methods of production and patterns of living—so as to get off the collision course on which we are moving with ever-increasing speed—we happily talk of unlimited progress along the beaten track, of "education for leisure" in the rich countries, and of "the transfer of technology" to the poor countries. (p. 4)

In two tables from his book, Schumacher (1993) showed population, fuel consumption (in million tons coal equivalent), and fuel consumption per head for "rich" and "poor" countries[6] in 1966 and a projected year, 2000. Malthus's geometric doubling readily appears in these tables: The world's population doubles (6,909/3,384), while the fuel consumption more than quadruples (23,156/5,509; pp. 13–14)—a striking result, borne out by recent figures showing the doubling of population growth and the quadrupling of fuel consumption since 1966. Furthermore, his data shows that 23% of the world was consuming 67% of the fuel in 2000, but that the "poor" nations of today are consuming more than the "rich" nations of only 34 years ago (7,568 million tons vs. 4,788 million tons coal equivalent), and in a greater percentage (1/7 or 1.43/9.64 in 2000 compared with 1/14 or 0.32/4.52 in 1966; pp. 13–14). What's more, China and India (considered poor in his survey) are now growing in Western ways,[7] and, thus, one must ask, is there enough fuel left to sustain first-world growth for another 34 years?

Another indication of energy use is the level of carbon emissions, which is also an area of increasing concern with regard to global warming. The American Museum of Natural History's Center for Biodiversity and Conservation stated that "people in the United States and Canada account for approximately 5.3% of the global population, yet they produce about 26% of global CO_2 emissions" ("Per Capita Resource Consumption," 2011), and, thus, if the rest of the world were to consume at the same rate per capita, the earth would need almost 5 times its current resources. The Center further stated that "Canadians, U.S. citizens, and European Union members are generally considered to have comparable standards of living, yet Europeans on average use 47.2% as many resources per person (in oil energy equivalent units) as their North American counterparts" ("Per Capita Resource Consumption," 2011). The International Energy Agency also predicted that carbon emissions will almost double in the next 20 years, three-quarters of which will come from China, India, and the Middle East (Moyo, 2011, p. 167). Other figures show that China now emits twice as much CO_2 from coal as all of Europe does (Hobbs, 2010, p. 105). What's more, the demand for resources is increasing.

But can current industrial resources sustain such growth? Recent growth has been spectacular, spurring on our over-consuming ways. Jackson (2009) noted about current supplies, "If the whole world consumed resources at only half the rate the U.S. does, for example, copper, tin, silver, chromium, zinc, and a number of other 'strategic metals' would be depleted in less than four decades" (p. 10). Furthermore, will there be enough food to sustain an ever-increasing population? If the population continues to grow at the current rate, a third more food will be required in 15 years (Zakaria, 2009b, p. 30) and 50% more by 2050 (Jackson, 2009, p. xvii).

According to Goldman Sachs, "the combined GDP of the four BRIC[8] economies—Brazil, Russia, India, China—could overtake the combined GDP of the G7 countries by 2035. These days, they say it could happen by 2027" (Zakaria, 2009b, p. xxii). To put the numbers in perspective, based on population alone, an increase in use of only 5% in Brazil, Russia, India, and China is equivalent to a 50% increase in the United States. If those in the West are already starting to see limits in available resources, imagine how much worse it can get given those figures.

Mason (2009) also recited the numbers:

> In 2007 global GDP growth was 5%—well above its historic trend for the fourth year in a row. Growth in the developing world averaged 8%; and in Asia it was 10%. Across the G7 countries it was 2.6%—slightly below the average for the 1990s. (p. 157)

To be sure, the growth has been staggering and without precedent. But as Jackson (2009) noted, "This extraordinary ramping of global economic activity . . . [is] totally at odds with our scientific knowledge of the finite resource base and the fragile ecology on which we depend for survival" (p. 13).

What's more, with regard to the effects such growth has on our economies, American economics professor Hyman Minsky noted that "the normal functioning of our economy leads to financial trauma and crises, inflation, currency depreciations, unemployment and poverty in the midst of what could be virtually universal affluence—in short . . . financially complex capitalism is inherently flawed" (Mason, 2009, p. 154).

And yet we are told that growth is the only model, unchecked until a correction occurs—often disastrously, as in the 2009 credit crunch and resultant worldwide recession,[9] which in some countries was in fact a depression. As noted by Morgan Stanley economist Stephen Roach, "A voracious appetite for economic growth lies at the heart of the boom that has now gone bust" (Mason, 2009, p. 157).

Of course, we should always examine the goals of industry and consider the effects such growth has on social patterns and sustainable living. John Stuart Mill reminded us that the ultimate aim of economics is toward "a stationary state of capital and wealth" and that the ever-increasing demands of an ever-expanding economy are simply incompatible with the ever-dwindling resources of a finite supply (Jackson, 2009, p. 122). As Jackson stated, "Economics—and macro-economics in particular—is ecologically illiterate" (p. 123).

Dwindling resources and increasing worldwide consumption have put us on a collision course between what we have and what we use, and while the West imports more and more oil, it is exporting more and more of its over-consuming ways—not to mention the increased waste and pollution from fossil fuel emissions and increased contributions to greenhouse gases and global warming. For a society based on a previously abundant supply of cheap oil, new strategies are needed—such as renewable energy and reduced use—and are imperative if we hope to continue our modern ways. The simplest mathematics of arithmetic and geometric progressions explain as much.

We seem to grow without any concern for sustainability, but a world that cannot reduce its over-consumptive ways may itself be consumed. A world that cannot reduce its waste may itself soon become waste. Just as in any failed doubling game.

Notes

1 **European paper sizes**: A4 (210 × 297 mm) = ½A3 = ½A2 = ½A1 = ½A0. An A0 poster (841 mm × 1189 mm) = 16A4.

2 **The ultimate Ponzi scheme**: Moyo (2011) doesn't pull any punches, stating, "Forget Bernie Madoff, forget Allan Stanford, the biggest Ponzi scheme has got to be the looming car crash that is Western pension funds. And like any well-run Ponzi game, its results will be devastating" (p. 79). She adds that 2008 American pension costs were $2.2 trillion or 15% of GDP, whereas in the United Kingdom they were an even more staggering $1.3 trillion or 64% of GDP (p. 79).

3 **Malthus and growth**: Zakaria (2009b) noted that Malthus's essay "is remembered today for its erroneous pessimism, but, in fact, many of Malthus' insights were highly intelligent" (p. 57). He further noted that "Malthus was wrong about Europe. His analysis, however, well described Asia and Europe" (p. 57).

4 **Malthus's influence**: Both early naturalists Alfred Wallace and Charles Darwin based the origin of their ideas about natural selection, where species must compete to survive, on Malthus's relation $2^x/x$. Herbert Spencer later applied the catch phrase "survival of the fittest."

5 **The end of oil?**: Hobbs (2010) noted that the estimated amount of oil remaining at the peak was 1,000 billion barrels, which amounts to less than 40 years left based on a current use of 35 billion barrels per year, and even less if demand from China and elsewhere increases (p. 36). Maass (2009) noted that "the advent of peak oil is yet another incentive to cut our dependency, because in the years ahead the price will only rise—skyrocket, really—if we fail to arrest our desires for it" (p. 223).

6 **Schumacher's rich and poor**: Schumacher (1993) defined a "rich" country as one with an average fuel consumption of more than one metric ton of coal equivalent (p. 13). He used United Nations figures throughout.

7 **Chinese and Indian energy consumption**: According to the United States Energy Information Administration (2011), world marketed energy consumption is expected to increase by 53% from 2008 to 2035, growing the most (117%) in non-OECD Asia (led by China and India). China's oil consumption increased by 6% to 8.3 million barrels per day (mbpd) (2008 to 2009) and its installed electric capacity more than 10% to about 800 GW (2007 to 2008; http://www.eia.doe.gov/EMEU/cabs/China/pdf.pdf). India's oil consumption is expected to increase by 3.2% to 3.1 mbpd in 2011 and its installed electric capacity by almost 50% to 232 GW (2007 to 2012; http://www.eia.doe.gov/EMEU/cabs/India/pdf.pdf). Note that almost 400 million people in India still have no access to electricity. The rank of the top seven oil consumers is USA (24%), China (8.7%), Japan (5.8%), Russia (3.2%), India (3.2%), Brazil (3.1%), and Germany (2.8%), totaling more than half the estimated 87 million barrels used per day (Hobbs, 2010, pp. 49–50).

8 **BRIC nations**: The BRIC nations (Brazil, Russia, India, and China) are a loose group of countries with similarly growing economic power. Recently, South Africa was added, giving the five-nation group the moniker BRICS. The nations all have large overall economies, but relatively poorer per capita incomes.

9 **Recession and depression**: A recession has been defined as a fall in GDP for two successive quarters and a depression as a fall in GDP for more than 3 years or a one-time drop of 10%. Neither is conclusively defined.

SHE TOLD TWO FRIENDS,
AND THEY TOLD TWO FRIENDS,
AND THEY TOLD TWO FRIENDS...

Carrying on with the doubling game, we can now more easily see how a pyramid scam works and why it ultimately can't sustain itself (and from that, the essential causes of the 1929 stock market failure and 2009 credit crisis, among other examples of such greed-fueled growth). Such scams are illegal in many jurisdictions, although they crop up in various guises from time to time. All prey on those who dream of future fortunes, seemingly made with only a modest investment and a few friends. How many friends? Well, let's see.

One particular scam (there are countless variations with different amounts of upfront money and various splits) made its way to Toronto in the 1980s and went by the name "The Mile High Club." Here, a "captain" was at the top with two "copilots," who each had two "flight attendants" (four total), who each had two "stewards" or "stewardesses" (eight total). To "propel" the plane, all each "stew" had to do was recruit two new "passengers" (16 total), who would each pay the captain a $1,000 initiation fee. The club would then split, with the captain pocketing $16,000 and everyone else moving up one rung. Passengers became stews, old stews became new flight attendants, old flight attendants became new copilots, and the old copilots got to fly their own planes, which were now in search of 16 new passengers each, or 32 total.

It seems too good to be true. Who wouldn't be impressed and eager to sign up? With only $1,000 and two friends, after four splits offering 2^4 or 16 times the money, one can go from passenger to stew to flight attendant to copilot to pilot in no time and pocket a cool $16,000. All one has to do is pay the $1,000 and wait for the money to roll in. One can start over and go through the loop again with two new friends, play a richer version with a bigger payoff, or play two at once with four new friends. One can play again and again, collecting a sixteenfold payoff each time . . . as long as one has two new friends. How can it go wrong?

Well, therein lies the problem. We all have friends—maybe 20 or so good friends, and maybe another 20 or so we could enlist in a surefire investment with a good sales pitch. But how long before our supply of friends starts to run thin? How long before the passenger pool dries up? How long before the fizz in the bubbly goes stale? *Only as long as one has two new friends.*

It doesn't take long for the numbers to run up or, in this case, to run out (of friends and friends of friends), exactly as in the chessboard doubling example. And since most of your friends will want to enlist their friends, too, and you likely have a lot of friends in common, pretty soon you'll be fighting over whose friends are whose. Eventually, the whole world will be fighting over whose friends are whose as they look for new friends and new friends of friends. At some point, there won't be any more friends—not one, not even in the extraordinarily expansive Facebook network (with its estimated average of 120 friends per user)— all because the doubling of a geometric progression is unsustainable after too many doubles.

As we saw in the chessboard doubling example, the math is straight-forward ($2^{32} = 4,294,967,296$), such that after 32 splits, more than 4 billion passengers are needed. Alas, by the 33rd split, in today's population, there will be no more people left (see Figure 2.1). Indeed, there are no more new friends and never can be.

Furthermore, why such pyramid scams fail or property bubbles burst so quickly (called Ponzi schemes, as we'll see below, after the original Florida swampland promoter) and can lead to a global credit crisis is seen in the last double. For example, if each split were to take 1 week, it would take 8 months to enlist half the world (the first 32 splits). But to keep the pyramid going at that same rate, the rest of the world would be needed *in one week*, an impossible feat—like a snowball that can no longer "snowball." In fact, such continuous doubling is all the more perilous the longer it continues.

Nonetheless, the game is tempting and the logic reasonable to the uninitiated. But the most insidious part of a doubling scam is that we don't think of the consequences—i.e., those who follow and aren't

| Figure 2.1 | Pyramid scam population growth |

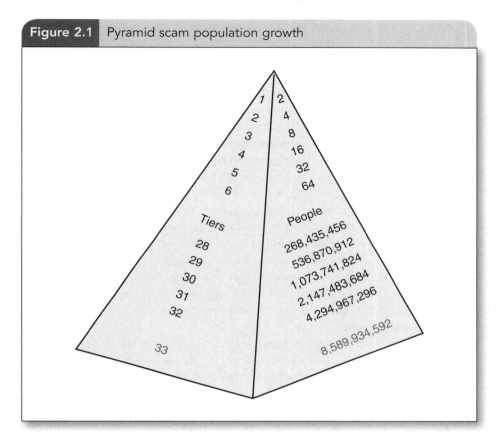

"smart" enough to get in and get out while the going is good, and are left holding the can. Pyramid scams are a perfect argument against the social Darwinist view that pursuing one's own good translates to pursuing the collective's good (more on that later).

But we also must understand that acting in a way that appears to be in one's own interest may not be, especially with the knowledge that our actions, although seemingly correct, can be responsible for someone else's misfortune, no matter how far down the line. Furthermore, we must not participate if we know that by our actions a particular endeavor will ultimately fail.

I suppose one can forgive not understanding the mathematics of doubles—who knew that doubling would eventually cause such pain, and so quickly once the cracks begin to show?—but the slimiest part of pyramids is that they are started by people who pretend to have played and, after recruiting a round of new blood with the requisite initiations, go elsewhere to start up another game *already at the top.*

In Ireland, one such scheme, called "Liberty," was set up by German con men, who were long gone before everything collapsed and the police were called in to sift through the wreckage. An offshoot, called "Speedball," also spread throughout Ireland, as did another variation among teenagers, who were duped out of their small savings. For those who think that pyramids have been around forever and are part of a valid investment landscape, they are started by cunning and deceitful con men who disappear before anyone is the wiser. For this reason, pyramids are illegal.

But we should also consider pyramids inherently immoral because of the built-in unsustainability (or instability) of doubling. And, indeed, it is not too far a stretch to the holding company whose sole purpose is to hold stock in other companies, including holding companies that hold stock in other holding companies—deemed by some to be the cause of the stock market crash of 1929, which triggered the Great Depression, or the 2009 credit crisis based on overleveraged, off-loaded, subprime debt, as we will see below.

Pyramid schemes do seem innocuous though, and we must be careful. We have all heard of or seen pyramid doubling in action in some way, such as in chain letters, Amway-like merchandising, or the "six degrees of Kevin Bacon" Internet game (which measures how closely an actor has worked with Kevin Bacon via their shared films), which is perhaps why it isn't seen as the scam it is. But when we don't think of the consequences of our actions, especially when we're removed in time and space from the direct consequences (i.e., the three or four tiers it takes for a pyramid scam to collapse), we don't see the harm.

2.I. Bernie Madoff and the Never-Losing Fraud Fund

A recent high-profile pyramid involving former NASDAQ stock exchange chairman Bernard Madoff was no different, using new investors to pay earlier ones, and was especially damaging, costing an estimated $50 billion in personal savings. *Newsweek* referred to Madoff's scam as an "affinity Ponzi" (Biggs, 2009) since it primarily targeted members of the same background, his well-off Jewish and Hollywood friends. More than 13,000 were duped, including Larry King, Sandy Koufax, and—in a case of life imitating art—the actor and Internet star Kevin Bacon. I doubt many are still Madoff's friends after losing millions and, by some accounts, hundreds of millions. New Jersey Senator Loretta Weinberg went so far as to refer to Madoff as a "sociopath with no sense of values" (Snow, 2009).

What's more, pyramid scams are becoming more common. In 2009, a Japanese businessman was arrested for tricking thousands of people out of more than $1 billion, promising annual returns of up to 36% in what was described as "a semi-Ponzi scheme" (Parry, 2009). Another highly dangerous example of pyramid doubling is the 1997 Albanian economy, endorsed by the government, which did not see the failed logic of endless doubling. Initially, Albanians were swimming in it to the tune of $1 billion, but as always, the demand for payoffs swamped the supply of new players, estimated to be two thirds of the population at its peak. In the end, 2,000 people were killed in the resultant rioting and the government was thrown out. At the height of the scam, the amount of pyramid money in the economy was almost half the Albanian GDP (Jarvis, 2000). In 1994, a Russian group promised 1,000% returns. After that scam failed, 50 bankrupted players committed suicide (Keeley, 2006).

Alarmingly, Robert Fitzpatrick, president of Pyramid Scheme Alert in the United States, believes the number of pyramid scams has doubled in 10 years: "Unless there is rigorous regulation, a lot of public education, people going to jail, I can't see it changing" (Rusche, 2009). Referring to another recent pyramid scam—the alleged $7 billion swindle of Texan Allen Stanford—FBI assistant director Kevin Perkins noted, "Economic crimes such as those alleged here today are unfortunately all too commonplace. These crimes strike at the heart of our economy and our quality of life" (Bone, Reid, & Spence, 2009).

The Enron fraud can also be likened to a pyramid scam, where debt was systematically hidden and share price artificially propped up. But as the share price increased, the ultimate peril increased; in the end, Enron was bankrupted in short order, with more than $30 billion in debt, although executives sold off up to $1 billion in the month prior to the bankruptcy (Gibney, Klot, & Motamed, 2004). At the height of Enron's arrogance, the company even shut down power plants to manipulate utility prices, while at the same time betting on the future share price—bottom-deck dealing and insider trading at their worst. Furthermore, ordinary Californians were cheated throughout, subjected to regular power shortages, and forced to pay inflated utility bills. Some even lost jobs. Enron's house of cards may even have been responsible for the election of Arnold Schwarzenegger, who benefited from voter unrest and budget deficits that arguably led to the recall of then-Governor Gray Davis (Gibney et al., 2004).

As for Madoff's high-profile scam, one that he eventually confessed to his sons was "just one big lie," the lawyers will be sifting through the wreckage for years to come. It took a credit crisis and panicked clients demanding their money back to topple the pyramid scam of all pyramid scams, which was particularly insidious given that Madoff hid it so well for so long, although hedge-fund manager Barton Biggs (2009) wondered how it could have gone so long undetected:

> The consultants never verified the books and records because they didn't exist. [Madoff] maintained it was highly sophisticated, tactical trading. Almost anyone in the investment business in their right mind knows that there is no such thing and even if there was, it could not have gone on for decades without even one losing quarter.

In a book about Madoff's scam, *No One Would Listen*, self-styled quant (a *quantitative* or numerical analyst in the financial industry[1]) Harry Markopolos (2010) noted that many European clients thought Madoff was "front-running" (think of the scam in *The Sting*, based on selling information before it was publically known), but since enormous gains kept coming in they didn't ask too many questions and assumed he was only front-running others. As Markopolos wrote, "They never bothered to look a little deeper to see if he was cheating other clients—like them for example. What they didn't understand was that a great crook cheats everybody." Of course, Madoff's was a simpler kind of scam, glossed up as a legitimate business, which Markopolos also noted was tied up in offshore funds, or tax havens: "It was not something that was spoken about. . . . While offshore funds certainly can be legitimate, to me it indicated that some of these funds were handling dirty money." No doubt, Madoff filched a few people who didn't take kindly to being filched.

Nonetheless, "Uncle Bernie"—or prisoner number 61727-054, as he is known to the American Federal Bureau of Prisons—was sentenced to 130 years in jail (at the age of 71). However, since he was not ordered to pay restitution, less than $1 billion has been recovered against a litany of compensation claims. Further charges were laid against an aide, an accountant, and two computer programmers who were said to have worked with "others known and unknown" to facilitate the scam. Further charges have also been laid against his long-time secretary and may be laid against others, including his wife, who withdrew $10 million on the day Madoff confessed.

2.2. More About Pyramids and Connected Growth

So what is it about exponential doubling? Do large numbers (and the indefinite time period) leave us uncertain about how such scams work, or are we reluctant to see our own greed in action? Of course, incentives are important—there must be a large enough perceived payoff to provide the lure.

Perhaps the reality of large numbers and the impossibility of maintaining stability can be seen in a pyramid in which I was once involved, and with which many likely have some passing experience. Tame by comparison, and without an overriding monetary incentive or the allure of large profits, it, too, had to fail nonetheless.

The pyramid I was recruited into was a recipe chain letter. When I was a university student, I got a letter from a friend with a list of eight names. The letter requested that I send a recipe to the name at the top, cross that name off the list, put my name at the bottom, copy the letter eight times, and send it to eight others, who were then to send their own recipes to the new name at the top of the list, cross that name off the top, put their names below mine, and so on. In a couple of months, or so the letter stated, I would be swimming in new recipes (8^8 or 16,777,216 to be exact).

So, I dutifully sent a recipe for my mother's Irish bread to the person at the top of the list and sent my crossed-out and copied list to six friends (at the time, I didn't know eight people who could cook), and as far as I know, only one of them (the one girl) sent her recipe to the new name at the top of the list. Twenty-five years later, I am still waiting for my first recipe. Oh well, I guess the perceived benefits of a smorgasbord of recipes weren't high enough. To be sure, money has its own unique lure.

Another interesting example of connected doubling (and, in this case, also harmless) can be found with the prolific Hungarian mathematician Paul Erdös, who was so sought after in his day that it was deemed highly prestigious to work on a mathematical paper with him. Those who published a paper with Erdös were given an "Erdös number" of 1, indicating the nearness of their collaboration. Those who published a paper with someone who had published a paper with Erdös were given an Erdös number of 2, and so on (Erdös himself, by definition, has an Erdös number of 0). Alas, just as in a pyramid scam, producing endlessly higher numbers is impossible (an Erdös number of infinity means only that one has never published a paper). Interestingly, as reported by Oakland University mathematics professor Jerry Grossman (2010) in his Erdös Number Project, five people have Erdös numbers as high as 13.

Similarly, in the popular Kevin Bacon connectedness game, if you have ever been in a film, even the smallest home production, you will be able to trace your way through the separation degrees from any actor to yourself; for example, from Kevin Bacon to Tom Cruise to Paul Newman to Tim Robbins to Rebecca Jenkins to Albert Schultz and, finally, to John White

(A Few Good Men, The Color of Money, The Hudsucker Proxy, Bob Roberts, Street Legal, a home movie).[2] Applying a similar numbering nomenclature to Kevin Bacon connectedness thus gives me a "Bacon number" of 6 from my connections above. Clearly, it doesn't take long on an exponential scale to run out of Hollywood stars if I can be linked so easily to Kevin Bacon.[3]

It is instructive that in the Erdös number world, the connectedness of doubles runs out after 13 tiers, or about 8,000 people, and in the movie world even faster (because of a smaller original-player pool). Interconnected web structures and genetic pools can also show the same limited connectedness, with shorter routes between nested tiers (so-called small-world phenomena).

A fascinating corollary of the interconnectedness of branched networks or binomial distributions is that we are all ultimately related to one another in some way. I have a fifth cousin in Ireland whom I know, but if I were to go back further, especially in the rather small and relatively isolated gene pool of Ireland, I would soon run out of unrelated ancestors (the population of Ireland is about 4 million, or 20 generations of doubles). Maybe I don't want to know how many great-great-grandparents are more related to one another than is perhaps wise. Furthermore, it is fairly easy to show that the actors above may also have smaller Bacon numbers along some shorter path (other than Tom Cruise, whose Bacon number is already 1).

Note that exponential growth isn't only by twos. A popular school movie from the 1960s and still available today[4] shows a person lying in the grass as the camera pans out from field to solar system to galaxy by tens (i.e., 10^x), as in the raised powers of scientific notation (10^1, 10^2, 10^3). But the principle is the same, where exponential growth is generic: 2^x is base 2, and 10^x base 10. What is most important with exponents is that things blow up quickly.

The bel also measures sound levels in powers of 10, where 2 is a whisper and 14 a jet plane (10^{12} times more sound intensity). The pH scale measures hydrogen ion concentration in solution, where $0.1\ H^+ = 1$, or acid, and $10^{-14}\ H^+ = 14$, or base (water equals 7 and is classified as neutral). The Richter scale also quantifies seismic energy in powers of 10, such that a 9 scale earthquake (which typically occurs once a year) is 10 times greater than an 8 scale earthquake (which typically occurs monthly).

Even dollars can be measured as such, as Silicon Valley executive Heide Roizen noted when Bill Gates gave away $30 billion, saying that his charity will surpass others "by many orders of magnitude" (Miller, 2009). To be sure, $30 billion is a lot of money, but not on an exponential scale ($30 billion = $3 x 10^{10}, or 10 orders of magnitude), highlighting once again the seemingly innocuous increases of exponential growth, which when compared with you or me giving away $1,000 is only seven orders different!

Interestingly, when probability decreases with size, as in the lower frequency of earthquakes at higher Richter numbers, a "power-law" distribution[5] exists, and is different from the probability distributions we are used to—such as the famous "normal" distribution, used for test scores, coin flips, and other binomial events (which we'll look at in Chapter 7). Power laws show how the probability of an outcome decreases as its size increases—for example, in such diverse systems as earthquakes, word count in books, or size of human settlements, to name but a few.

Similarly, the Pareto principle typically states that most of the wealth of a nation is held by relatively few people. Pareto expressed this in his "80:20 rule"—i.e., 80% of wealth is held by 20% of people, from which various corollaries abound: 80% of employees do 20% of the work, 80% of the time spent on a problem is on 20% of the solution, 80% of awards go to 20% of actors.[6] Power laws recognize the exponential relationship between two variables and are straight lines when plotted.

Ball (2005) noted that studies of small-world networks or friendship circles show "a high degree of clustering but [possess] numerous short cuts between clusters, creating the short average path lengths responsible for our 'six degrees of separation'" (p. 463). Clearly, we are more connected than we think. The Internet is similar, although because of a more intricate makeup, it shows more degrees of separation. For example, a study of the University of Notre Dame domain showed that any two webpages were separated by an average of only 19 links (p. 482). But because the Internet is inherently scalable (i.e., a power-law distribution), the number of links between pages will not prohibitively increase as the size of the Internet increases, which is just another restatement of how innocuously exponentials behave. As the Internet increases (say, a thousandfold), the number of extra links will increase by only one or two.

The upshot of all this is that you should be careful with exponents. They will trample you, and more so the higher you go. As for the appearance of small-world phenomena in the interconnectedness of doubling, one can circumvent the power of doubling through alternative routes, but only for a while until the doubling swamps such clustering.

The harder lesson, however, is that participating in such doubling ventures is wrong if one is hoping to find a sustainable future (movies, Erdös numbers, and recipes notwithstanding), no matter if one can get in and get out before the system collapses, because we contribute to the future instability *by playing*. In the same way a bad foundation can be stable to a point before breaking under an unintended design load, higher and higher pyramids will always fail, and all the more quickly and with more damage done as the growth continues. Furthermore, because such schemes are unsustainable by design, they are immoral, benefiting only those in the know, while those below lose all.

2.3. New Scam, Old Scam

Runaway expectations in exponentially based ventures are not confined to our time or even to the margins of legality. Perhaps the earliest modern pyramid failure was the tulip craze in Holland, a disastrous encounter with overhyped trading in the 1630s, when tulip prices went through the roof in a matter of weeks before falling after the bloom had wilted. In 1717, more than a few British investors also saw their dreams of earthly fortunes dashed when the South Sea Company collapsed. In exchange for acquiring most of the British national debt, the company had been granted a trade monopoly with Spain for, among other things, transporting slaves from North Africa to the New World. In this scheme, dubbed the South Sea Bubble for the speed at which it burst, stock prices bottomed out after relations between Spain and England soured and shares could no longer be propped up. Desire for unconfined exponential growth was the catalyst for both the tulip craze and the South Sea Bubble.

Land speculation in Florida in the 1920s was also likened to a pyramid or "Ponzi" scheme after transplanted Bostonian Charles Ponzi developed subdivisions at 23 lots to the acre in places next to nonexistent cities, the so-called Florida swampland, described by Harvard economist John Kenneth Galbraith (1992) "as repugnant to the people who bought it as to the passer-by" (p. 33). Citing American feeling prior to the 1929 stock market crash, Galbraith also noted that "the Florida boom was the first indication of the mood of the twenties and the conviction that God intended the American middle class to be rich" (p. 35).

Indeed, as money becomes exceedingly loose, everyone wants to get in on the action and greed plays a bigger part in the seductive nature of such ventures. Whether the stock market is a zero-sum game (I win, you lose) will be looked at later (see Chapter 11), but it is prudent to recount the reasons for the most abject failure ever of an investment structure, the 1920s stock market crash—a demise that can be traced to the headiest of financial chutzpah, that of "pyramid investing."

The story begins in the summer of 1929, the year prior to the crash, during the height of speculative giddiness—when many companies were buying heavily of their own security, hoping to influence stock prices.

Furthermore, the increase in "margin buying," where only a percentage of a stock price is surrendered as collateral, was encouraged at lower and lower percentages by banks keen to reap the advantages of the raging bull market (sound familiar?). But the most questionable practice of all was the growth of the holding company, a legal device created by the legendary Standard Oil lawyer C. T. Dodd to circumvent the antimonopolistic restrictions of the previously enacted Sherman Antitrust Act of 1890 (Boorstin, 2000, p. 419), which placed overbearing pressure on an already overmargined, poorly leveraged stock market and failed as soon as the panic selling began. As Galbraith (1992) noted with regard to the numerous precrash mergers,

> In the case of utilities the instrument for accomplishing this centralization of management and control was the holding company. These bought control of the operating companies. On occasion they bought control of other holding companies which controlled yet other holding companies, which in turn, directly or indirectly through yet other holding companies, controlled the operating companies. Everywhere local power, gas, and water companies passed into the possession of a holding-company system. (pp. 70–71)

To be sure, as long as prices keep rising—for tulips, slaves, houses, stocks, etc.—no one is any the wiser, but when the supply of those willing to buy falls short of those willing to sell, the bloom wilts, the bubble bursts, house prices fall, and the walls come tumbling down. In the case of the stock market crash, where the holding company was shown to be a sham and no more money was available to leverage or prop up the overvalued stock, three quarters of the market's value was wiped out in 2 years, precipitating a decade-long, worldwide depression and nearly 25% unemployment. As Galbraith (1992) noted,

> In 1929 the discovery of the wonders of the geometric series struck Wall Street with a force comparable to the invention of the wheel. There was a rush to sponsor investment trusts which would sponsor investment trusts, which would, in turn, sponsor investment trusts. The miracle of leverage, moreover, made this a relatively costless operation to the ultimate man behind all of the trusts. Having launched one trust and retained a share of the common stock, the capital gains from leverage made it relatively easy to swing a second and larger one which enhanced the gains and made possible a third and still bigger trust. (p. 83)

In the aftermath 3 years later, stocks previously valued at $100 could be got for around 50 cents (Galbraith, 1992, p. 161), while the overall stock market had dropped more than 80% from its peak (Moyo, 2011, p. 3). As Galbraith noted, "The fears of November 1929 that investment trusts might go to nothing had been largely realized" (p. 161). Mortgage defaults also increased to a rate of 1,000 per day (Boorstin, 2000, p. 284).

Do parallels exist in the excessive availability of loose money and sub-prime mortgages prior to the credit crunch from 2007 onward, facilitated by the Reagan administration's deregulation in the 1980s of economic safeguards put in place after the 1920s to prevent a recurrence of the Great Depression? You bet—an ever-expanding supply of players fueled by excessive greed and easy credit and kindled with lax government regulation is a powder keg waiting to explode.

By the start of the subprime mortgage market collapse, foreclo-sures had increased by 80% to 1.3 million, and by March 2008, an estimated 8.8 million[7] homeowners in the United States (almost 11%) were in negative equity, i.e., they owed more than their houses were worth (Moyo, 2011, pp. 69–72). As a result, Bear Sterns stock, which was tied to the American mortgage market, became worthless, while another traded at 9 cents on the dollar. What's more, when the sub-prime mortgage market collapsed, it triggered a meltdown in the hedge-fund market. Overleveraged funds couldn't double down any-more as margin calls took over and decreased their playing money, resulting in a frenzied selling of increasingly toxic subprime positions (Patterson, 2010, p. 223).

When the losses couldn't be covered, the end came in an instant, exactly as in any pyramid scam; the bad positions could no longer be hidden nor the losses ridden out. In the end, more than half the value of all global stocks was wiped out, dropping from $63 trillion to $31 trillion in only 10 months (compared with the 27 months it took the market to lose half its value after the 1929 crash; Mason, 2009, pp. 53, 187). The knock-on effects created a credit crisis and a worldwide recession as liquidity dried up around the world and exports slumped.

After the bubble burst on Ireland's decade-long property binge, accel-erated by the credit crisis, a staggering confirmation of short-sighted planning and unregulated growth economics was realized with the report that "there were up to 2 million homes and apartments but just 1.5 mil-lion households" (White, 2010), meaning that *one-quarter* of all homes were empty. By early 2010, half of all homeowners were in negative equity, and a year later, house prices had dropped by more than half from their peak 3 years earlier.

Of course, these crises were more about pumped-up credit than pumped-up assets,[8] which is the difference between a thick-walled bubble (assets only) and a thin-walled bubble (bank-leveraged debt). The former expands and contracts; the latter blows up, leaving an ugly mess, as has happened to banks all over the world. Such banking crises are the worst kind of pyramid scam.

George Soros (2008), one of the original hedge-fund managers from the 1960s, noted that boom-bust cycles are especially prone to leveraged debt rather than leveraged equity (p. 65), where everyone is getting in on the lending game and in greater amounts. In fact, lending becomes a free-for-all future-pay pyramid. But in the end, as always, something has to give. In this case, the correction came like a bang, unsurprising to anyone familiar with hyped expectations or pyramid scams or with a minimum understanding of history and the concept of a future beyond one's reach.

As to how such growth madness could continue unchecked, Zakaria (2009b) noted,

> [The naysayers] asked why packages of subprime mortgages should be rated as highly as bonds from General Electric. But each successive year ended with another eye-boggling earnings report or billion-dollar pay-day for the hedge-fund manager of the moment, the much-promised correction failed to materialize, and the naysayers grew quieter and quieter. (p. xv)

It would seem that the lessons of history are hard to learn, but we should at least know that any system built on greed and excessive credit cannot last. Perhaps the answer lies in the word *credit* itself, which from the Latin *credere* means "to believe." Alas, when there is no more belief, there is no more credit—but by then the bubble has already burst.

2.4. Scrooge or Faust? The Future Is in the Balance

One question asked in this book is whether we can sustain the unfettered growth economics of our world, or at least without any thought to the consequences. Is continuous growth a doomed model, where use continues unquestioned without regard to available resources or sustainability? Or are other models needed? What's more, why are we in such a hurry to have it all? It makes little sense to step forward into the unknown without considering our resources or how we ourselves enter the equation of discovery. We are no longer in frontier times.

Perhaps we think we can continuously correct ourselves when we stray too far, like a spring recoiling or a yo-yo always returning to hand. Such oscillatory systems are like an ongoing payback, a never-ending tit for tat. Literature is ripe with payback stories, seen in revenge plays such as *Hamlet* and *The Merchant of Venice,* or in any number of modern equivalents such as *Mad Max* and *Rambo*, where the revenge is usually a one-off with everyone dead at the end.

But "what goes up must come down" is not quite what is meant by an oscillatory system, because what goes up might go up only once and come back down only once (everybody dead), or decay as it returns (i.e., not quite the same on return, as in a metal-fatigued spring or magnetic hysteresis). Here, a bouncing ball is perhaps a better model—a ball that will eventually come to rest, the height decreasing a little more with each bounce, as in the exponentially damped oscillating system shown in Figure 2.2.

The earth is also an oscillatory system, revolving around the sun to produce the seasons and revolving on its axis to create day and night, but that system, too, will run out of steam eventually. Even the universe, according to the going physics, will end someday in a so-called "heat death,"[9] unable to continuously violate the first law of thermodynamics.[10] One day, the energy will be completely unusable, converted into entropy or a maximum state of disorder. Unfortunately, there is no such thing as a perpetual machine, not even for our own universe.

Figure 2.2 The damped oscillator

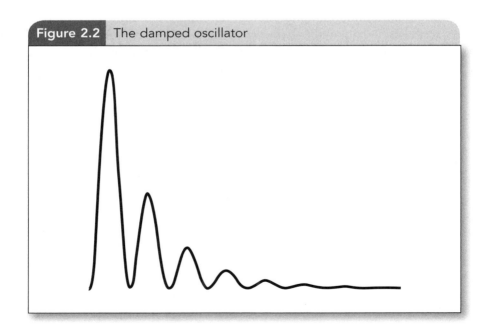

So what happens as we push ourselves to the limits? In 1957, Rachel Carson showed that we are killing our environment by polluting the landscape with pesticides and excessive mass-farming practices. Her advocacy—in particular, her landmark book *Silent Spring*—helped create the Environmental Protection Agency in the United States and other agencies like it around the world. In our era, prolonged use of fossil fuels is increasing CO_2 emissions, and global warming now poses a threat, perhaps in the same way that excessive and unregulated pesticide use caused untold damage to flora and fauna.

The implications for the future are serious. For example, developing the Athabasca oil sands[11] near Fort McMurray, Alberta, will produce 3 times the carbon dioxide as conventional drilling and could disastrously raise global temperatures (Webster, 2009b). Oil sands are particularly hard to develop since heavy crude oil on the surface is mined using energy-intensive methods (1 barrel used for every 5 produced, compared with 1 in 25 for traditional oil extraction methods; Stewart, 2010). As Stewart noted, "Getting it out comes at a price—a hell of a price." The response to the excesses of today's modern living and our own unbounded lifestyles may well determine our future.

Atwood (2008) wondered about such strategies, arguing that too much is bad because the balance is destroyed in the process—the balance between man and earth, man and man, creditor and debtor. She wrote in a retelling of Dickens's Scrooge story that because of an excessive focus on money, we are destroying the earth, with mono-crop farming, "which Nature has always disliked" (p. 188); overfishing, which is "really easy to do with megaships equipped with sonar for fast fish finding—and the eventual result is no fish" (p. 191); biofuels; soil depletion; rain forest destruction; and out-of-control carbon emissions, resulting in thawing tundra, superforce cyclones, and more. In Atwood's version of the story, Scrooge ignorantly pleads with the last of the spirits: "We approach things rationally now, what with science, and cost-benefit analysis, and the use of debt as a sophisticated investment vehicle" (p. 180). Add oil extraction to the mix, and we already may be using up what little is left, and at exceedingly alarming and harmful rates.

The Scrooge story is one metaphor for our time that might scare us into thinking we can change, even if our time is almost up. As Atwood (2008) wrote, recalling Schumacher's idea of "natural capital,"

Maybe it's time for us to think about it differently. Maybe we need to count things, and add things up, and measure things, in a different way. In fact, maybe we need to count and weigh and measure different things altogether.

Maybe we need to count the real costs of how we've been living, and of the natural resources we've been taking out of the biosphere. Is this likely to happen? Like the Spirit of Earth Day Future's, my best offer is Maybe. (p. 203)

Another story that Atwood (2008) investigated, however, is that of Faust, in which man makes a pact with the devil to live the high life on the layaway plan, living far beyond his means to enjoy material pleasures today at the expense of his soul tomorrow (p. 91). Faust is a much darker story, and as presented by Atwood, excellently highlights the giddiness of having too much.

So, are we being repaid because of our bargain with money, wanting to be richer than others while remaining seemingly indifferent to the world's destruction? In A Christmas Carol, a redeemed future is presented, with a chance to correct our past ways, whereas in The Tragical History of Doctor Faustus, only perpetual indebtedness exists. The choice is ours.

More important, can the world bounce back? Forgetting that all things must come to an end someday, the earth is pretty good at recovery. Like the Energizer Bunny, the earth keeps going, surviving an ice age, meteor strikes, and all sorts of man-made abuses, exponentially increasing or otherwise. There is nothing but a supernova standing between earth and a few more billion years. To be sure, our lack of concern for the future is alarming, as is our fascination with growth at any cost, but the earth will bounce back, spring will return for its annual rite, and the sun will continue to rise. Humans—well, that's another story.

Notes

1 **Quantitative traders**: A quantitative trader, or "quant," takes advantage of small, real-time inefficiencies in the market, especially in highly volatile times when second- and third-order effects (i.e., the first and second derivative of a stock price or derivative) can be more easily seen by a computer program. Many such programs, pioneered by mathematics professor Edward Thorp, use statistical arbitrage (or "stat arb") to break down stock market patterns. Patterson (2010) noted that "quants make their living juggling odds, searching for certainty, shimmering probabilities always receding into the edge of randomness" (p. 138).

2 **Six degrees of separation**: One can use any actor to construct a "connectedness" graph, although the connectedness between any Hollywood actor and me would still have a similar number of levels, highlighting the linear scaling property of exponentials.

3 **Erdős and Bacon numbers**: Interestingly, Erdős has a Bacon number of 3 when television and documentaries are included, whereas Bacon has an Erdős number of infinity (Harris, Hirst, & Mossinghoff, 2008, p. 30).

4 **Cosmic view**: The original book Cosmic View (Boeke, 1957) was the basis for the movies Powers of Ten by IBM and Cosmic View by the National Film Board of Canada

((http://aspire.cosmic-ray.org/labs/cosmic_zoom/index.htm and http://www.nfb.ca/film/cosmic_zoom/).

5 **Self-similar distributions**: Power-law distributions can also be thought of as "scale-free" or "self-organized," where the law remains despite the scale, illustrated most elegantly by Benoit Mandelbrot and his study of fractals.

6 **The 80:20 rule**: The ratio could be 90% and 10%; the important point is a lower probability at higher amounts.

7 **The negative equity blues**: By the middle of 2011, 10.8 million homeowners were in negative equity, about 22.5% of the total mortgage market ("CoreLogic," 2011).

8 **Bubble speak**: Moyo (2011) made the distinction between a "productive asset bubble" (equity values decline sharply, e.g., the 1995–2000 technology bubble) and an "unproductive asset bubble" (overleveraged bank loans must rapidly deleverage and recapitalize, e.g., the Japan real estate bubble of 1986–1990 that saw Japanese equities and land lose 60% from their peak; pp. 59–60).

9 **The heat death of the universe**: The German physicist Hermann von Helmholtz first postulated the idea of a "heat death" of the universe when he was studying heat flow and mechanical work in engines in 1854 and the idea of entropy or disorder (an irreversible process). If the universe were to expand forever, at some point it would be without energy, reaching a complete state of entropy, and thus dead. (See also "The big crunch.")

10 **The first law of thermodynamics**: Energy cannot be created or destroyed but can only be changed from one form to another.

11 **Athabasca oil sands**: Also known as the Alberta tar sands, although there is no tar. At 1.7 trillion barrels, the Athabasca oil sands have the second-largest proven oil reserves in the world, after Saudi Arabia (Webster, 2009b).

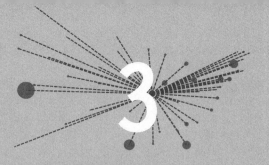

THE MATHEMATICS OF CHANGE

A New World Model

What do Aesop's fable about a race between a tortoise and a hare, the Dow Jones industrial average, and a Formula 1 car have in common? They all deal with change—in the distance run by two animals in a storybook race, in a stock market index (a measure of the performance of a widely held group of stocks[1]), and in the speed of a car. Fortunately, to analyze the data in all three cases, the mathematics is the same.

3.1. The Measure and the Change in the Measure

Perhaps the most fabled race in all folklore is that of the tortoise and the hare. Aesop gives little information about the race, but knowing only that the tortoise traveled at a constant speed (as tortoises do) and that the hare raced ahead and fell asleep (as hares seemingly do), we can plot their data in a distance versus time graph to see what conclusions can be reached (see Figure 3.1). This is not unlike how one might think through the various routes to or from work, factoring in traffic bottlenecks and speed traps along the way.

Since Aesop provides no length or time for the race, I have had to make some assumptions. First, I have assumed a 2-meter/minute tortoise, which is reasonable for a real-life tortoise and is equivalent to about 0.1 miles per hour.[2] I have also assumed a 200-meter course, which thus takes 100 minutes for our steady, 2-m/min tortoise to complete and also seems reasonable given how long an average hare might nap. Last, I have assumed that the hare runs as fast as the fastest human,[3] at about 600 m/min, and that he fell asleep at the midpoint, or 100 m.

| Figure 3.1 | Distance versus time for the tortoise and the hare |

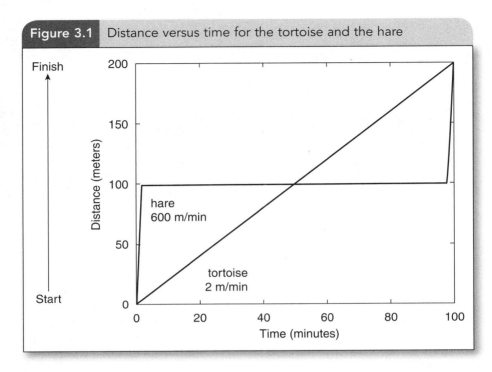

With the data at hand, we can now make a few important observations. I had always thought the race was a longer distance over a shorter time. I don't know if this was because of Disney's cartoon version, but if the race was, say, of marathon length (26 miles, 385 yards), a 2-m/min tortoise would take more than 2 weeks (without rest) to finish, and no hare would likely nap for that long (well, maybe a talking Disney cartoon hare would)—not to mention the patience the referee fox would need. Perhaps the hare hibernated during the race (Disney again?), happily dozing for 2 weeks, but that would only complicate the fable. As well, a faster race (say, the length of a cartoon) would require a faster tortoise, which is not only unrealistic but defeats the idea of the fable. Suffice it to say, the race was over a reasonable distance such that a steady tortoise could finish in a realistic time and a haughty hare could fall asleep and awake in time (well, almost in time).

It would also seem from the numbers that our hare was haughtier than originally supposed—he was so self-sure and cocky that he couldn't keep it together for 200 meters, or about 20 seconds. Perhaps our hare had a short attention span, one possible interpretation of the fable. Or perhaps he was more stupid than haughty, foolishly tiring himself out in a sprint when a slow but steady jog would have sufficed, given the unlikely competition. Perhaps he just overestimated his resources. Of course, we don't need exact numbers to understand a fable, but it does seem that retelling the story with reasonable numbers makes it even more of a cautionary tale about being haughty, distracted, and stupid.

But with such real-world numbers, we can now understand more about how data changes in time—for example, in an airplane flight data recorder, where the distance and speed data can reveal the cause of a crash; in inflation indices, where price instead of distance is measured; or in the stock market, where extreme changes in the price of a stock can result in huge windfalls or losses.

Let's start with the tortoise, since she is the easier to analyze. As seen in Figure 3.1, the tortoise's distance is as straight as an arrow: 20 meters after 10 minutes, 100 meters after 50 minutes, and 200 meters

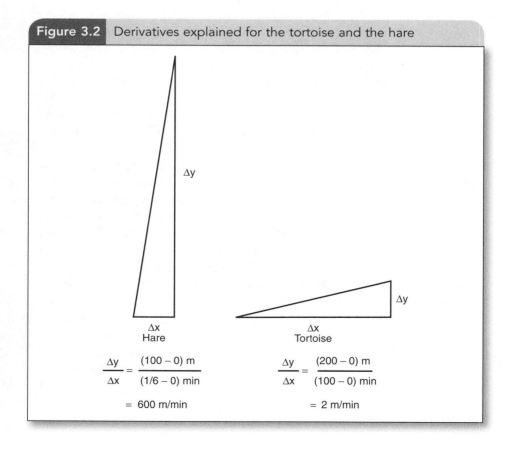

Figure 3.2 Derivatives explained for the tortoise and the hare

Δy

Δx
Hare

Δy

Δx
Tortoise

$$\frac{\Delta y}{\Delta x} = \frac{(100 - 0)\ \text{m}}{(1/6 - 0)\ \text{min}}$$

$$= 600\ \text{m/min}$$

$$\frac{\Delta y}{\Delta x} = \frac{(200 - 0)\ \text{m}}{(100 - 0)\ \text{min}}$$

$$= 2\ \text{m/min}$$

after 100 minutes, which is represented as $d = 2t$ (d is distance, and t is time). The hare's speed was also assumed constant, though in spurts, and is seen as two spikes in his d-t graph. The hare raced ahead (greater slope), slept (no slope), and madly dashed at the end to catch up (greater slope again). Since he raced his first 100 meters in 10 seconds (600 m/min), this is represented as $d = 600t$, i.e., 300 times faster than the tortoise.

Most of us are familiar with slopes (e.g., the change in height over the change in distance equals the slope of a hill, stairs, or a ladder, or distance versus time equals the speed of a car or plane). A slope shows the change of one variable with respect to another, as in a physical slope with respect to distance (the "rise over the run," as seen in Figure 3.2 on the previous page) or, more abstractly, versus time. Mathematically, a change over a change is called a derivative.[4]

As seen in Figure 3.3 in a plot of the Dow Jones index daily high over a monthly 50-year period (top graph) and the change in the index over the same period (bottom graph), slopes or derivatives can reveal seemingly hidden information. For example, spikes appear in the index data but are not nearly as pronounced as those in the derivative data. One can also see that the spikes are more frequent than those in the tortoise and the hare data, indicating a higher volatility. Furthermore, since 2000, the Dow Jones shows an increased spikiness, i.e., a much greater volatility. Note that the data is no different; only the presentation is—the change in the measure instead of the measure.

In auto racing, the same analysis applies, where speed versus time data is sampled using multiple high-frequency, onboard car sensors that record data in real time or at numerous sector points around the course. Figure 3.4 plots a Formula 1 car's speed versus time around the track at Monaco,[5] where the slopes show acceleration (positive slope) and deceleration (negative slope), marking where better pull-away speed and braking translate to better performance. Again, the spikiness is a direct measure of the change in speed as a car speeds up and slows down around the track.

In physics and economics, predicting a future trend (e.g., a ballistic trajectory or future price) is possible by quantifying the change in the "spikiness." Given the distance or price versus time, one calculates the car speed or price change by differentiation (calculating the derivatives

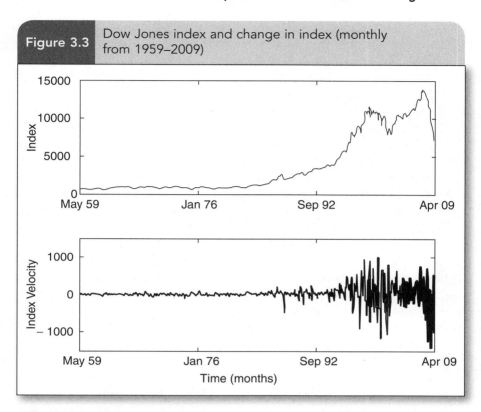

Figure 3.3 Dow Jones index and change in index (monthly from 1959–2009)

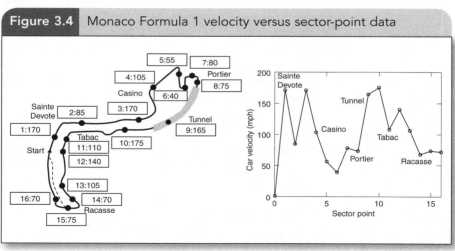

Figure 3.4 Monaco Formula 1 velocity versus sector-point data

or difference between two points[6]) and the acceleration by taking the derivative again. In the financial markets, this is how a dealer estimates whether a stock or index will continue to peak or plunge and by how much, where a low rate of change indicates a more prolonged trend or high resistance to change (inertia). Growth or contraction in the economy is also measured by such change.

Calculating differences is not hard—we just take the differences between two points. If the tortoise achieved her top speed of 1/30 m/s (i.e., 2 m/min) and the hare his breezy 10 m/s (i.e., 600 m/min) in 10 seconds, then the tortoise's acceleration was $1/30$ m/s^2 and the hare's 1 m/s^2.[7] The calculation is no different for the Formula 1 or Dow Jones data, although there are many more sample points—in this case, sector speeds or daily ticker prices.

What is most important with derivatives, however, is that one calculates by working backward, looking closely at how the data changes in time. In physics, the motion is exactly as prescribed by the applied force, as in gravity acting on an astronaut or an apple, which ultimately predicts a future trajectory. In economics, where the data doesn't necessarily follow a prescribed rule or law, the forces aren't as easily determined and future trends are harder to predict, but they can be found in the data.

As for the meaning of the change, one has to look at the cause. In physics, acceleration is associated with a force—recall James Bond's contorted face in *Moonraker* as the astronaut's centrifuge trainer was increased to more than 13 g[8] (without any force, there can be no *change* in the motion). The hare can thus be likened to a drag racer that burned out because of high stresses on the engine, related to a large change of speed in a short period of time. In hare terms, his body gave out. For the tortoise, although she was tired at the end of the race and "comfortably dozing after her fatigue," she didn't overly stress her body (no acceleration, no force) at her slow but steady tortoise pace of 2 m/min. As for racing, drivers accelerate and brake by applying pressure to the pedals while changing gears to improve performance. In the markets, however, the causes are much less clear and often colored by the politics of the day (which we'll look at later).

But change must always be considered—how fast, how slow, and at what rate—to understand the data as presented in any functional relationship, from sports standings to inflation, from everyday electrical usage to global temperatures, and from stock markets to races.

3.2. Distance Versus Speed, or Who to Bet on in the Crunch

The Greek letter for change is delta (Δ), and as we saw above, the change in a measure can be more important than the measure itself, more formally referred to in mathematics as a derivative ($\Delta y/\Delta x$).[9] Here, we look at an example from the world of sports that shows how discrete (or slowly changing) data—such as sports standings, rather than the more challenging continuous (or fast-changing) data of race cars and instant stock prices—is particularly dependent on change and the rate of change.

At the end of the 1992–1993 National Hockey League season, the New York Rangers led the New York Islanders by 3 points with 12 games to play as both teams battled for the coveted last playoff spot. The question is, could one tell from the data which team would go on to win? Or, if one was a bettor, which team would one bet on? The graph in Figure 3.5 shows both teams' performances through the year (top graph) and the

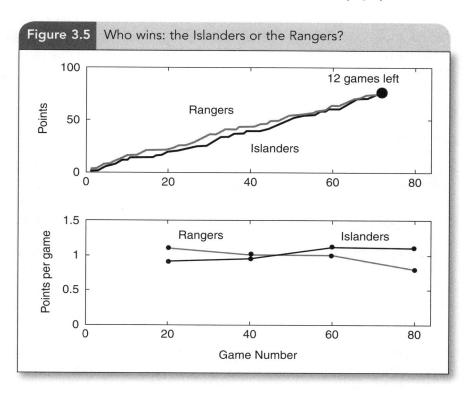

Figure 3.5 Who wins: the Islanders or the Rangers?

change in their performances over four 20-game periods (bottom graph). With the change in the data now plotted as shown, the question is asked again: Who would you bet on to advance—the Rangers or the Islanders?

The data is an almost static representation compared with the real-time race data or daily Dow Jones data, but the same analysis applies, that of measuring the change in the measure (points per game, velocity, inflation) rather than the measure (points, distance, price). As can be seen in the bottom graph in Figure 3.5, the Islanders steadily improved as the season advanced, whereas the Rangers got worse and, with only four games to play, were already in a tailspin despite having been ahead of their crosstown rivals for most of the season. As it turned out, the Rangers lost 11 of their last 12 games and the Islanders won by 7 points, a trend readily seen in the derivative data (points *per* game)—although one could just as easily say the Rangers lost it, as the derivative data prior to their meltdown shows.

Other teams have also managed to turn newfound "form" into success. The 1951 New York Giants beat the odds in the best-ever baseball pennant race by coming back to win after falling 13 1/2 games adrift of the Brooklyn Dodgers with 7 weeks to play. Over the final weeks, the Giants, with Willie Mays and Bobby Thomson, went 37 and 7 while the Dodgers won only 19 games, miraculously ending the season tied. The pennant winner was decided in the playoff by Bobby Thomson's famous "shot heard round the world." In 2011, the St. Louis Cardinals fashioned a similarly miraculous comeback, trailing by almost 10 games with 5 weeks to play. They won the National League wild card on the last day of the season and went on to win the World Series.

Analyzing sports scores over a limited number of games is a way of quantifying such form and can be a much better indicator of future success than league standings, as understood by the former head of President Obama's National Economic Council, Larry Summers, who as a schoolboy would compare baseball teams' midseason positions to their final positions to determine any telling correlation (Cohan, 2009). In the same way, a snooker player who is behind by 70 points can still win if the table is favorably positioned for a game-winning clearance, or a last-place race car with fresh tires can pass everyone and win. In each case, the change in the data is more revealing than the data itself.

Changing economic data can also show the strength of the economy, as reflected in the gross national product from one quarter to the next and as indicated by growth after a reduction rather than in annual figures (even better if presented monthly or weekly, as we will see in Chapter 5). From such simple examples, one easily sees how static data can be highly suspect and doesn't always represent the true state of affairs, as in the government statistic "economy grows by 3.4% in first quarter" or "unemployment is 2.5% less than last year," which may not represent current conditions, especially during times of economic turmoil.

Furthermore, if inflation measures or consumer price index data lag the real world, such data may be no better than common intuition gleaned from experience, such as department stores or restaurants offering more-than-usual reductions. Admittedly, change is hard to gauge and depends on the sample period (daily, weekly, monthly),[10] but we should at least know that the *change* in a measure holds more information than we think.

3.3. Inflation Decreases, So Why Don't Prices Go Down? It's All in the Delta

Another example from the world of economics highlights how change isn't properly understood and can be confusing even to the professionals. Prior to the 2009 economic downturn, business was booming in Ireland for more than a decade, with 7% average annual growth recorded since 1995. In times of great economic prosperity, housing prices and tax revenue can increase dramatically, as can inflation. In Ireland, particularly Dublin, inflation was on the rise.

One night during the peak, however, the news reported that inflation had decreased over a particular period, and a news reporter asked a representative of a national union to comment on the change. The union representative said the usual good things about the economy and that he welcomed the decrease in inflation, although he hadn't yet seen a corresponding decrease in prices.

At first thought, the union representative's remarks don't seem out of place, but if one pictures an analogous situation—say, that of a driver braking—the fallacy is apparent. Let's say Joe Economist is driving along, sees a cat crossing in front of his car, and slams on his brakes. He decelerates immediately, but of course the car still moves forward before eventually coming to a stop. How hard Joe Economist slams on the brakes (and how fast he was traveling) determines how quickly he will stop, but he will still move forward after braking (think of how long a jet or the shuttle takes to come to a stop after landing). In the same way, although inflation might decrease, *prices* will still rise. In fact, inflation can continue to decrease *forever* and prices will still *rise*, although by less and less.

A decreasing inflation means only that prices are increasing by a *lesser* amount, as shown in Figure 3.6, where the inflation decreases from 4% to 2% to 1% yet prices keep rising. A decrease in prices

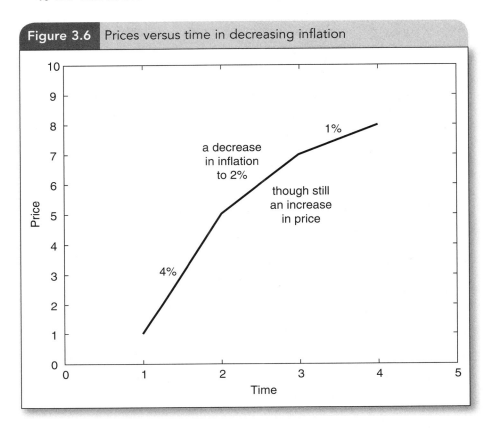

Figure 3.6 Prices versus time in decreasing inflation

requires *negative* inflation (or deflation), not a decrease in inflation—analogous to a car reversing, something that will never happen when one slams on the brakes. In a car, as in the economy, another gear is needed: reverse.

As we saw above, a force is a change in the change of distance versus time, felt when one decelerates or accelerates—think of the seatbelt restraining your forward motion as you brake or of your back pushing against the seat as you accelerate to pass. In economic terms, the increase in prices is inflation, the *rate* of which can decrease even though prices increase.

Alas, the decreasing inflation in Ireland didn't last long and prices began increasing again in their usual way—until the 2009 economic melt-down, which saw a dramatic decrease in rising prices followed by an actual turnaround, i.e., deflation (or negative inflation). Death and taxes aren't the only certainties in life; inflation isn't far off, economic melt-downs notwithstanding.

3.4. The Measure and the Change in the Measure: More on Slopes

All sorts of examples help us see the difference between a measure and the change in the measure, such as inflation with price or speed with distance. A country's GDP may be increasing (positive slope) but its rate decreasing (negative *change* in the slope), which will eventually translate to a decrease. It seems, however, that the mind is drawn to the instantaneous measure and not to the more important change in the measure, as the following example illustrates (McDermott, Rosenquist, & van Zee, 1987):

> Figure [3.7] shows a position versus time graph for the motions of two objects A and B that are moving along the same meter stick.
>
> At the instant t = 2 s, is the speed of the object A greater than, less than, or equal to the speed of object B? Explain your reasoning.
>
> Do objects A and B ever have the same speed? If so, at what times? Explain your reasoning.

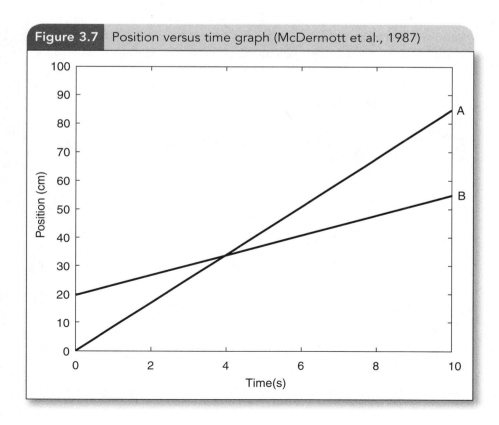

Figure 3.7 Position versus time graph (McDermott et al., 1987)

Many of us, however, confuse slope and height, thinking that at 2 s, B is going *faster* than A, when at 2 s, B is only *higher* than A. It is the slope (speed = distance/time) that matters and not the height (distance), yet most do not realize that B (although higher than A prior to 4 seconds) can *never* go faster than A. Recall the hare, who moved much faster than the tortoise but still lost the race (a distance, not a speed race).

As for the second part of the question, as convincing as it might seem that A and B are going the same speed at the point where they cross (4 s), they are not. Since both are straight lines, A is *always* moving at 8.5 cm/s and B is *always* moving at 3.5 cm/s (positively slug-like compared with the 600 m/s and 2 m/s of the hare and the tortoise). A and B may be temporarily at the same height (at 4 s) but will never be at the same speed once in motion.[11]

Similarly, as applied to markets, it does not necessarily follow that buying in a rising market is better than buying in a falling one. Here, the price matters more than the change in price. Thus, it is better to buy as close as possible to the bottom of the market, whether rising *or* falling, and sell as close as possible to the top (mathematically called a local minimum or local maximum). The old adage may have been "location, location, location," but the new adage is "timing, timing, timing." Timing is all about getting the change right.

The following teaser perhaps best highlights the difference between a measure and the change in a measure. If you can order the three ramps in Figure 3.8 from lowest to highest by height and by slope, then you are well on your way to mastering the mathematics of change.

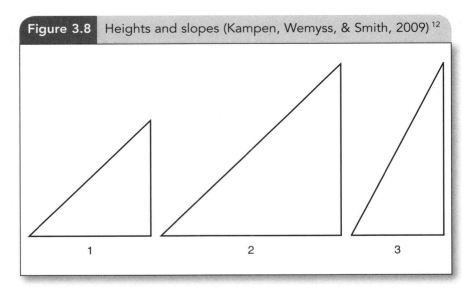

Figure 3.8 Heights and slopes (Kampen, Wemyss, & Smith, 2009) [12]

1   2   3

To be sure, numerical data is part of everyday life, from daily temperatures to hourly measures of our own health and well-being, which can be quantified by today's battery of health workers. But a *change* concerns us more often than the actual numbers. No one worries about a constant body temperature of 37° C (= 98.6° F), but when the thermometer starts to creep up or down at a perceptible rate, we become concerned, and all the more as the rate increases.

The same applies to trends in energy consumption or weather, as shown below in two sets of historical, or longitudinal, data (i.e., data versus time): electrical usage per hour over 7 days, as recorded by a utility provider (Figure 3.9), and global temperature per year from the mid-19th century to the present (Figure 3.10). Both sets of data show clear trends, which we can easily see and attempt to interpret now with a little more insight about change.

In Figure 3.9, we see a trend in daily usage—increasing steeply in the morning, continuing through the day, and peaking in the evening. The

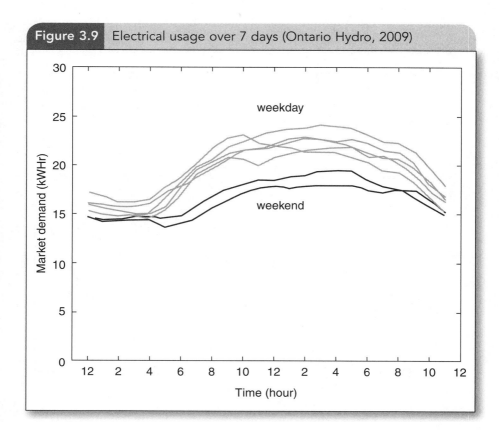

Figure 3.9 Electrical usage over 7 days (Ontario Hydro, 2009)

inference is clear: People get up, turn on appliances, work through the day, return home to cook, watch television, bathe children, etc., before turning off their lights to go to bed. There is also an increased load during the week (top five lines), suggesting that more electricity is consumed in the office than at home. Such data can inform power stations about how to balance electrical loads or give politicians ideas about encouraging peak-time reductions, staggered office hours, or public transit fare savings. Here, the plotted data helps us see a simple, obvious trend.

In Figure 3.10, the famous global warming "hockey stick" figure, the data shows a marked increase in temperature with time, plotted as relative "anomalies" to base years. The change in the data from one year to the next, however, is less predictable, as seen in the derivative data, but is certainly not as dramatic as the increase in the stock market data. And although cause and effect are not as easy to interpret—a subject we will look at later with the help of statistics and regression (see Chapter 6)—for

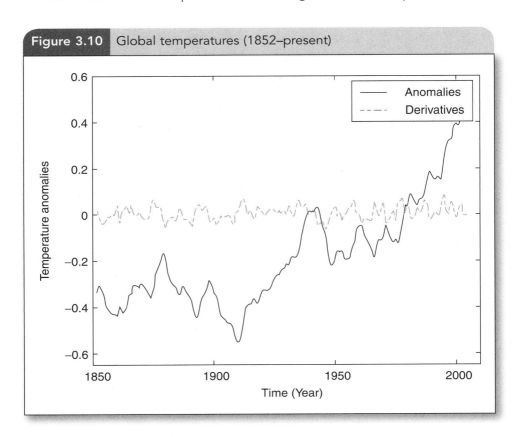

Figure 3.10 Global temperatures (1852–present)

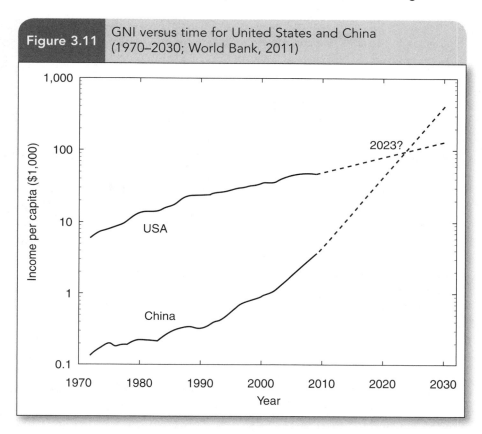

Figure 3.11 GNI versus time for United States and China (1970–2030; World Bank, 2011)

now, we can comment only on the appearance of a correlation that shows a steady increase.

Comparing increases in U.S. income per capita (gross national income, or GNI) to China's fast-growing economy can also show the extent to which China (and the rest of the world) is catching up. By plotting GNI versus time and extending the slope forward, one can even estimate a catch-up date, shown in Figure 3.11 to be in May 2023! But be careful— to achieve such a spectacular increase, China would have to continue growing as in the past 5 years—increasing annual income per person from $3,650 to almost $100,000—an unlikely if not impossible feat.[13] Nonetheless, if we could accurately model the decreasing increase in China's growth, we could come up with a more reasonable estimate, assuming everything stays the same.

So, can we predict the future from the past? That is the question one asks when showing data versus time. In some cases, the answer is yes, and in some cases no, but at least by plotting both the data *and* the change in the data, we can more easily see any changing trends and what might be the real cause if and when abrupt changes occur.

3.5. Turning Points: More Change and Illusive Growth

Formally, a derivative is the change of something with respect to something else, as we saw above in the stock market, Formula 1, and tortoise and the hare data. Speed is the change in distance versus time and acceleration is the change in speed versus time (first and second derivatives of distance), whereas inflation is the change in price (first derivative). We also saw that the spikiness is related to how much the data changes and that a slow change suggests a continuing trend and a fast change high volatility, as was especially seen in the stock market data after 2001, where the excessive spikiness in the index derivative gave a more explicit picture of the increased volatility. Although one can see that the index is changing, the effect is much clearer with the derivative.

Most of us know that the trajectory of a thrown ball follows a parabola (a quadratic function[14]), as seen when an outfielder throws out a runner at home plate or in canon fire on the high seas in an old war movie. The Dow Jones data looks similar, suggesting a similarly growing exponential relationship. But compared with the stock market data, one can see that the change in the Dow Jones has not been constant and has fluctuated dramatically in the past decade. The derivative data shows that the growth of the stock market has not continued unchecked.

Furthermore, the change was seen prior to the presumed global downturn of 2008—much sooner. John Casti (2009), whose research includes large-scale microsimulations of stock markets and road-traffic networks, stated that the financial market turnaround began in 2000, a result that can be readily verified by the derivative data. A snapshot of the Dow Jones data (see Figure 3.12) shows that the index leveled off well before the time generally assumed, which highlights the importance of the sampling period, where change appears less exaggerated over a shorter time.

The variation could be the result of noise (random fluctuations), and at first glance stock market data does include noise. For small time steps, the price variations (noise) follow a random walk, with varying step sizes corresponding to a power-law distribution (a so-called Levy flight, as noted by Benoît Mandelbrot). For large time steps, the noise is more Gaussian (think of the randomness of a series of coin flips).

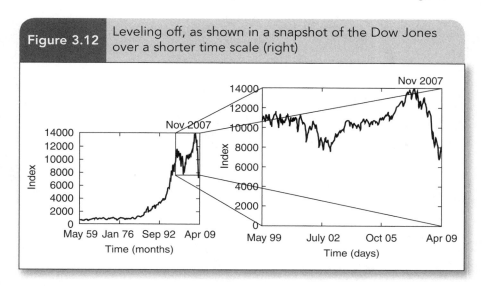

Figure 3.12 Leveling off, as shown in a snapshot of the Dow Jones over a shorter time scale (right)

For our purposes, however, any noise can be thought of as equally spread throughout the data (not depending on frequency). Such noise is called white noise (think of a hissing speaker) and can be subtracted by smoothing to reveal a truer story, for example, a quadratic fit.

Of course, there is no reason for the Dow Jones to follow a prescribed law of economics that dictates permanent constant growth, as today's investment brokers and politicians would have us believe. Indeed, the markets dropped from a peak of 1396 in November 2007 by almost 50% over 2 years before increasing again—hardly permanent growth. And what next, after the latest maelstrom abates—a single, double, or endlessly recurring dips?

To illustrate the difficulty of predicting economic trends, Figure 3.13 (top graph) shows three possible outcomes at point x, and the question is, does the measure (price, speed, index, winning percentage) go up (path 1), down (path 3), or stay the same (path 2)? Also shown in Figure 3.13 (bottom graph) is the seasonal change in temperature for New York City from 1999 to 2009 (using the same 10-year period as in the Dow Jones data above)—an oscillatory model, which we have already seen in electrical usage from daily work–home patterns. Here, one can clearly see the turning points as the sun makes its seasonal variation between the northern and southern tropes.[15]

Perhaps an oscillating model can better predict our economic future and better explain an ever-contracting, ever-expanding economy with

| Figure 3.13 | Where next? The stock market and New York temperatures |

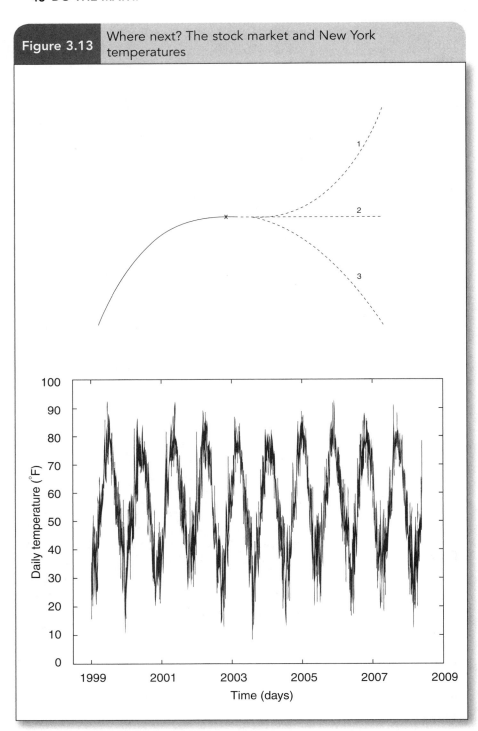

accompanying high-stress turnarounds—e.g., the Great Depression, Great Compression, Great Prosperity, and Great Expansion of the past 100 years, where the Great Compression decreased income inequality via an FDR-led New Deal after the Great Depression[16] and the Great Expansion increased income inequality via a Ronald Reagan-led dismantling of the New Deal after a prolonged post-World War II era of Great Prosperity. It is also believed that the universe is an oscillating system and that we are in the expanding phase, although the state of the universe may eventually turn around, contracting like a rubber band in a so-called big crunch.[17]

So, are we in for an up-and-down ride as our markets continue to fail—not just a so-called double-dip (or W-shaped[18]) recession, but an ongoing up-and-down or oscillating cycle of constant economic volatility? In reality, we have been experiencing permanently oscillating output since the beginning of growth-oriented economic strategies that fail when pushed beyond their limits. The stock market is now the bumpiest ride of all, like temperatures fluctuating between the uncomfortably hot and uncomfortably cold—as if following the same spiked strategies of the hare—where pensions, currency pricing, and national productivity have become more and more linked to the market.

Is this the cyclical model in which we are stuck—one of ongoing hiccups and permanent economic volatility, fully supported by the governments of today and composed of complicated, interconnected parts like a set of stacked dominoes, such that when one falls the whole system fails and no one is the wiser? One hopes not and that we can identify the causes in the most precarious of games. At the very least, by plotting the changes in the data as well as the data itself, we can better see the turns in the road as they approach—an essential part of any economic or political forecast—and not just comment on the changes after they have occurred.

Notes

1 **The Dow Jones**: The Dow Jones industrial average or index is a composite measure of 30 American companies, trading mostly on the New York Stock Exchange (traditionally industrial, transportation, and utility companies but now a variety of blue chip stocks). The original Dow Jones in 1896 was a true arithmetic mean of 12 stocks, but an index divisor is now used to account for stock splits (which change the price) and substitutions (e.g., in 2009, Travelers and Cisco replaced General Motors and Citigroup).

2 **0.1 mph**: 0.1 mph = 2 meters/minute × 60 minutes/hour × 1 km / 1,000 m = 0.12 kph / 1.6 miles/km = 0.075 miles per hour (~ 0.1 miles per hour).

3 **Fastest human**: The 200-meters world record is held by Usain Bolt of Jamaica at 19.19 s (622 meters/minute; as of August 2009). Bolt had previously won the 100-m (9.69 s)

and the 200-m (19.30 s) at the 2008 Beijing Olympics in world-record time, running one 200-meters faster than two 100-meters (19.19 < 2 × 9.69), highlighting the importance of the start on the time and his exceptional breakaway acceleration (i.e., continuously increasing speed).

4 **Financial derivatives**: In the fast-moving world of high finance, a derivative refers to a "derived" value but can also be thought of as the change in the price over time, used by investors to mitigate the risk of excessive change in their so-called forwards, futures, options, and swaps.

5 **Monaco Formula 1 sector velocity**: The data is recorded at each sector point, and, thus, the distances are unequal.

6 **Differentiation**: Differentiation calculates the rate of change of one variable with respect to another and is represented by the slope of a function at a given point. For example, if I walk 100 meters in 50 seconds at a constant pace, my rate of change of distance over time is 100 meters over 50 seconds or 2 m/s, which is the slope or derivative of the function that describes my walk ($x(t) = 2t$; x is distance and t is time). If I stop to tie my shoe or am blown forward by a gust of wind during my walk but still make the 100 meters in 50 seconds, my average slope is still 2 m/s, but my instantaneous slope will change accordingly as I stop and start.

7 **Acceleration from velocity**: For the tortoise, 1/30 m/s in 1 second = 1/30 m/s^2. For the hare, 10 m/s in 10 seconds = 1 m/s^2. In both cases, we assume they reached their top speeds uniformly (in mathematical terms, *linearly*). Note that physicists use /s^2 to mean per second per second, indicating the change in the change versus time.

8 **Force and acceleration**: The tortoise's acceleration was quite inconsequential (in the same way that we feel very little force when we start and stop walking), as was the hare's (about 1/10 the acceleration due to gravity of 9.8 m/s^2) compared to the force that keeps us on earth, although he likely reached his top speed even quicker—say, 10 m/s in about 5 seconds or 2 m/s^2. If he exploded out of the blocks to reach top speed in 1 second (10 m/s^2), he would feel a force of about 1 g.

9 **The derivative of a function**: A derivative of a function at any point is the instantaneous measure of the slope as Δx becomes infinitesimally small, i.e., dy / dx = lim ($\Delta x \to 0$) Δy / Δx.

10 **Changing data and the sampling period**: In the Rangers–Islanders example, different sample periods will show different results, but an analysis over different periods showed that the Islanders were the better team from the second half on. The question was whether they could catch the Rangers before running out of time. As it turned out, the Rangers were standing still by the end of the season and the Islanders blew past them.

11 **Calculated slopes**: A: (85 − 0 cm) / (10 − 0 s) = 8.5 cm/s. B: (55 − 20 cm) / (10 − 0 s) = 3.5 cm/s.

12 **Ramp slope and height answers**: height: 1, 2 = 3; slope: 1 = 2, 3.

13 **Asian dominance looks closer than it is**: Rosling (2009) used a similar method to calculate that the Indian economy would catch up to the United States and Europe by 2048. In his graph of income versus time, however, the income scale is logarithmic ($400, $4,000, and $40,000), and, thus, the intersection between the Asian and Western economies looks much closer than it is. The same is true for my U.S. and China GNI versus time graph.

14 **Parabola**: A thrown baseball follows a parabola or quadratic, $y = x^2$, whose derivatives are a sloped line ($y = 2x$) and a zero-slope line ($y =$ constant), i.e., all objects fall at a constant acceleration (9.8 m/s^2).

15 **Trope**: In Latin, *tropus* means "turn," which gives us the word *tropics*, the only region on earth where the sun can be directly overhead. The Tropic of Cancer is at 23.5° latitude (the sun directly above on June 21 at the northern hemisphere summer solstice), and the Tropic of Cancer is at –23.5° latitude (the sun directly above on December 21 at the northern hemisphere winter solstice).

16 **The Great Compression**: Economic historians Claudia Goldin and Robert Margo coined the phrase the "Great Compression," a narrowing of income inequality in the United States between the '20s and '50s (Krugman, 2009, pp. 38–39).

17 **The big crunch**: There has been much debate about the origin and evolution of the universe, such as whether it is infinite, closed, or oscillating, as well as about its composition (95% is still unexplained as dark energy or dark matter). We do know that the universe began with a "big bang," although not an explosion from a single location as is commonly thought but rather an expansion throughout all of space—as seen in galactic redshifts and 3K blackbody cosmic background radiation, leftover from a much earlier, much hotter state when photons and matter decoupled. Until recently, most cosmological models (based on whether the cosmic energy density was greater than a critical density such that gravity would stop the expansion) suggested a decreasing expansion or "big crunch" and, thus, an oscillatory universe (as Einstein had conjectured); however, recent observations from Type Ia supernovae (resulting from the explosion of white dwarf stars) suggest a constant expansion and, thus, a flat, infinite universe and the existence of repulsive dark energy. The debate continues. (See also "The heat death of the universe" in Chapter 2.)

18 **Recession shapes**: As their letter shapes imply, a V-shaped economy is a one-time reduction in growth, a U-shaped recession is a prolonged one-time reduction, and a W-shaped recession is a double dip.

FAIRNESS AND UNCERTAINTY

Who Watches the Watchers?

A measurement or decision cannot be precisely known, and, thus, uncertainty or "error" must always be taken into account. Here, error is not bad but a way of quantifying the unknown—what we don't know as well as what we can't know—and is a measure of our fallibility. Error is essential in measurement because of our incomplete knowledge or limited means. For example, in statistics, I simply can't measure the height of every man, woman, and child in a population, and so I must choose a sample statistic representative of the unknown population within some known error. As we will see, error is a fact of life.

In essence, error describes the fairness of our preferences, as seen in the board game Monopoly, phone-in television shows, and the competitiveness of North American versus European sports leagues, a few of the examples we look at in this chapter. Error also plays its part in the drama of the courtroom, where one must decide to err either on the side of compassion (in favor of the accused) or on that of security (against the accused). Both choices indicate a preferred way of living or bias that quantifies our uncertainty.

The same either/or dilemma applies to a number of seemingly disparate ideas, as we will see, such as the basics of statistical testing, the catastrophic decisions not to ground suspect planes or recall exploding cars, as well as Ralph Nader winning the 2000 U.S. presidential election for George Bush. All are subject to the error or bias in our decision making, which puts the mathematics of the unknown squarely in the middle of the debate about how we behave.

4.1. Random or Systematic Error: *Economically Advantaged American Redial Idol*

In a measurement, a scale is only as good as its finest division, excluding a little visual interpolation, thus making it impossible to measure something exactly. And, thus, the height of a person, width of a piece of wood, or weight of a block of cheese, for example, is measured as something *plus or minus* its error—say, 1.23 pounds ± 0.005 pounds, but not 1.233 pounds exactly (see Figure 4.1).

It's a bit like irrational numbers never ending or the seemingly endless divisibility of matter, yet 3.14 suitably represents π in most engineering calculations (the error decreases as more decimals are added), just as protons, electrons, and neutrons sufficiently model an atom in most atomic theories (without the need for quantum mechanics, particle theory, or string theory).

Exact measurements don't exist, and, thus, all measurements have an associated plus/minus, just as in the uncertainty of a poll, where the square root of the number of people polled is used as an estimate of the uncertainty. Obviously, the spread depends on the number tested, with a larger sample providing more certainty. As such, the size of a sample is the origin of the ± error tagged on the end of a population estimate.

Errors that fluctuate—sometimes plus, sometimes minus—are called *random* errors and should, if they are random, occur with more or less equal frequency. Perhaps a grocer reads the scale at a different angle each time, the flow rate in a gas pump changes as the temperature fluctuates up and down, or a clock sometimes runs fast, sometimes slow. The direction of a random error cannot be predicted. If you like, a random error is a fair error.

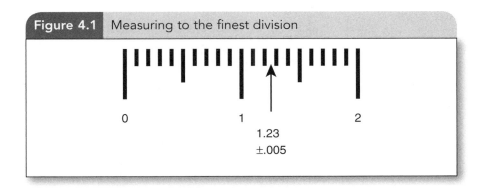

Figure 4.1 Measuring to the finest division

0   1   2

1.23
±.005

However, not all errors are fair. What if the scale is not zeroed properly and reads 0.1 pounds with no weight on it? What if the sun heats the gas pump and increases the flow rate or one sets the clock fast in hopes of being less tardy? The error is then *systematic* or unfair—that is, always in the same direction. There may be no intended bias, as in a faulty factory scale, and could just as easily go against the grocer—as in a −0.1 zero reading—but is nonetheless always in the same direction. Or there may well be an intended bias, such as a grocer who purposely puts his thumb on the scale, jacking up the price the same amount for every purchase— say, 15 cents. A pound of cheese at $5/pound would thus cost $5.15, whereas 2 pounds would cost $10.15, the grocer collecting 15 cents extra for every weighing, large or small.

Such errors in the weights of cheese over a significant number of purchases—if they are, in fact, random—should follow the so-called "normal" distribution (aka the Gaussian or bell curve), and any deviation within known tolerances can be taken as an indication of a purposeful or systematic error.[1] The same applies to cheese weights, population heights, and even the chest widths of Scottish soldiers, as Flemish mathematician Adolphe Quételet noticed when he was looking through a past issue of the *Edinburgh Medical and Surgical Journal* (Mlodinow, 2008, p. 155).

In fact, Quételet used such a method to show that would-be French soldiers were lying about their height to get out of serving in Napoleon's army, finding 2,200 extra "short men" below the cut-off height of 5 feet, 2 inches (Mlodinow, 2008, p. 155). The same analysis uncovered fraud in French bread weights (discovered by another French mathematician, Henri Poincairé), backdated stock options matched to price dips to maximize value at maturity, and American college basketball point-spread shaving scams, where a Wharton School economist found evidence of fraud in up to 70,000 games (Mlodinow, 2008, pp. 156–157).

In the board game Monopoly, the ups and downs of Chance and Community Chest cards are also seemingly random, sometimes for and sometimes against the player. Yet a simple analysis shows a bias that is *not* random, where 67% of Chance cards are *for* the player and 33% *against*. The results of Community Chest cards benefit the player even more, with 76% in favor—rising to 87% when weighted by their associated dollar values (see Table 4.1). Indeed, in Monopoly, a marked bias *favors* the player,[2] unlike in real banks, where errors mostly go against the customer. That real bank errors so often go against the customer suggests a *systematic* and not a *random* bias.

Tests and polls also provide good examples of unfairness and are rampant with systematic error. There is little point in asking the same IQ

Table 4.1	Chance and Community Chest plus/minus						
	Chance				Community Chest			
	Plus	Minus	Redirect	Repairs	Plus	Minus	Redirect	Repairs
#	4	2	9	1	10	3	2	1
Value	$103.75	–$50			$650	–$100		

question in New York and California if the question is culturally biased, because what is understood as commonplace in one area or demographic may not be in another.

A famous culturally biased test is the Dove Counterbalance General Intelligence Test, or so-called Chitling Test, created by sociologist Adrian Dove (1968) to show that exposure to African American culture would score a student more points, as in the sample questions in Figure 4.2. How different life in the United States would be if the Chitling Test were used as the measure for entrance, say, to Harvard University.

To be sure, bias crops up in any test or survey. Obviously, if one were to ask 100 farmers who they would vote for in an American presidential election, the result would be of little surprise—and the same for asking delegates their preference outside a Democratic convention. The validity of a statistic is only as valid as its sample, especially with smaller sample sizes.

It would be no surprise to find that fewer voters in Orange County than in Brooklyn preferred Barack Obama to John McCain, because of the right–left divide, or preferred John Kerry to George Bush.[3] Or who do you think was voted player of the game in a baseball game between the

Figure 4.2 The Chitling Test

1. A "handkerchief head" is:

   a) a cool cat, b) a porter, c) an Uncle Tom, d) a hoddi, e) a preacher.

7. "Hully Gully" came from:

   a) East Oakland, b) Fillmore, c) Watts, d) Harlem, e) Motor City.

9. Cheap chitlings (not the kind you purchase at a frozen food counter) will taste rubbery unless they are cooked long enough. How soon can you quit cooking them to eat and enjoy them?

   a) 45 minutes, b) 2 hours, c) 24 hours, d) 1 week (on a low flame), e) 1 hour.

SOURCE: Dove, A. (1968, July 15). Taking the Chitling Test. Newsweek.

Baltimore Orioles and the visiting Toronto Blue Jays, won 10–1 by Toronto: Blue Jays shortstop Yunel Escobar, who went three for four and hit a grand slam; Blue Jays starting pitcher Shawn Marcum, who pitched five innings and gave up one run; or Oriole right fielder Nick Markakis, who had two doubles and a single in five at bats? Yep—the hometown favorite Markakis, with 8 times as many votes as the other two.

Another example shows the bias in a Connecticut firefighters' exam, the results of which were overturned in the midst of Supreme Court appointee Sonia Sotomayor's confirmation proceedings. Sotomayor had allowed the nullification of the exam by the city of New Haven, presumably on the grounds that the test was biased, since no African Americans scored high enough to warrant promotion. Eighteen white firefighters, however, who had scored high enough but were not promoted, filed suit, believing they had been discriminated against because of the nullification. Writing in the 5-to-4 ruling, Justice Anthony Kennedy stated, "The city rejected the test results solely because the higher scoring candidates were white" (Liptak, 2009). Here, the bias seemed real given the sample size (18), a result unlikely to have occurred by chance.

Getting the sample right is essential. Football audiences are mostly male, so if polled as to their preferences for blue or pink, the answer would be obvious, as would the likelihood that viewers of a religious channel believe in God. Television channels also have a bias, presumably reflected in their viewers, and given that a television poll is a poll of *television* viewers, that sample is also inherently biased.

The whole subject of statistics is predicated on getting the sample right and ensuring that a sample *statistic* is representative of the larger *population*, a fact wholly ignored in one famous telephone poll that went horribly wrong. The poll, conducted in 1936, predicted a landslide presidential win for the Republican candidate, Alf Landon. The problem was that two thirds of Americans, most of whom were poor, didn't have telephones in 1936, and ended up voting for the Democratic candidate, Franklin Roosevelt (Huff, 1954/1993, p. 20). In fact, Roosevelt won the presidency in the most lopsided Electoral College result ever with 98.5% of the vote!

Huff (1954/1993) succinctly addressed the problems of selection bias and choosing the right sample:

> If your sample is large enough and selected properly, it will represent the whole well enough for most purposes. If it is not, it may be far less accurate than an intelligent guess and have nothing to recommend it but a spurious air of scientific precision. It is sad truth that conclusions from such samples, biased or too small or both, lie behind much of what we read or think we know. (p. 13)

A representative, large-enough sample size is essential, such that "2 out of 3 dentists" recommending the latest tartar-fighting, plaque-removing toothpaste is meaningless if the sample size is unknown. It could be 3 dentists, it could be 300. And not only is the sample size indeterminate, but which 2 out of which 3 dentists? Dentists paid to do the test? How many of those 2 out of 3 dentists would recommend another competing product, which they may recommend more favorably? And why did 1/3 of all dentists polled not recommend the advertised toothpaste? Is there something wrong with a toothpaste that 1/3 of dentists did not recommend? Getting the sample size right is not easy and depends on a number of factors—the margin of error (say, 3%), the confidence level (within a probable spread of data, say, 95%), and the sampling method, as well as careful planning.[4]

Huff (1954/1993) also used the average income of Yale graduates to show how surveys can mislead because they reflect intentions and not reality—for example, what Yale graduates said they earned (p. 12)—but don't include valid results from those who have not done so well, such as "clerks, mechanics, tramps, unemployed alcoholics, barely surviving writers and artists" (p. 15). Casti (2009) believes people will say anything, noting that "what they do in the ballot box may be very different than a survey."

Poll questions can also be misleading, depending on how they are posed. For example, in a Ross Perot poll, 80% polled said yes to the question, "Should laws be passed to eliminate all possibilities of special interests giving huge sums of money to candidates?" But when the question was asked slightly differently, "Should laws be passed to prohibit interest groups from contributing to campaigns, or do groups have a right to contribute to the candidate they support?" only 40% said yes (Paulos, 1996, p. 15). In another CBS/New York Times survey about permitting gays to serve openly in the military, the numbers showed a marked decrease when "homosexual" was used instead of "gay men and lesbians" (from 70% to 59% in a 1,084-person survey; Sussman, 2010).

TV polls can also show a marked bias, especially if a charge is applied, which then reflects the views of those who would call a phone-in number and can afford to do so. On American Idol, X Factor, and any number of network cash-grab talent shows, it is patently unfair to charge a fee to vote. Publishing intermediate results can also influence a decision. What's more, how many so-called votes are from different numbers? If repeats are allowed, the result becomes a measure not of what the public thinks but of the likes and dislikes of the more affluent and their ability to press redial as well as those with better access to voting procedures.

Polling is not easy, where in essence the real question being asked is, what is the public and to whom do the words *audience* and *community* apply? In many cases where incorrect polling methods are used, the answer is in the poll itself. One vote, one person does not always apply. In all honesty, *American Idol* should be called *Economically Advantaged American Redial Idol*.

What's more, the numbers are no small peanuts. More than 63 million votes were cast in one *American Idol* season finale, which host Ryan Seacrest noted was more "than any president in the history of our country has ever received" (Stanley, 2006). In Britain, 4 times as many people under 25 voted in the first *Pop Idol* final as in the previous general election, with profits from telephone charges totaling £11 million (Sherwin, 2002). Its successor, *X Factor,* and *Britain's Got Talent* are the most popular shows on British television, with average viewing numbers of more than 11 million per week, generating advertising and other revenues in excess of £6 million per show (Sabbagh, 2009). The advertising revenues on *American Idol* are estimated at $15 million per hour (Lewis, 2009).

But should we be concerned that the results for such shows are greatly biased, a practice encouraged by the networks, which make millions from the "voting" public? For that matter, should we be concerned that more people want to vote for best singer or dancer than for best political leader?

Interestingly, in an attempt to stop the 2009 *X Factor* winner from also winning the Christmas No. 1 spot in the British music charts, a Facebook campaign was launched to "vote in" Rage Against the Machine's *Killing in the Name*. Decried by *X Factor's* pop mentor Simon Cowell as cynical, the Rage Against the Machine antivote campaign is nonetheless an excellent example of using influence and persuasion to manipulate a result—showing its own bias, in this case, against engineered consumerism.

As for who is right, one could easily argue that Cowell's machinery is the more cynical, timed perfectly to take advantage of the Christmas buying season and free publicity, and in so doing defining its own standard. Regardless, both the Cowell and anti-Cowell campaigns are motivated by their own biases.

It is also interesting that a poll of *American Idol* viewers showed that 35% thought their votes counted as much as in an election, revealing another bias, toward pop culture—to which the ever-self-effacing impresario Cowell wryly suggested, "What if, and I'll throw this challenge out . . . , we do *President Idol*. A couple of songs, a bit of dancing, and I'll judge it. Maybe that's how we should decide the next president" (Ayres, 2006). Cowell's suggestion might save time and money but, nevertheless, still includes its own inherent bias.

One need only look at the simplest of examples to see how important bias is: a typical presidential election in a high school with a more-or-less equal distribution of boys and girls. Who do you think would win between three equally able candidates, two of which are boys and one a girl? As Randy Jackson might say, "Is it me? Check it out, dude."

4.2. Bias and Unfairness: From Politics and Maps to Search Engines and Sports

A better mathematical understanding of error reveals the biases of media outlets and how they reflect their own agendas, typically those of their owners. Fox News doesn't present news in the same way as PBS, or the BBC in the same way as ITV. Political apologists such as Glenn Beck and Jon Stewart do not speak for all Americans but, rather, for their franchise, their fraternity. Paul Krugman writes from a different perspective for *The New York Times* than does George Will for *The Washington Post*. As standard-bearers for their political sides of the spectrum, they have an agenda, a vested interest, a bias. Indeed, who among us does not?

In politics, bias is built right into the system because of income and can even direct government policy. As Krugman (2009) noted, higher earners pay more attention to politics and, thus, are more likely to vote: "As a result, the typical *voter* has a substantially higher income than the typical *person*, which is one reason politicians tend to design their policies with the relatively affluent in mind" (p. 70). Galbraith (1958/1999) noted that "present laws are notably favorable to the person who has wealth as opposed to the individual who is only earning it" (p. 70). Even bad weather on election days favors the more well-off, who can more easily ferry themselves to and from the polling station.

Maps show a bias because of the problem of projecting the surface of the earth onto a rectangular grid, which in today's typical classroom map artificially weights the mid-northern hemisphere regions of Europe, Asia, and the United States compared with those nearer the equator. Such maps are based on the original 16th-century Mercator projection and show Europe as larger than South America, when it is actually about half the size. Alaska is shown as bigger than Australia, Greenland looks bigger than Africa, and India appears as a blip. In many cases, the midpoint is also misrepresented toward the middle of Europe, where Mercator was from.

To more realistically represent countries and continents, the Gall-Peters map projects spherical dimensions onto a grid using equal areas (see Figure 4.3). Although the appearance of the Gall-Peters map[5] is jarring at

| Figure 4.3 | Mercator projection (left) and Gall-Peters projection (right) |

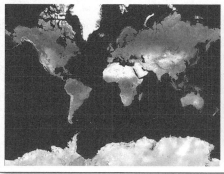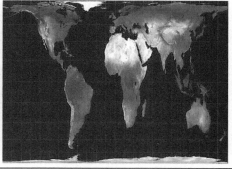

SOURCES: NASA Earth Observatory/MODIS Science Team; NASA

first, such a projection more realistically weights land masses than does the Mercator projection and can help us understand not only the geography of our world but the increased political interconnectedness of globalization (e.g., why Africa's resources are so important). Of course, all 2-D maps misrepresent the 3-D world, emphasizing what is important to the mapmaker—whether distance for ease of navigation (Mercator), area for size of land mass (Gall-Peters), or even demographic fidelity (area cartogram), which weights political divisions by population (Monmonier, 1996, pp. 17–18, 99).[6]

Bias appears even in search engine results. Google (or Yahoo or Bing) calculates its page rankings based on the number of links from other pages, and, thus, a standalone page that has no links is ranked at the bottom, whereas a page with numerous links—such as Wikipedia, with its thousands of internal and external links—is ranked at the top.[7] Of course, Wikipedia is designed to score top billing in Internet searches, but does that mean it should or that a particular top-billed Wikipedia page is the most informative, the most authoritative, or the most appropriate to one's interest? A top-ranked page may be all those things, but according to the search engine ranking criteria, all we know is that it is the most linked to by other pages, which may represent the so-called Internet community's collective authority—but then again, it may not. At the very least, internally linked-to pages should be excluded from any ranking criteria.[8] What's more, it is not easy to know if a ranking has been fairly established by the programmed spider machine and its unknown criteria.

So what can be done to counter bias? To begin with, all error must be quantified, i.e., in which direction and by how much. We must know how level (or unlevel) a playing field is in any game or system if we hope to understand the underlying reality. Furthermore, any numbers quoted should at least come with a full statistical provenance—where, when, who, and how many. Barring that, the number surveyed should be included, the average and standard deviation stated, and due effort afforded to ensure that a sample is representative, with no repeats (more on that in Chapter 5). No one should be permitted to report "2 out of 3" anything without stating the total number in the sample or suitable references.

The more important questions, however, are whether equality is missing from society in general and whether a representative spread of views is being fairly voiced and heard. We might think we live in a fair world— unbiased and error-free—but in reality, fairness may be of little concern. At the very least, we should be aware of the presence of bias and attempt to examine its influence.

The Truth About Sports Leagues: Owner, Not Player Competitiveness

The New York Yankees win more often because of bias based on revenue, translated to higher player salaries (the average starting player earns more than $135,000 per week, tops in the sporting-team world; Harris, 2010). They win more because they have more money, which attracts better players year in and year out, whereas the poorer, small-market teams have less chance to be competitive. Come October, temperatures cool, the leaves fall, and more often than not, the Yankees are fighting it out with whichever team some other owner forked out the most money for that year.

In European soccer as well, the transfer market is ruled by those with the deepest pockets, providing a systematic bias in the league based on money. Michel Platini, president of the Union of European Football Associations (UEFA), openly lamented the spiraling imbalance in the European soccer league transfer market after Christiano Ronaldo's £80 million, world-record transfer from Manchester United to Real Madrid in 2009. Platini stated that the fee was a "serious challenge to the idea of fair play," adding that he wanted to "clean up the system" (Ducker, 2009), presumably by restoring some sense of equality to the pay structure.

It is interesting that sports leagues in the United States and Canada are more fair-minded than those in Europe (Major League Baseball and the New York Yankees notwithstanding), with salary caps and reverse-order player drafts the norm for North American leagues. Because of parity-centered practices in the National Football League (NFL), small-market teams such as Green Bay and Baltimore have as much chance of winning the Super Bowl as do the marquee teams in New York and Los Angeles. In the NFL, the playing field is releveled each year and helps promote a true national sporting interest that supports fair-mindedness, thus contributing to its broad-based popularity. If you like, the NFL chooses to err on the side of fairness, based on *team* competiveness rather than *owner* competitiveness.

In English football, however, the competiveness of any team is directly related to the depth of the owner's pockets, and no team outside the big four (Manchester United, Liverpool, Arsenal, Chelsea) has much chance of winning. In the past 30 years, only four teams other than those in the big four have won the league title, whereas the big four have won a total of 26 times.

The main problem in English football stems from the size of the owners' balance sheets and the increasing debt loads to finance players' salaries—almost all lose money (Veseth, 2005, p. 100). Manchester United owes £720 million (Marlowe & Rushe, 2010), whereas billionaire Chelsea owner Roman Abramovich routinely clears his team's annual losses, amounting to £44.4 million in 2009 (Hughes, 2009). What's more, the entire Premier League is almost £2 billion in debt (more than half of all European football league debt), costing £150 million in annual interest charges alone (Kay, 2010). Platini has, in fact, tried to ban indebted Premier League teams from the UEFA Champions League because they have an unfair financial advantage, although Abramovich has found a way around this by converting a £340 million loan to equity (Hughes, 2009).

Interestingly, television revenues in the English Premier League, German Bundesliga, and the French Ligue 1 are shared, which helps the lower-ranked teams, although insufficiently without further funds (which is one reason they are lower ranked). In Spain, however, teams can negotiate their own television deals, prompting higher-ranked teams such as Real Madrid and Barcelona to try to form their own "elite" league in the hopes of divvying up the lucrative revenues among themselves.

When money is at the core of a league's success, the question is not one of sport but of how elite (or small) the league is. As Veseth (2005) noted, "The fundamental problem, questionable accounting aside, is that the super clubs form a winner-take-all market that benefits the top clubs disproportionately" (p. 100), something that also applies to economic markets, as we'll see later. In Scotland, the competition is so skewed that only Glasgow Celtic and Glasgow Rangers ever win (no others since the league's most recent remake in 1998 and rarely before that). Highlighting the inequality, only once in the Scottish Premier League has another team finished first or second, which, to be fair, can hardly be called a league.

Fans in the United States should take note, however, as money begins to dominate league competition unduly (which we'll calculate statistically using correlative regression in Chapter 6). Whenever such bias rules, the game is not played for the enjoyment of fans but rather for that of the team owners.

4.3. Statistical Error: Who Makes the Choice?

In a quantum mechanical world, error is built into the system, because of the uncertainty of measuring particles with less energy than the light used to detect them. To compare protons, electrons, and other subatomic particles to our classical world, a simple example suffices: If you turn on a light in a room, the chairs and tables don't bounce around, yet that's what happens in the quantum world—when you observe atoms, the energy of the light is greater than the energy of the atoms, and they whiz off in unknown directions. Just by observing the system, one changes it—a major problem of measurement at the atomic level, as stated in the Heisenberg Uncertainty Principle, where the position *and* the velocity of a particle cannot be known at the same time.[9] Indeed, modern physics reached a crossroad when the results showed that the experimenter was part of the experiment and could observe only a system that was being observed and not the system itself—a seemingly paradoxical dilemma about the notion of uncertainty in the material world. We just can't know.

In our world—the everyday slow-speed and large-mass world of tortoises and hares—uncertainty also prevails—paradoxically, it seems, since we are being observed all the more now, causing a growing concern for privacy issues with mobile phone tracking, Internet clicking, Google mapping, and the ubiquitous eye-in-the-sky security camera, or e-police. More police officers and cameras on the street may make it harder for the artful dodgers of the world to pick a pocket or two, but a perfectly policed e-world doesn't guarantee a perfect public.

The issue, again, is one of error, where even the most irrefutable evidence can be wrong, such as a grainy photo, a smudged fingerprint, tainted or planted DNA, or human parallax that gives us simple differences in the social construction of the "facts." Even in an increasingly more policed society, we can't have infallible witnesses with 20/20 e-eyes to e-police our e-world, and, thus, we have only our imperfect selves and the fallibility of our own devices. Besides, who would watch the watchers, or police the e-police, if we could construct a perfect policeman?

A trial provides the perfect example of imperfection and uncertainty in our decisions. In any trial, the evidence is presented, testimony is given under oath, and the lawyers make their summations. As in any episode of *Law and Order*, however, the verdict is uncertain even as it is about to be read out—as we the television viewers guess which way the judge or jury has decided, no doubt an indication of our own views or interpretation of the facts. It is hoped that the verdict is correct, but, to be sure, it can be wrong. Given that a judgment can be wrong, the million-dollar question is, which error would you rather make—hanging an innocent man or free-ing a guilty man?

The same logic can be applied to many situations—for example, whether to close the Thames River barrier to protect against flooding from North Sea tidal surges. Is it better to close the barrier when not needed than not to close it when needed, knowing that closing the bar-rier costs money (~£5,000); increases the wear and tear on the ten, 3,500-ton steel gates; and prevents shipping (Llewellyn, 2008)? The decision to close or not close (it is closed about 10 times a year) comes down to how we prioritize our choices and which possible error is worse or deemed worse. The same is true for the Mississippi River: Do we purposely break a levy upriver to allow flooding, with resultant damage to farmland and livelihoods, or allow the river to swell and possibly burst its banks else-where at much greater cost?

Many arguments can be couched in the same way: Do we lock our doors and live securely but less freely? Do we permit gun ownership, which may protect property and restrict government control but increases gun deaths? Do we reduce safety regulations in the oil industry and risk spills or apply more stringent measures and reduce exploration? Do we ground planes after volcanic eruptions to protect jet engines or fly unre-stricted but risk passenger safety? These questions are the same as that in the murder trial: If one is on trial under penalty of death, should we err on the side of caution and hang him though innocent or on the side of compassion and free him though guilty?

In most cases, the verdict is not much in doubt. Usually, the judge or jury can determine guilt or innocence from the evidence and testimony and make a just pronouncement. In most cases, no error is made. But in those cases where there is doubt, the onus of "beyond reasonable doubt" takes up the slack to make it harder to hang the innocent. In our society, we acknowledge that mistakes can be (and are) made, and so we choose to err on the side of compassion and put the burden of proof on the state. It is a cornerstone of our society to err on the side of compassion, i.e., the presumption of innocence until proven guilty.

Of course we hope the guilty are punished and the innocent freed, but the value of a civilized society is that it recognizes its own fallibility, and so we must lean in one direction to ensure against the other, worse error. As such, we choose to let the guilty man go free, for we deem it more serious to execute an innocent man than to let a killer off the hook, however worrisome that may be. This is not a question of justice but of fallibility, which errs on the side of compassion and demands that we free more O. J. Simpsons rather than incarcerate more Hurricane Carters. No matter how hard we try to judge correctly, we must choose which error is the lesser evil.

In past societies, we erred on the side of vigilance, for which we suffered the Star Chamber and the Inquisition—strong deterrents, but not very effective for the wrongly convicted. In Islamic law, too, harsh penalties act as a strong deterrent but are of little consolation to the wronged innocents. The same can be said about death row in the United States. In fact, Illinois Governor George Ryan put an end to executions in his state after 13 death row inmates were found to have been not guilty, a particularly horrific example of the either/or error.

Enemy combatants in the Guantanamo Bay detention facility, who are neither war prisoners protected by the Geneva Conventions nor criminals entitled to due process of law, have also been victimized in a rather nightmarish, Catch-22 version of either/or error. As suspected terrorists, they have been subjected to many inhumane practices before trial or even before any formal charges have been brought. As a result, President Obama issued an executive order at the start of his presidency to close the extrajudicial jail, citing the cornerstone rule of law that presumes innocence or at least demands that the burden of proof be on the state—although the order was not ultimately followed.[10]

Statistical testing can also be thought of as determining the innocence or guilt of a defendant, where the defendant on trial is a batch of product. Manufacturing processes are fraught with inconsistencies and unknown precision. Here, a *null hypothesis* is given as a statement one wishes to prove (or disprove)—for example, "This batch of widgets works." The test of the hypothesis has four outcomes, as in the criminal trial: acquit the innocent, convict the guilty (both correct outcomes), convict the innocent, or free the guilty (both errors). In this case, the good outcomes are to ship a good batch or reject a bad batch and the bad outcomes are to reject a good batch or ship a bad batch (see Table 4.2).

The two types of error in a statistical test are called Type I and Type II errors, or "rejecting the null hypothesis when it is true" and "accepting

Table 4.2	Four possible decision outcomes	
	Innocent	*Guilty*
Didn't do it	Acquit	Error I (reject good)
Did it	Error II (ship bad)	Convict

the null hypothesis when it is false"—similar to the medical equivalent of a "false positive" or "false negative." Let's say that for a batch of light bulbs, we will ship if the bulbs are good to 10%. A test of 20 is performed, and if 4 fail (2 in 10), the whole batch is rejected, saving the time of testing the entire lot. The partial statistic (20% failure ± 10%, or between 10–30%) is deemed representative of the entire batch. If only one bulb had failed, the batch would be accepted because the statistic would be 10% ± 10%, which includes 0% failure. That's not to say that there aren't errors in some bulbs, but to the best of our knowledge, the error has been minimized.

We could, of course, change the stringency of the criterion depending on which error was deemed worse. Depending on the product, it may be worse to ship a bad batch and have to pay for recalls than to reject a good batch and lose sales. Nonetheless, statistical testing recognizes that not only is nothing made perfectly but, when repeatedly manufactured by the millions, the reliability of a widget, light bulb, or assembly line product cannot be precisely known for the whole batch. As Boorstin (2000) noted about the advent of acceptable tolerances in early machine manufacturing in the United States, "The old adage about striving for 'perfection' had to be revised" (p. 196).

To be sure, no measurement—in product testing or in a trial—can be known with perfect certainty, whether we like it or not. What's more, our fallibility is paramount in the either/or decision-making process.

4.4. Cost–Benefit, Either/Or Error Decisions

Many cost-benefit decisions are based on the two types of error we saw above, often with disastrous results. In 1989, Boeing 747 United Airlines Flight 811 bound from Honolulu to Sydney experienced an explosive decompression, resulting in a cargo door opening midflight and the subsequent deaths of nine people. A company investigation blamed the ground crew, although considerable evidence pointed to a design flaw in the door. Nonetheless, a decision had to be made: Would it be worse to take the plane and others like it out of service to fix the flaw, costing millions of dollars in lost revenue per plane, or to let the planes continue flying, risking the occurrence of similar accidents? The likelihood of another accident may have been low but was certainly real given that one such accident had already occurred. Preferring to err on the side of business over safety, however, United chose not to ground any flights, and Boeing didn't issue any alerts (Higgins, 2003).

This is the classic error dilemma: better to fly with a suspected flaw than ground the plane and lose millions. That's the logic. The morality is choosing between the devil you know and the devil you don't. The devil you know is the high cost of downing hundreds of planes, which isn't good for the bottom line. The devil you don't know is whether more planes will crash. As stated by one industry analyst, United chose to "err on the side of economics rather than safety," and the reality about not fixing costly flaws is "one airplane every ten years, one airplane every five years, 200 or 300 people, the cost of doing business" (Higgins, 2003).

A similar choice was presented after an Airbus Air France flight from Brazil to Paris crashed in 2009, costing 228 lives. Faulty speed sensors were cited as the cause of the crash—the same problem discovered 13 years previously when two Boeing 757s crashed, resulting in hundreds of lives lost—yet no planes were grounded to make any changes (Swain, 2009). As the head of aviation for one law firm stated regarding flawed Boeing 777 Rolls Royce engines not being withdrawn for repairs,

> The regulatory authorities are reluctant to ground aircraft, and the harsh economic and political reality is that it will take a fatal 777 crash for that to happen. Just hope that you are not a passenger or crew on that flight. (Webster, 2009a)

It would seem that, in the airline industry, keeping the costs down is more important than safety, or at least the preference is to err on the side of saving money instead of lives. And, thus, the real question is, what does one value more—people or profits? Furthermore, how many lives are worth the money saved by not downing the planes, and can or should such a value be determined? Should we, as in the criminal trial, prefer to err in hanging the innocent or freeing the guilty?

Another example of choosing reduced costs over increased safety is the Deep Horizon accident and resultant oil spill in the Gulf of Mexico—the worst environmental catastrophe in American history—where crucial warning systems had been disabled (Goddard, 2010) and almost 400 maintenance tasks were more than a month behind schedule (Forston, 2010).

In fact, profit motives were at the core of numerous violations over the years, including BP "asking regulators to approve three successive changes to the well over the 24 hours before the disaster" (Frean, 2010b). The resulting billion-dollar mess will take decades to clean up.

The Japanese car manufacturer Toyota also chose high-speed growth and ambitious cost cutting over tried-and-true financial prudence in its lust for more market share, which ultimately led to quality lapses and well-publicized recalls. Pushing for 10% cheaper parts, vehicle quality suffered at the expense of customer safety (Ohnsman, Green, & Inoue, 2010). Although the auto market is highly competitive, pushing for too much can have dire consequences in the cost-benefit equation.

The either/or dilemma is also the basis for the infamous Ford Pinto case, where gas tanks exploded on rear impact. Here, Ford initially chose not to recall any cars, which they estimated would cost $11 each or $137 million in total to fix compared with projected settlements of $49.5 million.[11] In this case, however, there is no doubt on which side Ford erred, a choice that nonetheless backfired when a $128 million lawsuit was awarded against them, forcing a recall of 1.5 million suspect cars ("Ford Pinto," 2006).

In the case of United Airlines and Air France, they had no prior cost-benefit calculations from which to figure out the profit/loss value of their decisions not to ground planes, although they knew they were risking lives. The same was true with BP's operation of the Deep Horizon oil rig and others like it around the world, although they played hard and fast with regulations for the sake of the bottom line. With the Pintos, however, Ford knew, having calculated the cost to them of a human life before any crashes, as determined by their bean counters, compared with the cost of litigation following crashes. Using one estimate that 27 people died because of a Pinto rear-end-crash fire and the projected settlement costs, Ford's cost for a human life works out to $1.8 million. As noted by a Ford engineer, the sad reality is that "this company is run by salesmen, not engineers; so the priority is styling, not safety" ("Ford Pinto," 2006).

Styling versus safety, profits versus people, erring in one direction or the other—it all comes down to which side of the either/or equation you prefer.

Ralph Nader and How Cost-Benefit Analysis Elected President George W. Bush

One of the biggest crusaders for consumer causes is Ralph Nader, who championed automobile safety, better drinking water, and reliable consumer products, among others in his long career. You could call him the father of modern consumer advocacy. In his 1965 book *Unsafe at Any Speed*, he argued that the Corvair killed people because it was not fitted with seatbelts, and his advocacy helped change the law such that mandatory seatbelts were introduced in most states by 1985, a straight-forward example of how choosing to err in one direction (lower speed) affects another (more safety).

Interestingly, Ralph Nader is mostly remembered today as a candidate in the 2000 presidential election, one whose entry in the race changed the outcome, all because of an either/or preference. Indeed, Nader poached more Gore votes than Bush votes because his politics were closer to those of the Democrats (or less like those of the Republicans), regardless of whether they were, as he called them, "Tweedledum and Tweedledummer." Commendably, he weighed the effect of his running—that inner profit/loss reckoning of one's own actions—and decided it was better to try to get out the message about corporate greed and con-sumer rights than to do nothing and watch nothing change, which none-theless helped George Bush's numbers in the process. Nader's conscience spoke, and no one can argue with that.

As it turned out, Nader received 2.7% of the national vote, including almost 100,000 in the all-important state of Florida, with its 25 Electoral College votes, which ultimately siphoned off enough Electoral College votes from Gore to give Bush the election (won 271 to 266). Forget the hanging chads; it was down to Nader, where the Florida results clearly tipped the balance (Bush 2,912,790; Gore 2,912,253; Nader 97,488).

However, in another state, Nader's total votes were also more than Bush's tally less Gore's in a Bush-won state. Which state? The seemingly unimportant state of New Hampshire with its lowly four Electoral College votes, which nonetheless tipped the results in Bush's favor (Bush 273,559; Gore 266,348; Nader 22,198). In fact, had Nader's New Hampshire votes all gone to Gore, the Democrats would have won the White House 270 to 267 despite the Florida result.

Noting the effect on the outcome of the 2000 election, some Republicans even gave money to support Nader's campaign in 2004. And although his numbers were reduced to about 1% (about 400,000 votes nationally), his presence still favored Bush, no matter how the urban/rural vote split in Ohio or how the red/blue states segregated. In just the same way, in 1992, Ross Perot cost George Bush Senior his reelection to Bill Clinton by dividing the Republican vote. After the fact, one has to marvel at how some elections are won, all because of the statistics of error.

In fact, profit/loss strategies and the statistics of error are more common than we think, from our own everyday biased choices to the simplest of strategic thinking. For example, game shows can be played to take advantage of error—that is, if the contestant doesn't squirm too much when faced with the loss of hundreds of thousands of dollars on a single question. On *Who Wants to Be a Millionaire*, the standard advice is to reach $32,000 (if you can), have a go at the free $64,000 question, and stop there if you win. The prevailing wisdom is not to be beguiled into trying for $125,000, since you might lose $32,000. Here, however, the theory of profit/loss strategies helps: If the payout percentage is greater than the odds, you should go for it.

In this case, $125,000 / $32,000 (3.91 to 1) is slightly less than 4 to 1, and, thus, you will lose more often than you will win and shouldn't bet. However, if you know that one of the four answers is wrong, or better yet two—say, because of the 50-50 lifeline—then your payout percentage (3.91 to 1) becomes higher than the reduced 2-to-1 odds, and you should go for it. Easier said than done when you're under the gun and only get one go, but such a strategy does work in the long run—for example, not paying a parking meter when the odds are 5 to 1 of getting a ticket but the fine is less than 5 times the cost, feeling confident of success in committing a crime when security is lax (or unworried if the punishment is not sufficient), or comparing the cost of litigation versus possible settlement claims, as we saw with the Ford Pinto.

Another excellent example of a cost-benefit analysis occurred in a 2010 World Cup Football quarter-final knockout game between Uruguay and Ghana, the last remaining African team in the tournament and sentimental favorite. In the dying seconds of the game, a Uruguayan player used his hand to stop a sure goal, which would have put Ghana through to the semifinals had it gone in. Although a penalty kick was awarded (and missed) and the Uruguayan player was red-carded for the foul—ejecting him from the rest of the game (meaningless given the time) as well as the next game—the punishment did not outweigh the benefit of the action (especially after Uruguay went on to win on penalties). As such, the Uruguayan player made his decision based on the cost and benefit to him and his team, saving a sure loss. And who can blame him? But had the rules stipulated a harsher penalty (e.g., a long suspension and large fine), he would no doubt have acted differently. According to a more stringent cost-benefit scenario, he might have tried instead to head the ball off the line, a more sportsmanlike action and one that would have led to a fairer and less controversial outcome.

The same is true in any cost-benefit situation, where greater punishments will reduce wrongdoing, resulting in fewer plane crashes, rear-end deaths, or oil-rig fires. Profit/loss or cost-benefit analysis may well be how anyone (aka agent, automaton, boid, or peoploid in game theory) plans his or her actions, whether paying pot stakes in poker when the odds are favorable or making a financial deal over time—even in the evolution of a species. To the favorable profit/loss go the spoils, in the long run anyway.

In some instances, however, a better strategy can be counterintuitive, as explained by James Heckman, University of Chicago economist and 2000 Nobel laureate in economic sciences. According to Heckman (2006a), it is better to spend money on disadvantaged students at the *earliest* age possible, because a later, albeit well-intentioned intervention (e.g., lower tuitions or job training programs) is much less effective. He further noted that "skills beget skills" and "motivation begets motivation," with results showing a tenfold return for every dollar invested on the very young, from newborn to 3 years old. But perhaps that's obvious, as in the adage "an ounce of prevention is worth a pound of cure," although as Heckman further noted, "later interventions are less effective at current levels of expenditure and this is especially true for less able and less motivated children."

Despite the uncertainty in our choices and the direction of our preferences, good strategy is a gift that keeps on giving. Bad strategy, on the other hand, can depend entirely on which side of the either/or equation we choose.

4.5. Haggling About Uncertainty: The Real Price of Goods

Everyone has a price, or so they say. In the film *Indecent Proposal*, the price was $1 million for Demi Moore to spend a night with Robert Redford, a man who was not her husband. This modern morality tale is not unlike that attributed to George Bernard Shaw. As the story goes, Shaw was bored at a dinner party and asked a woman if she would sleep with him for £1,000,000, to which she replied, "What woman would not?" Shaw then asked if she would sleep with him for sixpence, to which she replied, "Who do you think I am?" Shaw's oft-quoted answer was, "We've already established that. We're now haggling about the price." Is Shaw's story an indecent proposal, or is one man's indecency just another woman's price?

Asking the price is something we do every day. Would you work at your job for half what you are currently paid? How much is it worth to clean toilets or collect garbage? What is the real price of a basket of goods? Supply and demand are generally believed to dictate the worth of a job, the price of a food basket, or the amount of one's honor, which translates into a spectrum of prices, low- to high-paying jobs, and low- to high-end stores. If sixpence is too low and one million pounds too high, as in Shaw's poignant story, somewhere in between lies an agreed-on value for exchanged goods or services.

Using the commonly taught economic logic, price determination is based on Alfred Marshall's ideas of supply and demand and marginal utility (aka the law of diminishing returns), where a producer adjusts supply to meet consumer demand and a natural price appears, from jobs to food to honors (which also implies that in a perfect marketplace, profits can't exist because price is set to maximize demand). Assumed in the logic is Adam Smith's invisible hand,[12] where everyone—whether supplier or consumer—seeks his or her own natural good, and somehow everything works out. As Smith wrote in his landmark book *An Inquiry Into the Nature and Causes of the Wealth of Nations* (1776), which assumes a free-market economy:

> Every individual necessarily labours to render the annual revenue of the society as great as he can. He generally neither intends to promote the public interest, nor knows how much he is promoting it. . . . He intends only his own gain, and he is in this, as in many other cases, led by an invisible hand to promote an end which was no part of his intention. Nor is it always the worse for society that it was no part of his intention. By pursuing his own interest he frequently promotes that of the society more effectually than when he really intends to promote it.

No one can deny that in a perfect economic world, a bumper crop yields lower prices or a shortage means higher prices—international tariffs, OPEC cartels, agricultural subsidies, diamonds, airport food, and other monopolies notwithstanding, which little reflect supply and demand or a scarcity-based price. But we don't live in a perfect economic world. We don't live in a world where perfect exchanges create perfect prices, where haggling and uncertainty is inherent in the process.

Interestingly, the idea of a fixed price did not become entrenched in daily life until the arrival of the department store and its advertised cash-and-carry products. As noted by Boorstin (2000), "'Price lining'—the production of items to sell at a predetermined price—expressed a new way of thinking" (p. 114) and occurred during the rise of the Five and Ten

Cent stores in the 1880s to guarantee a common and repeatable consumer experience. Furthermore, salesmen didn't need to be as skilled at haggling or, as it were, as trustworthy.

However, everyone knows that suppliers manipulate prices and that the valuing of a good or service changes for reasons other than free-market economics. Such artificial price determination can especially be seen in the ongoing battle over Chinese currency manipulations, where the renminbi and, hence, Chinese foreign exchange rates are held artificially low, resulting in cheaper imports and affecting trade balances and public debt to the tune of trillions of dollars. The *real* price is meant to float between the highs and lows of a *free* market, in an ongoing dialect of economic bartering. But does it?

Is such price determination any different from how individuals estimate their worth? If I say to someone that I am 6 feet tall (when I am actually 5 feet 7), I exaggerate my height and the person I am trying to convince will clear his throat and politely remark that I look more like 5 feet 8. The same goes for someone who lies about his or her age, such that whatever final point is agreed on, the truth doesn't matter.

Many hedge the truth in normal discourse. I have a friend who says his house is 20 minutes away when he wants to meet me and 40 minutes away when he doesn't. Speaking in terms of time may reflect the amount of traffic and other factors at a particular time of day or particular day of the week but is not always a fair assessment. It more reflects his *wanting* to come and surreptitiously conveys his desire by the length of time he quotes. But wouldn't it be better to say that he lives 10 miles away and let others judge the time? No hedging.

All sorts of hedges fit the bill. In Spain, a rampant black market economy produces lower declared house sale prices to avoid tax. On Internet dating sites, people of both sexes routinely lie about their age, marital status, and presumed appeal. Some think health warnings and inoculations for global viruses are excessive since the rates are lower than predicted, although one can easily argue that rates are low *because* of the warnings and inoculations. In most cases, constant evaluations and reevaluations are made, where haggling indicates competing preferences.

If someone drops a piece of litter on a bus, no one will likely take notice, but if someone drops a scarf, most will call it to their attention or pick it up. Some might even run after the person, wanting more than just to return the scarf. What if someone dropped $10 or a wallet? Again, most would return it. But what if someone dropped £1,000,000? Can we be our own witness, or do outside factors always enter into our decisions? Are we just haggling over the price as we decide on what is right? What is the price of our honesty, and can any of us be a man or woman for all seasons?

More Haggling: An Everyday Fish Tale

The same applies to the persuasive methods we use to present a result, particularly when numbers are involved and especially with politicians. A good example of such hedging occurred in 1999, when Northern Ireland and the Republic of Ireland held separate referenda as part of the Good Friday Agreement. The question in the North was, "Do you support the Agreement reached at the multi-party talks on Northern Ireland and set out in Command Paper 3883?" and in the South, "Do you approve of the proposal to amend the Constitution contained in the undermentioned Bill? (Nineteenth Amendment of the Constitution Bill, 1998)."[13] The result was 71.1% voting in favor in the North and 94.4% in favor in the South.[14]

As soon as the votes were in, however, the fun began—i.e., the hedging according to who was doing the reporting, which has important knock-on effects to this day. Certainly, a majority voted in favor—more so in the South—and various words were used to describe the outcome. In the North, descriptions included "large majority," "strongly in favor," and "swamped" referring to the "Yes" side (71.1%) and "significant part" referring to the "No" side (28.9%).

For the result in the South, "massive majority" was commonly used. *An Phoblacht*, the newsletter of the Republican party Sinn Fein, stated that "the vast majority of the Irish people endorsed the Agreement in referendum" (Ferris, 1999). To be sure, many more voted in favor than did not—71.1% in the North and 94.4% in the South—but is that a *vast* majority?

The *Éire-Ireland: Journal of Irish Studies* went so far as to say, "The arrangements were overwhelmingly approved by referendum in both the Republic and Northern Ireland" (McAuley, 2004, p. 193)—a misleading remark given that 71.1% in the North is hardly "overwhelming." Furthermore, the referendum itself was referred to by the Irish press as "popular," although in the South only 55.6% of the electorate voted—not so popular, according to the numbers.

The problem may lie in how a number is characterized. If *An Phoblacht* was referring to results in both the North and South, as one would expect from Sinn Fein, "vast" is not all that vast. Properly weighted, the "Yes" vote was 85.7%—the average of the two results weighted by the number of voters, $(971{,}745 \times 71.1\% + 1{,}428{,}331 \times 94.4\%) / 2{,}380{,}076 = 85.7\%$. Impressive but not quite so vast, overwhelming, or massive.

A particularly egregious numerical misrepresentation came from then-Secretary of Northern Ireland Mo Mowlam, who stated that the overall margin was 3 to 1, which is not equal to 71.1%. Of course, she was possibly unaware of what she was saying, but the error more favorably characterized her government's position—that is, she hedged, and, as is so often the case, she hedged in her favor.

Some might say that 3 to 1 is the simplest whole number representation of 71.1%, but what's wrong with saying 71% or 7 to 3? What is most worrying about her hedge, however, is that few may have noticed the difference, believing that 75% voted in favor of the Agreement.

Furthermore, in the North, an estimated 96% Catholics and 52% Protestants made up the 71.1% vote, and, thus, misrepresenting one number hides another—i.e., that almost *half* of Protestants in the North were opposed to the Agreement, despite reports claiming overwhelming or vast approval in the North and South. Is it any wonder that continued mistrust and political instability exists in Northern Ireland when official claims contradict the real numbers? Is it any wonder that Unionist support is seemingly at odds with the wider perception? When so many decisions are made based on the smallest of differences, a hedge is a hedge: 71.1% is not 75%.

Numbers can always be used to mean different things when qualitatively referenced. There are numerous examples (perhaps an extraordinary number!). But what is wrong with using only the numbers, without misrepresentative adjectives? How big is a *big* fish? What are *massive* savings? Furthermore, smaller hedges create an atmosphere of impunity where small hedges become big hedges and big hedges become bigger hedges, such that the truth is no longer important.

To make a case for invading Iraq in 2003, the British prime minister Tony Blair stated that Saddam Hussein could deliver weapons of mass destruction in 45 minutes, when in fact he couldn't. Not only were there no such weapons, there was no such means to deploy them on any time scale near 45 minutes. So how did the bomb counters arrive at 45 minutes? Is 45 minutes the amount of time that defines imminent danger and stirs public support? Would 60 minutes have seemed less dire? Two hours? Or was it all a hedge? Worse still, as Blair later confirmed, the existence of weapons of mass destruction did not contribute to his decision to go to war, thus revealing that he had misled the House of Commons from the beginning and led a country to war under false pretenses.

Of course, a case for war had to be made, despite the Americans and British having previously supported Hussein during the 1980–1987 Iraq–Iran War—even after he gassed his own people in the Kurdish town of Halabja, repeatedly cited as a precursor to the 2003 U.S.-led occupation of Iraq (Fisk, 2006, pp. 260–262). In fact, the West—opposed to a Shiite Islamic theocracy forming across the Middle East—was instrumental in its support of Iraq prior to Iraq's occupation of Kuwait, providing Hussein with chemical and conventional weapons throughout that war. But the perception had to change, and a different spin or bias had to be presented.

And what of President George W. Bush saying that Iraq was seeking uranium in Niger when it was not and the resultant willful leak by a Bush assistant of the covert identity of a CIA operative, whose husband had broken the story? These are no longer hedges but, rather, lies used to manipulate intelligence to make the case for invasion, timed, as some have said, to distract voters from Bush's failing domestic policies and plunging popularity—lies for which impeachment proceedings were recommended in the House. As Mason (2009) noted, "A new bull market took off on practically the very day American tanks went into Iraq" (p. 74), suggesting that the real story lay elsewhere.

Such acts are a cover for the truth, such as the Americans and British invading Iraq to ensure cheap oil; the Russian excursion into Georgia, ostensibly to protect South Ossetia for the same reasons; or the duplicity of selling arms to everyone, as is routine in the Middle East—a recent, $60 billion U.S. arms sale to Saudi Arabia was the largest ever of its kind and "will increase unease about a gathering arms race in the Middle East" (Spillius, 2010).

If politicians are so often spinning facts, twisting the truth, and double-dealing, is it any wonder we have become numb to their claims, particularly those with numbers? And like the boy who cried wolf, if we are so often conned, will we pay attention when real danger comes?

It is true that numbers can mislead, just as words do, and qualitative descriptions can be interpreted differently—where one man's rebel is another's insurgent. But haggling with the truth is something too often done and is not, as some would say, an intermediate point or invisibly agreed-on value between two extremes, i.e., those of the seller and the buyer. No, it is a systematic misrepresentation designed to favor one's own side. In fact, price and truth determination is anything but free and, instead, reveals the inherent biases in the process and the system.

4.6. Testing Error: Do We Test the Student, the Teacher, or the Test?

A similar uncertainty or error in how a test is applied reveals the same inherent bias in the education system, where either/or decisions and cost-benefit analysis ultimately determine who passes and who fails. Most of us believe that the purpose of testing is to assign a grade to each student, representing their mastery of the subject. But how good is a mark of 82 on one test versus 64 on another if the standards don't compare? Do we, in fact, test to rank, i.e., to hand out places regardless of score with an attached spread to the data?

In the United States and Canada, a 64 on a test (out of 100) is considered poor, where below 50 is a fail. The letter equivalent is a "D," not the kind of grade one hopes to bring home to Mom and Dad. In the United Kingdom and Ireland (as elsewhere in Europe) the same 64, however, is not such a bad result (a "C"), where below 40 is a fail. Thus, given that the standards are different, it is not how good one is but how good one is *relative* to everyone else that determines rank. Indeed, telling a prospective employer that you finished in the top of the class is much better than saying you got a 64, even if 64 was near the top.

Of course, to compare different groups, we must apply a standard, but standards are notoriously hard to determine. Some think that the North American education system is better than the European and others the reverse, but the truth is that both are comparable and both produce as many good students rated for population. Nonetheless, the scales don't match, and like inches and centimeters or Fahrenheit and Celsius, one student's "D" is another's "C."

If a test does not reflect the course material, however, it is unlikely that the scores will properly represent any student's understanding. For example, if the test is set too hard, the average will be low, which disadvantages the better students, who thus score as poorly as the others. If the test is set too easy, the better students are again disadvantaged because the test pulls up the weaker students. In both cases, the marks don't show enough variance or spread.

To be sure, testing is not as simple as we think. How a test overlaps with the material is as important as one's understanding and mastery of the subject and can reveal flaws in the test rather than in the students. Here, the "error" is how we choose to overlap the perceived standard, where testing becomes as much about matching the test to the class as it is about a spread of marks (which themselves can have an associated error or uncertainty depending on the marker, say, by as much as 5%). The biggest challenge in any test is getting the match right, which is as much an art as a science.

An apocryphal story about two medical students illustrates this point. In the story, the two students don't have enough time to study for an exam and so each studies half the course. The first studies the beginning, the second the end, and one passes while the other fails. But who is the better student—the one who got the higher mark having studied the *test* material or the one who may have known more but studied what wasn't tested? We all prioritize our lives and decision making in similar ways, but which student would you rather have as a doctor?

How questions overlap with knowledge can be seen in simple game shows, such as *Who Wants to Be a Millionaire?* or *The Weakest Link*, where one contestant might get all the hard questions, which she answers incorrectly, whereas another might be asked only easy questions, which he easily gets right. But who is smarter? Or more important for the purpose of the game, who will do better in future rounds? Here, past questions may not appropriately measure future success.

Paulos (1996) noted that the various criteria used to judge something affect the outcome of the judging and can thus misrepresent the characteristic being judged—such as well-known celebrities who perennially make the worst-dressed lists but are surely better dressed than a homeless person, and the usual world city rankings that appear in the newspaper from time to time, which are based on numerous yet vague measures (p. 193). New York and London are great cities, but as Paulos succinctly noted, "Someone with the artistic sensibility of Edward Hopper might give high marks to a Midwestern town with many deserted streets, dreary bars, and empty diners" (p. 194). You can't judge a book or a city by its cover.

Here, set theory, a visual representation of the mathematics, can help highlight the logic. Two sets (A and B) are shown in a Venn diagram in Figure 4.4 (see page 80), where A is all the things a class was taught and B is all the things a class learned. What the students learned from what was taught lies in the intersection (A *and* B) or overlap of the two sets. What they knew of the test material (C) is shown in the figure on the right as a smaller intersection of the three sets (A *and* B *and* C).

One can see that the students learned less than what was taught (B < A), as well as some of what was not taught, say, from outside reading or at home (B *and not* A). Furthermore, the test overlapped mostly with what was taught but not with what was learned, and significantly with introduced material that wasn't taught or learned. From such powerful diagrams, other concepts can more easily be explained, such as conditional probability—i.e., A *given* B.[15]

Conditional Probabilities: What Was the Intent?

Conditional probabilities show the importance of matching intent with result, such as the two very different scenarios in the following

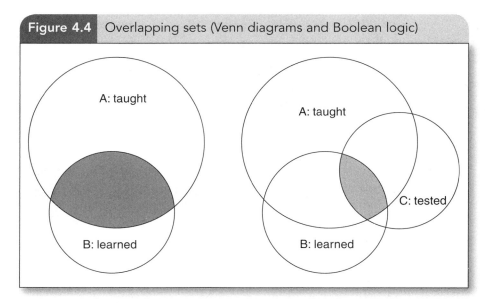

Figure 4.4 Overlapping sets (Venn diagrams and Boolean logic)

Venn diagrams (see Figure 4.5), which can and do come up in testing and in life. On the left, the test did not represent much of what was taught, although the students knew the course material well (perhaps the test was too hard). On the right, the course was taught to the exam, but the students didn't learn much other than the exam, which could be the result of bias or cheating by a teacher who wants to parade high-performance results as evidence of good students (and good teaching).

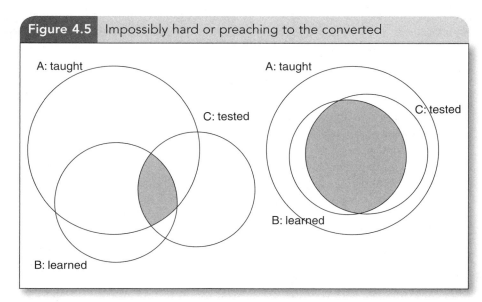

Figure 4.5 Impossibly hard or preaching to the converted

Such diagrams help one understand the intended audience that every politician or advertiser hopes to target. In the wake of spiraling road deaths, the prime minister of Ireland called for drivers to slow down, with the heartfelt plea, "I hope that just gets through to the psyche of the community" (Heatley, 2006). But was he speaking to the right community? Generally, it is the teenagers or young adults in the "community" who drink and drive or drive at high speeds and don't use seatbelts, but teenagers don't watch the news or read newspapers as much as the general public does and typically tune out the marketing and moralizing machine of their elders as they test untried waters. Here, the overlap, as shown by the intersection in our diagram, is not large enough to get the message across to the required or intended audience, and other methods are needed (e.g., a zero-tolerance policy on alcohol consumption by drivers under 21, or stricter penalties, such as a revoked license for a first offence).

Getting the audience right applies in many areas. For example, airport screening doesn't stop the real baddies, those who know how to blow up planes. Neither do most border or building entrance checks. Such measures deter the use of simple means but not the more complex means at the disposal of a professional terrorist. As such, the public is subjected to removing shoes, belts, and jackets for no good reason. Better profiling (i.e., getting the audience right) is needed—despite the intrusion into areas once considered private—and security forces are beginning to introduce such measures through message scanning, facial recognition techniques, and global security cooperation.

In reality, profiling is no more than filtering our judgments into obvious categories—something we all do, from judging the intent of a person approaching (friend, foe, neutral) to job applications, where one must scan countless pages by some identifiable and representative means (neatness, brevity, appearance, academic qualifications, grades, experience). We all do it as we attempt to identify the good and the bad among us, whether in testing, criminality, or life. But the means we use may ultimately determine the result.

The same applies to preaching to the converted or speaking over people's heads. There is no point to a Green party member reading the party manifesto at an environmentalist meeting or an academic citing facts and figures to a subliterate teenage audience. Knowing the audience, who construct meaning from our words through the filter of experience, is essential to getting the intended message across.

Getting the balance right is a challenge not only in testing but in life, where overlapping understanding and overlapping uncertainty are an everyday reality. Uncertainty is a part of life and represents a measure of our fallibility, but knowing what we don't know and how to quantify what we don't know—the uncertainty in our measures and choices and, indeed, the instruments of our analysis—help us understand the biases in daily life and in the choices we make. Knowing the context in our data or that of our audience is paramount.

Notes

1 **Normal curves and power laws**: A normal curve shows equal errors above and below the mean, whereas a power law shows a declining probability with increased size.

2 **Chance and Community Chest bias**: The net windfall is about $20/card for Chance and $35/card for Community Chest. Redirect cards and street repairs cards are not considered, since it is not possible to gauge their benefit as both depend on the stage of the game. The plus/minus payoffs from redirects and penalties from street repairs increase as play continues (more houses/hotels).

3 **Orange and Brooklyn County vote**: In 2008, 48% voted for Obama in Orange County and 80% in Brooklyn compared with 50% and 20%, respectively, for John McCain (Cable News Network, 2009a, 2009b). In 2004, the results showed the same Republican–Democratic divide, where John Kerry outpolled George Bush in Brooklyn County 75% to 25% yet received only 40% in Orange County to Bush's 60% ("Election 2004," 2004a, 2004b). Note that California as a whole voted 55% in favor of Kerry and 61% for Obama ("Election 2004," 2004a, 2004b).

4 **Sample size determination**: Numerous tests exist to determine the significance of statistical measures (e.g., sample means, proportions, and coefficients of correlation), for which a valid sample size is required, but they are too detailed to discuss here. Burns and Burns (2008) discussed sample size determination in much more detail.

5 **Spherical projection**: In 1974, German filmmaker Arno Peters created an equal-area projection map similar to an earlier version proposed in 1885 by the Scottish clergyman James Gall.

6 **Earth projections**: NASA's Earth Observatory is an excellent repository of global projections using composite satellite images. Its *Blue Marble Next Generation* collection especially shows monthly global changes (see http://earthobservatory.nasa.gov/Features/BlueMarble/).

7 **Internet rankings**: Spam link farms are removed from the rankings since they are created solely to make the most of this ranking criterion by linking to one another.

8 **Wikipedia's rankings**: Wikipedia does not confer its own high ranking (which it receives from external pages) to other externally linked-to pages (using the NOFOLLOW tag), supposedly to stop spam.

9 **Conjugate variables**: Prior to the methods of Heisenberg, Schrödinger, and Dirac, quantum mechanics used Fourier transforms, which explicitly relates the error of two conjugate variables—for example, bandwidth and the number of superimposed component frequencies. As applied to quantum mechanics, the same either/or dilemma is to err on the side of position and know nothing about velocity (i.e., momentum) or err on the side of velocity and know nothing about position.

10 **Guantanamo Bay detention facility closure**: Barack Obama did not keep his promise to close Guantanamo Bay (what he described as "a sad chapter in American history")

within a year of being elected president. Instead, military commissions were established, although prisoners were eventually granted habeas corpus.

11 **Ford's Pinto rear-end crash death estimates**: 180 burn deaths at $200,000 per death, 180 serious burn injuries at $67,000 per injury, and 2,100 burned vehicles at $700 per vehicle each year ("Ford Pinto," 2006).

12 **The invisible hand** (or the first fundamental theorem of welfare economics): Competitive markets are the most efficient, naturally producing winners and losers, resulting in a "Pareto efficient" allocation of resources—i.e., the rich get richer and the poor get poorer in a competitive market left to its own devices. Thomas Friedman (1999), however, noted that perfect competition is a fallacy in today's world: "The hidden hand of the market will never work without a hidden fist—McDonald's cannot flourish without McDonnell Douglas, the builder of the F-15. And the hidden fist that keeps the world safe for Silicon Valley's technologies is called the United States Army, Air Force, Navy and Marine Corps."

13 **Nineteenth Amendment of the Constitution Bill, 1998**: As part of the Belfast Agreement, the Republic of Ireland agreed to a referendum calling on the removal of two articles in its constitution (Articles 2 and 3), which ascribed territorial and jurisdictional authority over the entire island of Ireland.

14 **Belfast Agreement vote breakdown**: 71.1% in favor in the North (951,745 votes or 81.1% of the electorate) and 94.4% in favor in the South (1,428,331 votes or 55.6% of the electorate).

15 **Monty Hall and conditional probability**: Conditional (or contextual) probability is widely misunderstood, but with a little clarification, even the famous Monty Hall *Let's Make A Deal* problem is easily cracked, a problem that has challenged the highest tier of professional mathematicians. Here, a game show contestant chooses between three prize doors (one good, two bad) and after being shown one bad door by the host is asked whether he or she would like to switch. Contrary to intuition, one should switch, since the odds are now 1 in 2 that the other door is right, yet still 1 in 3 that the original door was right, a seemingly counterintuitive result that has nonetheless been corroborated by computer simulation and more than 30 years of door picking on *Let's Make A Deal* (Mlodinow, 2008, pp. 43–45). However, such a result is obvious if one imagines holding 1 of 10 million lottery tickets, all with a 1-in-10-million chance to win. If 999,998 were opened by the lotto host without a win, leaving yours and only one other, would you switch? Of course you would, since the odds are almost certain that the other is the winner, while your original ticket is still the same 1-in-10-million shot as before. Monty Hall makes all the difference. Now, had the other 999,998 been opened *randomly*, and somehow you were still in the running, it would make no difference whether you stayed or switched as the odds are the same. Of course, switching and losing, having originally picked a 1-in-3 winning door (or a 1-in-10-million winning lottery ticket!), may be harder to take than staying with one's original pick and losing, despite the better odds of switching—no matter what the mathematics say.

STATISTICS MADE SIMPLE

The Science of Weighting

As the joke goes, the world has two kinds of people: those who divide the world into two kinds of people and those who don't. Advertisers are those who do, spending billions of dollars each year selling anything and everything, day in and day out. Armed with the latest statistics on age, gender, hair color, etc., advertisers know whether one is more likely to buy brand X than brand Y. The cost of all this selling?—more than half a trillion dollars per year (Coen, 2007).[1] That's a 5 followed by eight zeroes and preceded by a dollar sign.

As statistical fodder in the advertiser's cannon, people can now be divided in scores of ways, from Freud's psychological "normal" and "perverse" to the 16 Myers-Briggs modified Jungian archetypes[2] to the usual social measures of education, income, and religion. Male or female? Tall or short? High or low IQ? Fair of face or full of grace? In advertising, one is "not loving and giving," one is *so much* loving and *so much* giving. To counter such practices, we start with the basics of statistics—the mean and standard deviation—not only to learn about the advertiser's world but to keep from becoming statistics ourselves. From there, we move on to correlative methods used to compare different statistics.

Most of us are familiar with the mean,[3] which gets bandied about as the so-called "average," but just because one is of average height, weight, or income doesn't mean one is average—although the mean is the simplest measure used to estimate a population. In effect, as we will see, the mean is no more than a balanced weighting. The standard deviation (a measure of the error) and regression (which quantifies correlation) can also be thought of as simple weightings.

Of course, we must always be wary of how data is used—of how one can lie with statistics—and some concrete examples can keep us on our toes. As Huff (1954/1993) noted, "The fact is that, despite its mathematical base, statistics is as much an art as it is a science. A great many manipulations and even distortions are possible within the bounds of propriety" (p. 120). As such, in this chapter, we look at some statistical tricks—from the American Electoral College and the stock market to no-money-down loans and inflation measures. As we do, we are mindful that a result doesn't follow just because a statistic says so.

5.I. Today's Information Age: The Art of Numbers

One could say that mass-produced statistical measures began with Herman Hollerith and his International Tabulating Company. Build a better mousetrap and the world will beat a path to your door; build a better people-counting machine and the future is yours. In 1890, Hollerith built the best people-counting machine ever for the American census using punch cards (aka Hollerith cards), which are still in use today more than 100 years later (mostly in dreaded student multiple-choice tests) and were the standard computer input method before personal computers.[4]

Hollerith used his punch-card invention (see Figure 5.1) to count electronically every single person in the United States, as well as to gather personal data such as sex, age, marital status, "whether a soldier, sailor, or marine during the civil war; able to read; able to write; whether defective in mind, sight, hearing, or speech" (U.S. Census Bureau, 1890), and more. In total, he counted 63 million Americans. More than 120 years later, the American population is numbered at more than 300 million, 78 million of which are in the so-called baby-boomer demographic born between 1946 and 1964 (Elliott, 2009b).

Of course, advertisers have access to all the latest data and know the market like the back of their hands. Watch any children's show on Saturday morning, and you'll see the G.I. Joe and Barbie ads. Watch *Oprah* or *Dr. Phil*, and during the breaks you'll hear the constant blather on lotteries, horoscopes, and psychic lines: "Life, love, happiness. It's in the stars."

To be sure, the advertisers have done their homework, relying on instant, computer-generated demographics to sell to today's masses—e.g., the boomers drink 60% of the beer and soft drinks and eat more than half the candy (Elliott, 2009b), and Gen Yers, born between 1979 and 1994, represent the "biggest discretionary spending power of any teen demographic in history" (Barber, 2007, p. 168), according to Saatchi & Saatchi. Anything can be marketed, from blanket ads on sheep to

Figure 5.1 The Hollerith computer punch card

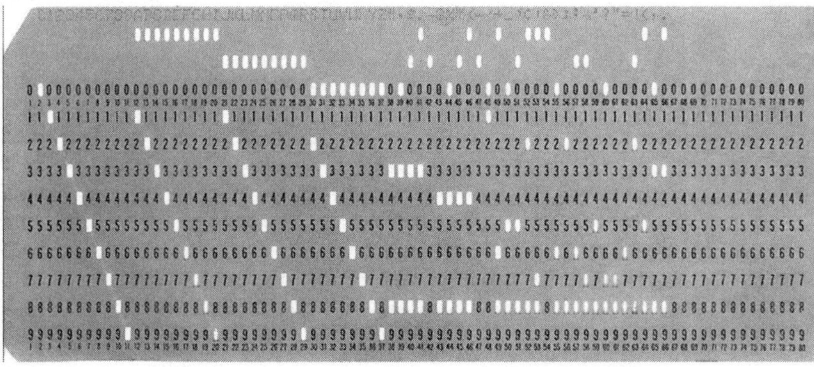

SOURCE: http://commons.wikimedia.org/wiki/File:Blue-punch-card-front.png

"Bumvertising" on the backsides of the homeless (pp. 182, 227)—even God has been advertised by that rainbow-haired, John 3:16 guy who stood behind the green at golf tournaments and behind the end zone during televised football games.

The proliferation of brand products in today's media extends to every corner of our lives, from sandwich boards to newspaper inserts to billboards to that big red *American Idol* Coke cup, somehow always strategically positioned toward the camera. But if what we are told is true, then we have to accept the theory of scrubbing bubbles, quicker picker-uppers, hold in its purest form, 65% lash life, ocean-fresh skin, pore-target technology, supreme cleaning power, dangerous washing machine limescale, rebellious hair, the goddess within us, the savings revolution, etc., etc. Given the moronic lengths to which today's advertisers will go—with their arsenal of dubious statistics, air-brushed lies, and fine print so small it couldn't be read in Lilliput—one could be excused for thinking that today's culture is no more substantial than the fizz in a pop drink gone stale.

Sadly, building a better mousetrap is no longer the aspired-to maxim that promotes ingenuity and entrepreneurship but, rather, a means to sell to an ever-increasing customer base. Numbers have replaced people. And so, to keep from becoming statistics, we must arm ourselves with the same tools the sellers wield—that is, if we want to keep from being ruled solely as consumers.

What became of Herman Hollerith? After running the first modern census, he turned his company into the computing giant International

Business Machines, from which we have inherited an astounding number-crunching legacy. On that day in 1890 when Hollerith's people-counting machine counted 63 million Americans, the information age began.

5.2. Electoral College Weighting: How Some Votes Are More Equal Than Others

The American Electoral College provides a useful introduction to statistics and how distributions are weighted, i.e., how some votes are more equal than others. Based on the Centurial Assembly of the Roman Republic, where citizens were divided into groups of 100 and given one vote each, the American Electoral College system was created to avoid problems of political parties, alleviate the difficulties of national campaigns over a diverse and large nation, balance state and federal interests, and represent both large- and small-population states more equitably. As such, states are given electors relative to their populations—about 1 elector per 700,000 citizens—as well as a minimum of two electors each to represent smaller states.

The idea is basically one of scale. In a time before electronic tabulation, it was hard to poll an entire nation; so 538 "electors" were given the job of choosing a president every 4 years, their numbers calculated relative to the population of each state (1 per 700,000, plus a minimum of 2). As such, California (the largest state) has 55 electors and Texas (the next largest) 34, whereas seven states with populations of fewer than 1 million have only 3 electors. In effect, the Electoral College is a proxy, subject to a "weighting"—just as one might weight the opinion of an expert or friend over that of a stranger.

The Electoral College has worked well in satisfying the voting wishes of a large, diverse country, and on only two occasions has the weighted vote of the Electoral College not reflected the overall national vote—first, in the 1888 presidential election (when Benjamin Harrison won small in 20 large-population states while Grover Cleveland won big in 18 small-population states, giving the Republican Harrison a majority of Electoral College votes yet polling about 1% less votes nationwide) and, second, in the 2000 Bush-Gore election. In the famously disputed 2000 election, the Electoral College weighting gave George Bush the win, although he polled fewer overall votes. In fact, Al Gore polled 48.4% of the vote to George Bush's 47.9%, yet Bush won the Electoral College 271 to 266.[5]

To see how this can happen, a histogram of percentage population votes minus Electoral College votes is shown in Figure 5.2, illustrating

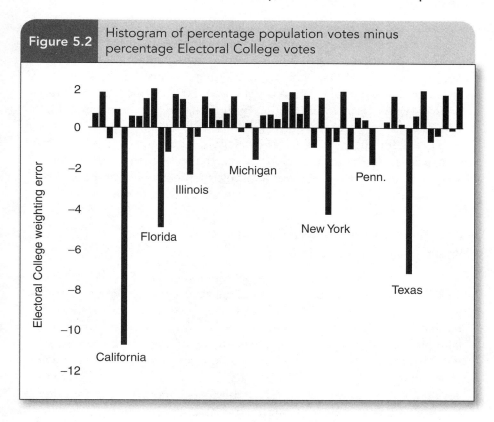

Figure 5.2 Histogram of percentage population votes minus percentage Electoral College votes

how Congress overweighted the Electoral College for 34 states and underweighted it for 17 others (the total includes the District of Columbia). And although most people remember Florida winning the presidency for George Bush—requiring a Supreme Court ruling to halt disputed recounts—a number of states contributed, because the Electoral College votes did not reflect the overall popular vote.

Here, one can see how many small states add up to relatively few big ones, including the largest underweighted states: California (–11), Texas (–7), Florida (–5), New York (–4), Illinois (–2), Pennsylvania (–2), and Michigan (–2)—all of which were won by the Democrats, except George Bush's Texas and Jeb Bush's Florida. Note that the underrepresented states all have large populations, which reflects the intention of the Electoral College, that small-population states be weighted relatively higher. However, had the Electoral College represented the actual state population and not included the minimum two votes per state, the result would have been different by 10 votes, with Bush *losing* 261 to 276.

As populations get out of whack with Census data, because of changing birthrates, migrant workers, immigration, etc., the allotment of Electoral College votes can also misrepresent in greater or lesser proportion the actual state population. Had current Census data been used to calculate the Electoral College weighting in 2000, the result would have been a swing of four votes favoring Bush, resulting in a more comfortable 275 to 264 win.

It is extraordinary that a simple variation in how the data is weighted (in this case, via a disproportionate proxy) can make such a big difference. To be sure, every vote is important, but when elections can be and are won by a single vote, how the votes are weighted is crucial.

Other Electoral Weightings: First-Past-the-Post Versus Proportional Representation

A similar disproportionate weighting applies in first-past-the-post (FPTP) voting systems, where it is possible to receive the highest percentage of votes in an election and yet receive no seats—for example, by finishing second in every constituency behind two other split parties (although not as likely in a predominantly two-party political system). In the 2010 British election, such a weighting significantly affected the electoral fortunes of the Liberal Democrats, a party that received 23% of the vote but ended up with only 8.5% of the seats,[6] and greatly changed the balance of power in determining the eventual government.

Transferable voting systems (one for first preference, two for second, etc., where those with the fewest votes are removed in rounds and the next preferences transferred up the chain until a majority is reached) attempt to deal with the inadequacies of FPTP weighting and, as such, were high on the negotiation agenda of the Lib Dems in their eventual coalition with the Conservatives.[7] Nonetheless, weighting of another kind is at play and brings its own unique peculiarities to the mix. For example, a candidate not preferred by a large majority of voters can be elected and smaller parties with no plurality in any single riding can win numerous seats, as regularly occurs in Ireland's proportional representation electoral system, typically producing a broader representation to the parliament than would an FPTP system.

To try to limit minority winners in diverse, multiparty systems, where majority rule can prove unmanageable, the French use a runoff between the top two candidates in their presidential elections. Nonetheless, this method backfired in 2002 when many left-wing parties divided the vote, leading to a runoff between the two main right-wing parties of Jacques

Chirac and Jean-Marie Le Pen. The extremes of political weighting are also seen in a system that pits each candidate against the others, where the candidate with the most individual "wins" is elected—called a Condorcet winner in choice theory. Here, the result depends on head-to-head comparisons, akin to how people interact in social clusters (Ball, 2005, p. 381).

Each of these systems is devised to represent its constituency as fairly as possible but, nonetheless, can ultimately determine the winner. Paulos (1996) constructed such a scenario, where a different candidate wins a state caucus depending on the voting system. In his fudged election, which takes place in the made-up state of Nebrarkamasscalowa, five different candidates end up winning under five different systems: (1) standard plurality (FPTP); (2) a runoff between the top two placers, decided by who is more preferred (FPTP and Condorcet); (3) transferable voting (proportional representation based); (4) weighted voting (5-4-3-2-1); and (5) man-to-man (Condorcet). The various systems resulted in wins for Tsongas, Clinton, Brown, Kerrey, and Harkin (pp. 104–106).

We may seem to get the politicians we deserve, but the choice of system can also dictate the government. Of course, other factors must be considered, such as the ease of implementation (a Condorcet system takes more time because of the increased number of candidates and individual interactions, as does a proportional system with its many transfer allocations during vote counting), although computers are making electoral accounting easier. The best system might be to have everyone vote at the same time with an identifiable handheld device, such as a phone, which would at least remove excessive bias from outside sources—say, in larger countries, where the result in the east can be known before voting has finished in the west—as well as limiting inherent clustering in small-world systems.

The moral of the story, however, is to beware—not all numbers are what they seem, especially when weighted. Knowing how and why weighting is applied and by whom is essential.

5.3. Numerical Distributions: A Question of Balance

In a distribution of data, from elections to test scores, weighting can also be used to explain the mean (average) and standard deviation (error), the calculation of which is no different from that of balancing weights on a central point or fulcrum. The average of a distribution is like a playground teeter-totter, where a parent sits closer to the middle to balance the lighter weight of a child on the other end. The middle, or fulcrum, is the point at which the product of the weight *and* the distance left equals the product of the weight *and* distance right (i.e., all add to zero), such that the teeter-totter is in perfect balance. As more kids are added, the parent moves farther out to keep the balance. In the same way, a stalemated arm-wrestling match (two points), a tied tug-of-war (*N* points), and a thrown wrench (any number of integrated points that rotate around the center of mass) are all in balance.

Mathematically, one weights the data by frequency. For example, with 10 test scores (9, 4, 7, 6, 8, 4, 9, 7, 8, 8), we could add the individual scores $9 + 4 + 7 + 6 + 8 + 4 + 9 + 7 + 8 + 8$ and divide by 10 to get a mean of 7.0. Or we could add $(2 \times 4) + (1 \times 6) + (2 \times 7) + (3 \times 8) + (2 \times 9)$, divide by 10, and get the same 7.0 mean (see Figure 5.3).

The idea of this kind of weighting around the midpoint is shown in Table 5.1. Here, the only difference is that we used the frequency of each score, a somewhat more involved construction but one that explicitly includes the idea of weighting and will greatly simplify our understanding of the standard deviation and other higher-order moments that tell us much about a distribution and the effect of outliers or rare occurrences in the data (as we will see later).

Paulos (1996) told a tale about three statisticians on a duck hunt, where the average is worked out in this way: The first statistician misses high by 6 inches, the second low by 6 inches, and the third declares, "We got it!" (p. 4). The same judgment could be applied to my dart-throwing ability: I am guaranteed to miss the center with all three darts, but if I average my misses (in both the *x* and *y* directions), my mean position might well be in the middle—bull's-eye!

Both stories also highlight the concept of accuracy and precision, which is another way of looking at the two types of error (random and

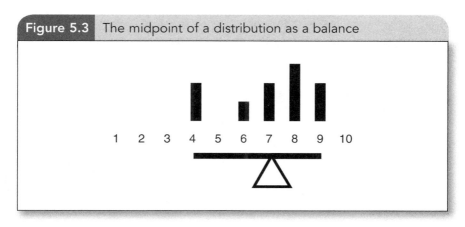

Figure 5.3 The midpoint of a distribution as a balance

Table 5.1	The first moment of a distribution: the average $(2 \times 3 + 1 \times 1 = 3 \times 1 + 2 \times 2)$			
Frequency $f(x)$	Score x	Product $f(x) \cdot x$	Distance Δx	$f(x) \cdot \Delta x$
0	0	0		
0	1	0		
0	2	0		
0	3	0		
2	4	8	−3	−6
0	5	0	−2	0
1	6	6	−1	−1
2	7	14	0	0
3	8	24	1	3
2	9	18	2	4
0	10	0	3	
$\Sigma = 10$	Total: $\Sigma = 70$ Average: 70/10 = 7.0			$\Sigma = 0$

systematic). One can be accurate but not precise (random error) or pre-cise but not accurate (systematic error) as seen in Figure 5.4. Nonetheless, on the left, the three throws *average* to three perfect bull's-eyes.

Figure 5.4	Accuracy (random error) and precision (systematic error)

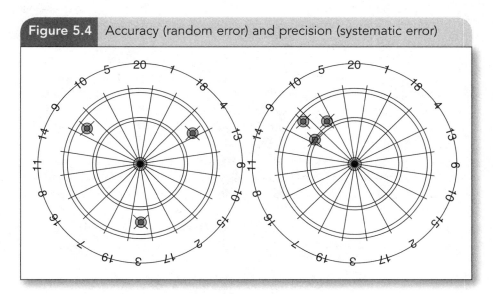

Weighting regularly occurs in the sporting world—for example, at the Olympics, where the medals are weighted to construct an overall points table. In the 2010 Winter Olympics in Vancouver, a gold medal weighting gave the host nation, Canada, the points lead with 14 gold medals, followed by Germany with 10 and USA with 9. But when weighted by total medals, USA topped the table with 37, followed by Germany with 30 and Canada with 26. Using a 3-2-1 weighting, the result was USA 70 (9, 15, 13), Germany 63 (10, 13, 7), and Canada 61 (14, 7, 5), while a 5-3-1 weighting bumped the Canadians to joint second with the Germans because of their extra gold medals. In this case, the Americans were still on top, but the margins depended on the weighting.

One could also argue that the weighting should consider the number of competitors—it seems unfair to compare Ireland to Canada or the United Kingdom to the United States. Perhaps medals per capita is a fairer comparison between unequally represented nations, in which case Norway would be the winner with its 23 medals (9, 8, 6) for a population of just less than 5 million.

Weighting also applies to ratios. For example, stating that the Dow Jones or FTSE 100 rose by 400 points is misleading without also providing the base amount of each index. Referring to an increase by percentage makes more sense, where the weighting is easily compared to other percentages. It is better to say that the Dow Jones rose by 4% (e.g., 400/10,000) or the FTSE by 8% (e.g., 400/5,000), rather than using the context-dependent "400 points." However, reductions given as percentages, such as an advertised "70% less fat" in potato chips, are no good if the amount is unknown (in this case, of fat), because it could still be beyond what is harmless.

Similarly, the ubiquitous post-Christmas, "up to 50% off all stock" sign could mean anything because of the "up to" restriction (typically written in a considerably smaller font). Here, "up to" could mean what the seller wants us to think—i.e., all stock 50% off—or, more likely, a restricted number of items at 50% off and the rest at a much smaller percentage off.

Weighting numbers can also reveal more about polls that state an either/or preference rather than a spread of data. As one *Washington Post*/ABC News poll showed, George W. Bush's approval/disapproval rating of 40% to 60% little reflects reality if those who disapproved did so *strongly*. The same goes for a poll that states only a 60% faith in Barack Obama when a mean approval weighting would be more useful, and even more so if quantified by the error (as we will see below). As Paulos (1996) noted, "Survey or poll results that don't include confidence intervals or margins of error are often misleading" (p. 153).

Another example shows explicitly how weighting can get it wrong. In 1981, both the Cincinnati Reds and St. Louis Cardinals lost out on post-season play because of weighting. That year, baseball players went on strike for 9 weeks, but instead of totaling the winning percentage over all games played, the owners decided to have a playoff between the winners of two split seasons. Both the Reds and the Cardinals, who won the most games in their respective divisions over the whole season, lost out, not having finished first in either of the split seasons. On the other hand, the Kansas City Royals, who lost more games than they won (50 and 53), fared well enough in the second half (30 and 23) to make up for their poor first-half performance (20 and 30) and advanced.

Sticking with baseball, the following example is similar and also seems counterintuitive, but is no different when presented in terms of weighting. In baseball, averages are used to compare players with different numbers of at bats (Ave = Hits/At Bats), where 3 hits per 10 at bats (3/10 = .300) is considered a high average. But what if the sampling period is divided into two halves—say, before and after the All-Star break? As Table 5.2 shows, the numbers tell a different story, but who is the better player?

As the data shows, Player 1 has a higher average in both the first and second halves but a lower *overall* average at the end of the year—a seemingly impossible result borne out because of the associated weightings.

As we saw in the American Electoral College system, weighting can hide the real numbers. For presidential elections, the weighting can mean the difference between winning and losing, since two electors are minimally appointed for small states, thus misrepresenting the population—as

Table 5.2 Split-term average: Who is the better player?

	Player 1	Player 2
First half	20 AB, Ave .500	80 AB, Ave .400
Second half	80 AB, Ave .300	20 AB, Ave .200
Total	100 AB, Ave .340	100 AB, Ave .360

does electing two senators for all states, big or small, where the influence in Congress of the smallest state (Wyoming: pop: ~500,000) equals that of the largest state (California: pop: ~37 million).

The board game Scrabble provides an excellent example of the importance of weighting and the associated frequency, $f(x)$, of each data point, x (in this case, letter). To see who goes first, players pick a letter from the Scrabble bag, with the person drawing closest to the start of the alphabet playing first (a "blank" is considered before A). Say, in a two-player game, I draw first and draw an M; what, then, are the odds that my opponent will go first? Well, if the letters are not weighted by frequency, my opponent has an equal chance of drawing a letter before or after M (13 letters from "blank" to L and 13 letters from N to Z), resulting in an even, 50/50 chance of going first.

But if we include the frequency of letters, the odds change. Scrabble has 100 tiles, ranging in frequency from a high of 12 for E to a low of 1 for J, K, Q, X, and Z (see Figure 5.5). Therefore, there are 53 ways of choosing a letter before M (2 + 9 + 2 + 2 + 4 + 12 + 2 + 3 + 2 + 9 + 1 + 1 + 4) and 45 after M (6 + 8 + 2 + 1 + 6 + 4 + 6 + 4 + 2 + 2 + 1 + 2 + 1), giving a 54% chance (53/99) that my opponent will go first[8]—all because of the frequency of letters in the distribution of Scrabble tiles, i.e., the weighting.

Figure 5.5		Scrabble weighting by frequency, $f(x)$: Who goes first when I draw an M?							
		A_1	A_1	A_1	A_1	A_1	A_1	A_1	A_1
A_1	B_2	B_2	C_3	C_3	D_2	D_2	D_2	D_2	E_1
E_1	E_1	E_1	E_1	E_1	E_1	E_1	E_1	E_1	E_1
E_1	F_4	F_4	G_2	G_2	G_2	H_4	H_4	I_1	I_1
I_1	I_1	I_1	I_1	I_1	I_1	I_1	J_8	K_5	L_1
L_1	L_1	L_1	M_3	M_3	N_1	N_1	N_1	N_1	N_1
N_1	O_1	O_1	O_1	O_1	O_1	O_1	O_1	O_1	P_3
P_3	Q_{10}	R_1	R_1	R_1	R_1	R_1	R_1	S_1	S_1
S_1	S_1	T_1	T_1	T_1	T_1	T_1	T_1	U_1	U_1
U_1	U_1	V_4	V_4	W_4	W_4	X_8	Y_4	Y_4	Z_{10}

5.4. The Mean as a Picture

The following example especially shows the importance of weighting, using the concept of a weighted average, where the "population middle" of a country is calculated and physically represented on a map and can provide valuable information about economic growth, political boundaries, or where to build future infrastructure.

To start, we plot all American state capitals by longitude, latitude, and circle with a radius proportional to state population (see Figure 5.6). Such a presentation also helps illustrate 3-D data (e.g., longitude, latitude, population) on a 2-D page, where the "weighting" of California (big population, big circle) and numerous small-population eastern states can more easily be seen compared with a table or spreadsheet (one could use bars of different heights or colors, but the bars might obscure each other and we would need 50 colors).

To highlight the effect of the *weighted* mean, however, the population-weighted middle is calculated by multiplying the population (the weighting) by distance (longitude and latitude) in the same way we averaged our three missed darts to get a bull's-eye. As such, assuming that an entire state's population resides in its capital (e.g., all California's almost 36 million people live in Sacramento)—a good first guess for a simple 50-point analysis—the population-weighted midpoint of the United

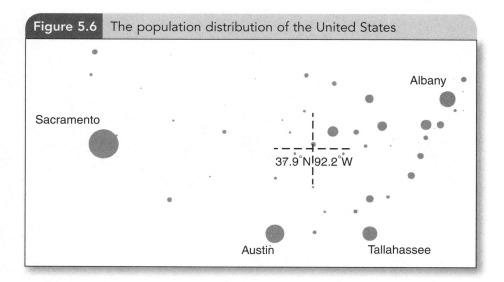

| Figure 5.6 | The population distribution of the United States |

States is 37.9° N, 92.2° W, or about 40 miles north of Springfield, Missouri. That is, as many people live north as do south and as many people live east as do west of the population-weighted midpoint near Springfield (see Figure 5.6). Note that Lincoln, Nebraska, could be considered the "unweighted" mean, showing how population weighting moves the mean south and east, as one might expect from an understanding of American demographics, i.e., where most people live.

As can be seen in Figure 5.7, where a line is drawn from the midpoint to each capital, the midpoint acts as the point of balance, just as in a teeter-totter—although here the balance is two dimensional (north–south *and* east–west). In fact, we can have any number of dimensions, where *distance* becomes a measure of the *difference* between two variables, such as country GNPs, Facebook user ages, basketball player heights, or actual distances.

In a similar analysis for Canada, the population middle is 46.7° N and 85.6° W, which, as it turns out, is in the United States (north of Deer Park, Michigan, in Lake Superior)! Such a calculation highlights how the mean is sometimes a meaningless statistic, where in this case the "average" Canadian lives in the United States, primarily because of

| Figure 5.7 | The population-weighted midpoint of the United States |

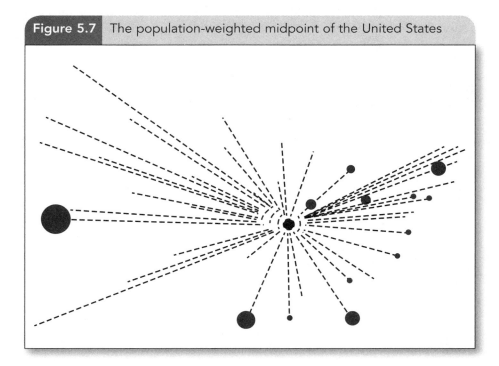

a nonsymmetrically shaped and unevenly populated country. The analysis also highlights the effect of using a small sample set (13 capitals), where if Montreal, the largest city in Quebec, were used instead of its capital, Quebec City, the population-weighted mean would be 46.4° N and 83.6° W, which at least is within the borders of Canada (in Ontario, north of the Trans-Canada Highway, between Blind River and Sault Ste. Marie). To get an exact result, we would have to include every city, mathematically integrating the population density at every point—a doable but highly laborious calculation.[9]

Here, we also see how a distribution can be dominated by a single data point, where the midpoint of Canada is close to its largest city (Toronto; see Figure 5.8), although we must be careful because many small but distant data points also affect the average. As we saw in the United States, the most populous state, California, is a long way from the middle but is balanced by a number of less-populous, distant eastern states, especially New York and Florida. In fact, if the outliers (large deviations from the mean) Alaska and Hawaii were excluded, the weighted mean of the United States would move south by 0.2° and east by 1.0° (about 15 miles), practically to the middle of Dillard Mill State Park.[10]

Nonetheless, such pictures show in a straightforward way how a country's population is distributed and how such demographics can affect governance, infrastructure, and even culture. We can certainly use such statistical pictures to compare different distributions (in this case, country demographics), one of the main reasons for using statistics to represent large data sets.

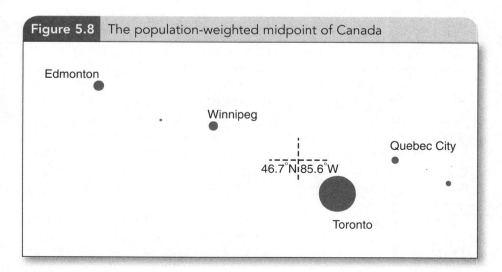

Figure 5.8 The population-weighted midpoint of Canada

In what other ways can such data help? Well, if I wanted to open a business with only one warehouse and I had to truck my product to stores around the country, I would want the warehouse to be in the population middle, not the geographic middle (e.g., Springfield instead of Lincoln). The sum of all my trips to distant small cities and close large cities would be balanced by the warehouse's midpoint location, like the parent on the playground teeter-totter moving outward to balance more children. The population-weighted center tells me where the middle is such that it is closest *on average* to all the data. Suffice it to say that the weighted center is useful for the home-delivery business in cities, states, and countries.

In fact, mail-order businesses helped shape the economic landscape of 19th-century North America, providing a working model for distributing goods across great distances, for which a detailed knowledge of the population distribution was essential. In today's changing world, deciding where to build new power plants in the national grid, erect cell phone masts, or create new high-speed rail lines also depends on population-weighted distributions. In each case, money is saved by knowing where one's customers are and how close they are to the source, where efficiency equals speed.[11]

Of course, I could always include other data and weighting criteria to increase the sophistication of the model. Let's assume that the roads are better in the east and electricity cheaper. It might make sense to move the warehouse east to take advantage of cheaper shipping and office costs, but perhaps labor is more expensive in the east. Using more cities would also improve the result, although at some point the effect of smaller cities will be insignificant (the diminishing effect itself is an interesting statistic). I can refine my model to include as much data as available, weighting each accordingly, where more degrees of sophistication improve the model but take more work.

In just the same way, one could weight geographic data by income to get the money middle of a country, which doesn't make as much sense since money doesn't move in the same way as goods do but could show there is more money in the south and west or highlight a changing trend in the distribution of money over time, an important statistic if one wants to invest in large-scale house construction or relocate a business, an ever-broadening concern in the more globalized landscape of today's international business world.

Using state populations to compare results to other countries with different politically administrated systems, however, is not recommended. Using the most populous cities—say, the 10 largest—is more appropriate and also an easier calculation. The population midpoint of the United States calculated in this way is 36.8° N, 93.1° W—roughly the same as before but about 75 miles closer to St. Louis.

Comparing different countries is now straightforward. For example, results for Ireland and the United Kingdom show unique differences. In Ireland, the population middle is 53.4° N, 6.7° W (near Navan); in the United Kingdom, it is at 52.4° N, 1.0° W (near Rugby). Here, however, the results are subtly different: Navan lies between the two most populous cities, Dublin and Belfast, roughly according to population (3.75–1), but in the United Kingdom, large outliers affect the result more. Although Rugby lies between the two largest cities (London and Birmingham), the more northern cities of Glasgow and Edinburgh shift the average farther north than one might think (London is 7 times more populous than Birmingham) and again highlights how the mean acts as a "balance" (see Figure 5.9).

As shown in Figure 5.9,[12] Ireland also appears more centrally administered than the United Kingdom, although relatively speaking, it has more populous cities.[13] One could also anticipate further expansion with increased infrastructure between Dublin and Belfast compared with Dublin and Cork, whereas the UK midlands should see growth, ultimately reducing urban sprawl in London.

In a similar analysis applied to the United States, Europe, China, and India, one can readily see the differences in each country, from which important conclusions can be made about how they could be better administered (see Figure 5.10). In Europe, where political and monetary union is continually being questioned and redefined, all roads seem to lead to Stuttgart, Germany. (Prior to expanding in 2007 to include Bulgaria and Romania and in 2004 to include Poland, the population-weighted midpoint of the European Union was just outside of Lyon, France.)

In China, future growth is expected to expand out from Shanghai, and in India, from Mumbai—both cities already located near their population-weighted midpoints. In fact, both are already expanding predominantly along a north–south corridor, with China building high-speed rail lines to accommodate further growth, projected to account for 50% of world capacity by 2020 and operating more than 18,000 kilometers of track (Moyo, 2011, p. 144). The geography is a bit different in the United States, but from the weighted demographics, St. Louis—an advantageous, population-weighted midpoint—would be the ideal hub from which to build an interstate, high-speed train system.

Interestingly, all the distributions show the same balance, conjuring up the image of a nodal network or even an abstract painting. In effect, the

| Figure 5.9 | The population-weighted (10 largest cities) means of Ireland and the United Kingdom |

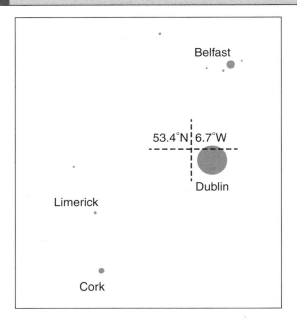

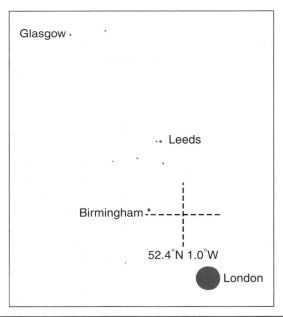

Figure 5.10 The population-weighted (10 largest cities) means of the United States (top left), Europe (top right), India (bottom left), and China (bottom right)

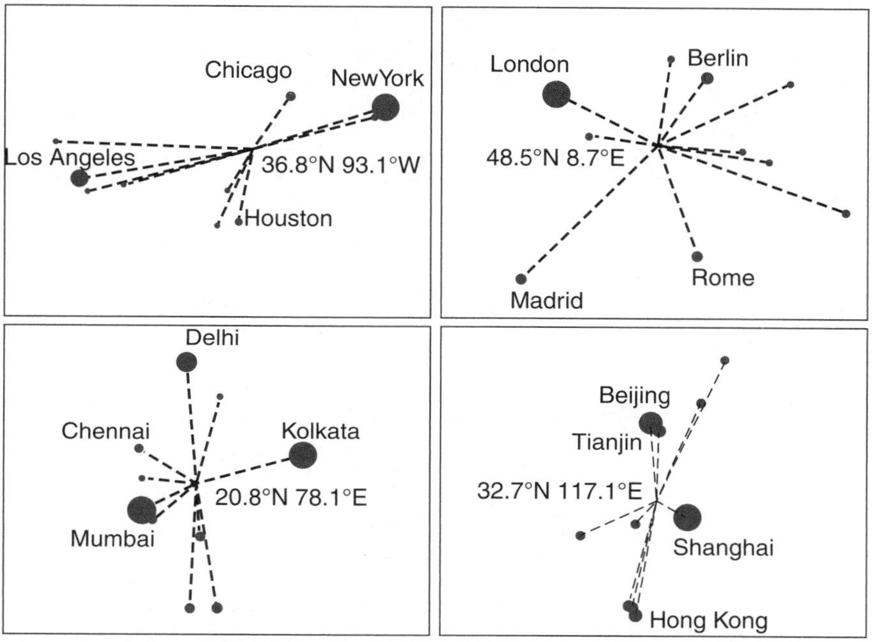

average is just that—a balance of all points weighted about the midpoint, from which one can calculate the nearness of any relationship between points or clusters, regardless of the content. Furthemore, the relationship between points above and below or left and right of the mean is easily seen.

The same is true when we plot any set of 2-D data. In Figure 5.11, life expectancy versus income per capita is plotted for 220 countries, using weighted circles for population. The average divides the chart into equal top–bottom and left–right halves, in this case, indicating average life expectancy (69.3 years) and average wealth ($14,400).

Data can also be shown versus time, as in an electronic flip chart, to illustrate four-dimensional information, an impossible feat on the 2-D page. Rosling (2009) displayed a fascinating plot of life expectancy and income data that comes to life when animated in time, showing the rise and fall of economies, the effects of wars and credit collapses, and the

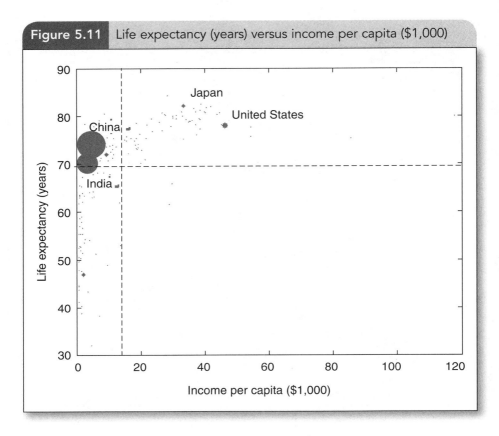

Figure 5.11 Life expectancy (years) versus income per capita ($1,000)

growth of populations in an ever-changing geopolitical world, as if in a time-lapse film. A similar plot was used to calculate graphically the intersection of the U.S. and European economies with those of China and India, predicted for 2048 (Rosling, 2009).

Other weighted data can also be used to similar effect—for example, in sports, where the weighted average can show where a batter likes to hit, a wide receiver's pass pattern on third and long, or where a soccer striker scores most of his goals. An analysis of bank slips from different ATMs can paint a picture of where a particular group likes to hang out, informing an enterprising businessperson where to locate a new coffee shop or a local advertiser where to start a new campaign. Income as a function of age can show that the best group to pitch a new drink to is the 21-to-25 recreational softball crowd. From such statistics, we begin to see why marketers want to know about our habits, what we eat for breakfast, lunch, dinner, and every snack in between.

In choosing a hospital, a nearer, busier hospital may be worse than a more distant yet better-equipped and less busy one when the data is appropriately weighted. Weighted data can also show a high occurrence of cancer in a particular region and possibly indicate a correlation between the incidence of cancer and high levels of toxins, a nearby chemical plant, or the screening practices of a particular hospital. As explained by Gigerenzer (2002),

> Too often in health care, "geography is destiny." For instance, 8 percent of the children in one community in Vermont had their tonsils removed, but in another 70 percent had. In Maine, the proportion of women who had a hysterectomy by the age of 70 varies between communities from less than 20 percent to more than 70 percent. In Iowa, the proportion of men who have undergone prostate surgery by age 85 ranges from 15 percent to more than 60 percent. (p. 101)

According to physician David Eddy, the differences come down to doctors following the "local pack" (Gigerenzer, 2002, p. 101). Although why (cause and effect) is not as easily determined (which we'll look at in the next chapter), it would seem that a second opinion is a good idea—preferably, in this case, from outside one's area. To be sure, how data is weighted and what the average and variance in the average are (which we look at next) can be a matter of life and death.

5.5. The Standard Deviation or Error in the Mean

To simplify a distribution of data, the mean is the most often quoted measure but is not the only useful statistic—the standard deviation (or spread in the data) also tells us much. Using the concept of moments, the standard deviation is no more complicated than the mean, which weights data about a central point (think again of physical weights on a teeter-totter). All we have to do is square the data first before adding, since it doesn't matter which side we're on but only how far away we are from the middle.

As we have seen, all measurements have an associated error—nothing can be known with perfect certainty—and in statistics the error is related to the second moment. In our distribution of 10 scores earlier, the average was 7.0, which was calculated from the product of the frequency and scores divided by the number of data points. To get the standard deviation, we do the same, but this time we *square* each score before multiplying by the frequency, and then sum and divide by the number of points.

As such, instead of the average (the first moment), we get the variance (the second moment; see Table 5.3).

But the variance isn't much use if we want to relate it to the mean, and so we have a little tidying up to do to get an error with the same dimension as the mean. To do this, we subtract the average squared (to center the moment) and then take the square root, and voilá, we have the standard deviation. Note that squaring and square-rooting is not mathematical trickery but ensures that the sum doesn't equal zero[14]—in essence, it is the distance from the mean that matters, not the direction.

Essentially, the standard deviation is a plus/minus on the mean (or the width of the error), showing how tightly or loosely packed a distribution is, and is a more often quoted measure than the variance. Furthermore, with the standard deviation, we can now easily compare different distributions[15] (always important, as we saw above with the mean). For example, bank withdrawals vary—some account holders withdraw a little and some withdraw a lot, e.g., $20, $40, $90 from Bank A and $49, $50, $51 from Bank B. Both have means of $50, but their spreads are clearly different, which when quantified tells us about the withdrawal habits of their respective customers.

Table 5.3 Weighted distributions (second moment, the variance)

Frequency $f(x)$	Score x	Product $f(x) \cdot x^2$	Distance^2 Δx^2	Moment $f(x) \cdot \Delta x^2$
0	0	0		
0	1	0		
0	2	0		
0	3	0		
2	4	32	9	18
0	5	0	4	0
1	6	36	1	1
2	7	98	0	0
3	8	192	1	3
2	9	162	4	8
0	10	0	9	0
$\Sigma = 10$	Total: $\Sigma = 520$ Variance: $520/10 = 52$			$\Sigma = 30$ $30/10 = 3$
	Standard deviation: $(52-49)^{1/2} = 1.73$			$3^{1/2} = 1.73$

Using the moment calculation, we get a spread of 29.4 for Bank A and 0.82 for Bank B (see below), and, thus, we can infer not only that Bank A has more wealthy account holders but that Bank B customers are more alike in their habits.[16] Most important, the average can now be explained more informatively: For Bank A, $50 ± $29.4 indicates a *spread* of withdrawals from $20.60 to $79.40, whereas Bank B has a much more narrow spread, $50 ± 0.82, i.e., ranging from $49.18 to $50.82.[17]

Bank A spread: $((20^2 + 40^2 + 90^2) / 3 - 50^2)^{.5} = 29.4$
Bank B spread: $((49^2 + 50^2 + 51^2) / 3 - 50^2)^{.5} = 0.82$

Even better, by standardizing the error, we can also compare completely different distributions, whether bank withdrawals and dart-throwing ability, heights and weights, or oil production and GNP. In each case, a measure of the spread allows for a general analysis regardless of the content. In fact, the study of error began in the 18th century when the Dutch-Swiss mathematician Daniel Bernoulli first compared bull's-eye errors in archery to aberrations in astronomical measurements and determined that the distributions were the same (Mlodinow, 2008, pp. 135–136), which led to the later formulation of the famous normal distribution by De Moivre and Gauss.

As can be seen in Figure 5.12, the normal distribution shows a spread of data about the mean (the first moment, μ) with a width (the square root of the second moment, σ). Here, the two distributions overlap ($\mu_1 = \mu_2 = \mu$), where the first distribution is narrower than the second ($\sigma_1 < \sigma_2$), not unlike our different bank withdrawal amounts. What is most important, however, is that any such distribution can be represented this way, no matter how small or large, by only two measures, μ and σ—whether heights of a five-player basketball team or ages of a 300-million-strong country.

5.6. Real–World Statistics and Advertising Abuse

Can such statistics help combat the daily abuses we are subjected to as consumers? One hopes so given the barrage of numbers and myriad unsubstantiated claims that abound today, a sample of which is cited next from but a cursory look at today's advertising world.

For example, for an advertiser to claim that "31 women over 12 weeks noticed an improvement" is absurd. What kind of women? How much of

Figure 5.12	The normal or Gaussian curve (mean, μ, and standard deviations, $\sigma_1 < \sigma_2$)

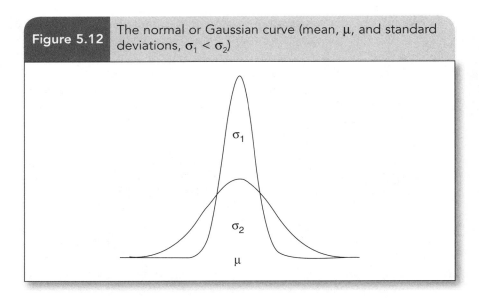

improvement? How many women did not notice an improvement? How many noticed a worsening? Was a group of similar women given a control treatment to compare against the women given the manufacturer's treatment to see if other factors accounted for the "improvement"? More important, what was the variance (or standard deviation) in the improvement? Here, if the statistical provenance were included, we could quantify how much of an improvement was noticed or how many women noticed a worsening and by how much.

Stating that "out of 166 women, 73% agree" fares little better, although the sample is almost 5 times larger and the data includes a count of women who don't agree (presumably 27%). But how much do they agree or disagree (a lot or a little)? Did some women strongly disagree? How strongly did they agree (weighting), and what is the uncertainty (plus/minus or standard deviation) in their agreement?

Another study of 292 people, claiming that "after 44 days, 82% felt better," is also better because of the larger sample but doesn't give any test conditions. The survey was possibly done in spring, when everyone feels better. Perhaps, after 44 days, the participants were happy to spend the money they were paid to do the study.

Unfortunately, the statistical provenance of such studies is typically unknown, without which the quoted numbers are of little use to us as consumers. As Goldacre (2009) noted regarding abuse of numbers and misleading newspaper reports about heart attacks,

I want to know who you're talking about (e.g., men in their fifties); I want to know what the baseline risk is (e.g., four men out of a hundred will have a heart attack over ten years); and I want to know what the increase in risk is, as a natural frequency (two extra men out of that hundred will have a heart attack over ten years). (p. 259)

Alas, some advertisers purposely flout statistics for effect, such as one soft drink advertiser that included a fake asterisk in the statement, "Makes you 73%* more attractive to the opposite sex (*Department of Completely Fictitious Statistics)," or the deodorant ad that touts, "More spray, more pull."

Even worse are the not-so-obvious yet equally dubious falsehoods found on many supermarket items[18] today, such as "50% extra free" or "buy one, get one free." But "extra for the same price" does not mean the same as "free," only that the price has been lowered, i.e., a price reduction of 50%—not something for nothing as the seller would have you believe. The so-called "drip pricing" on many online purchases today also hides the real costs to consumers, a practice most evident in the purchase of airline tickets. Such hidden costs make it harder for consumers to make effective price comparisons.

One wonders how so little regulation is placed on using numbers and statistics in advertising, given the moronic lengths to which "the truth" is stretched. Without appropriate weightings applied to the data, little inference can be made to determine the validity of an advertiser's claim. But including the sample size, mean, and standard deviation in any cited statistical study is easy. Without such measures, no consumer can be assured that the numbers aren't completely made-up.

5.7. The Law of Large Numbers:
A Simple Measure of Uncertainty

It is interesting to look at how the average changes as the sample size increases and how we can become more certain about future data points, which has applications in many fields of data analysis, from the stock market and other probability-related games to music encoding and filtering methods. The analysis can be applied to anything from test scores to the stock market, from the number of occupants in a vehicle passing a certain point at a given time to the number of strikeouts per game by a pitcher over a season. The question is whether one is more certain about a statistic as the sample size increases.

Figure 5.13 shows the sample test scores from before (top), illustrating how the average tends toward 7.0 (the calculated average) as the number of points is increased from 1 to 10. The other two data sets have been randomly generated (all numbers from 0 to 10) and also show the tendency toward the mean as the number of data points increases. An average of 5 is expected, seen with greater certainty as the number of points increases from 100 (middle) to 1,000 (bottom). This is referred to as the "Law of Large Numbers" or Bernoulli's theorem, after Daniel's uncle Jacob Bernoulli, who first proved it.[19]

In essence, large swings early on will affect a running average, but not as much as the run continues. Similarly, coin flips will tend toward 50:50 as the number of flips increases, the average dice throw will tend toward 3.5 with more throws ((1 + 2 + 3 + 4 + 5 + 6) / 6), and all casinos will ultimately win 2.7 cents for every dollar bet (see Chapter 7).

Figure 5.13	Average over 10 (top), 100 (middle), and 1,000 (bottom) points

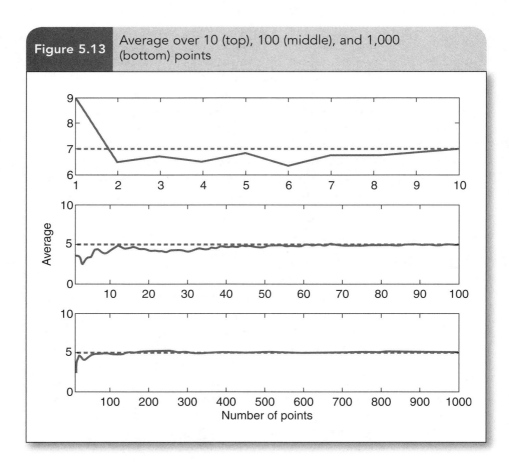

Hedge-fund managers use this principle to great advantage, when losing positions need to be held until favorable sell conditions exist. Deep pockets are essential to outlast other investors, although such strategies can fail from time to time with disastrous results when the price strays too far from the mean (i.e., a greater-than-expected deviation) and the bankers come calling. Patterson (2010) noted a record hedge-fund meltdown that lost $5 billion in a single week in 2006, as well as extreme fluctuations leading up to the 2008 credit crisis, which saw a swing of 25 standard deviations every day for 3 days wipe out hundreds of billions of dollars from paper balances (pp. 155, 238). In reality, however, such "never-to-be-seen" deviations are due to failed models, not a failure in the law of large numbers.[20]

A running average is also similar to "smoothing" or "flattening" data to make a distribution more manageable. It is much easier to understand a distribution of data points reduced to a single mean and standard deviation (e.g., our 10 test scores: 7.0 ± 1.7), but a running average also helps reveal any trends that may exist in the data. For example, in the stock market, running or rolling averages are especially used to smooth the data to see how a stock is performing over the long term. As shown in Figure 5.14 (see page 112), a set of 1,000 data points (top) is smoothed by a rolling average of 10 points (middle) and 50 points (bottom), making the data much easier to interpret.

Smoothing can also be applied to large data sets to reduce the size by selecting filtered points (e.g., every second or tenth point), from precise geographical vector boundaries to digitally encoded megabyte music files. But we must be careful because highly variable functions can lose data when sampled, depending on the frequency (number of points sampled), and can skip important information (or sounds in music files). Figure 5.15 (see page 113), shows the changing quality in a sampled signal as the number of points decreases from 1,000 (top) to 11 (middle) and then to 4 (bottom).

Clearly, in the reduced file (low-frequency sampling), all information has been lost, which is another way of stating that the certainty of a measure decreases with a smaller number of points or that the validity of a sample decreases the smaller the sample size.

5.8. Skewness and Recognizing Outliers: The Third Moment

Moments were used earlier to show how the mean and variance can be easily calculated (the first two moments, where the data is squared to

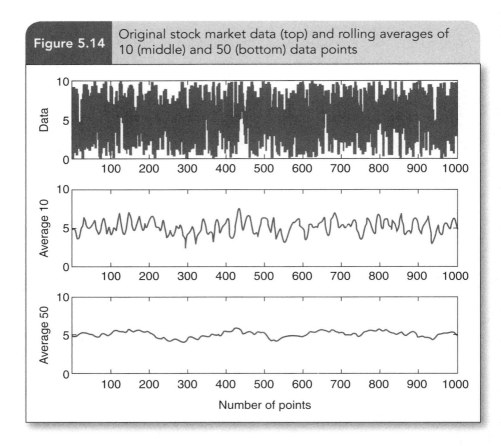

| Figure 5.14 | Original stock market data (top) and rolling averages of 10 (middle) and 50 (bottom) data points |

calculate the second moment or variance).[21] Higher-order moments, such as skewness (the third moment, where the data is cubed), are not any different and can reveal hidden information about highly varied distributions.

The third moment, skewness, measures the *asymmetry* in a distribution, which arises when the mean is not equal to the median, i.e., an unequal weighting left and right of the mean. The mean may balance opposing weights, as in our test scores example above ($2 \times 3 + 1 \times 1 = 3 \times 1 + 2 \times 2$), but such a distribution is clearly not symmetric (think again of one parent balancing three or four children).

Taleb (2004) applies this logic to the stock market, where he noted that it doesn't matter if a trader is right or wrong *on average*, only by how much he is right or wrong (pp. 98–103). For example, a trader could be wrong 90% of the time, but if he makes a 200% profit when right and loses only 10% when wrong, he will be up a great deal more than he is down in the long run. The skewness is a measure of this.

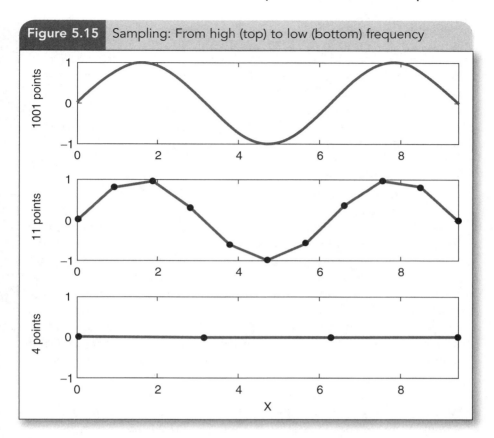

Figure 5.15 Sampling: From high (top) to low (bottom) frequency

In fact, Taleb (2004) especially aimed to take advantage of the rareness of such skewed events in his strategy as a stock trader, stating that because the past is not an indication of future performance (p. 108), one should instead plan for the rare event or outlier in a distribution: "There is a category of traders who . . . lose money frequently, but in small amounts, and make money rarely, but in large amounts. I call them crisis hunters. I am happy to be one of them" (p. 112). Interestingly, however, he suggested that a loss hurts more than a gain helps (p. 193), which doesn't follow mathematically speaking. All that can be said is that a loss and a gain are on opposite sides of the mean. How far from the mean (distance) quantifies the variance, and for particularly varied distributions, the skewness.

The same applies in blackjack, where occasional large bets are made with the hope of a big payout, for example, a blackjack in a card-counting system when a potential stacking of high cards accrues, as opposed to losing small amounts in a prior, small, flat betting method—a strategy used by a group of highly successful MIT gamblers, as depicted in the popular book and movie *21: Bringing Down the House*[22] (but can also attract the attention of suspicious pit bosses and so must be disguised). In soccer, long goal kicks also produce asymmetrical results since most turn over possession immediately to the opponent, a seemingly wasted strategy, but not if 1 in 10 results in a goal, a straightforward example of the importance of skewness. In football, a "hail Mary" pass is just that, a desperate nod to an unlikely but well-rewarded outcome.

Scientific data, particularly in the medical field, is also subject to skewness, where positive results are almost 20 times more likely to be published than negative results (Goldacre, 2009, p. 213), a clear example of an asymmetric reporting or publication bias. According to Goldacre, "This is more than cheeky. Doctors need reliable information if they are to make helpful and safe decisions about prescribing drugs to their patients. Depriving them of this information, and deceiving them, is a major moral crime" (pp. 215–216). Furthermore, positive results are often dressed up in different guises and republished many times over to appear more important, whereas negative results (e.g., the reporting of side effects) are hidden and researchers, who are funded by the pharmaceutical companies, bullied (pp. 215–216).

There is also an inherent asymmetry in the risk and reward associated with being long or short in the stock market (Soros, 2008, p. 166). For example, if one owns a stock (i.e., being long) one can, at most, lose the price of the stock (if it goes to zero), but if one agrees to sell a stock one doesn't have (i.e., being short)—say, sell now for $2 and buy back tomorrow for $1—and the price shoots through the roof, one can theoretically lose everything. As Soros noted, this creates an asymmetric market that discourages short selling (p. 166). However, the exact opposite asymmetry is the basis of a credit default swap (CDS[23]), which limits liability for traders going short on bonds and is at the core of current problems with the euro, where CDSs are being used to speculate *against* countries with poor credit ratings, picking them off one by one by betting against their future solvency. As well as being particularly toxic (p. 169), such practices put pressure on a country's credit worthiness, hampering its ability to issue further bonds at manageable rates. This has occurred in Greece, Ireland, and Portugal, where bond prices have become prohibitively expensive and where, as a result, the governments are now fighting solvency issues because of increasing national debts.

How this can happen and whether it should be permitted, including the effect of CDSs on the collapse of Lehman Brothers and the origins of the 2009 global economic meltdown, will be examined in Chapter 12. For now, suffice it to say that a complicated financial derivative—unregulated at that—is being used to encourage casino-style bond raiders because of an asymmetric risk/reward that favors short-selling credit derivatives. We can also wonder, as Soros (2008) did, why "the issuance of stock is closely regulated by the SEC, [but] not the issuance of derivatives and other synthetic instruments?" (p. 169).

5.9. The Science of Longitudinal Weighting: A Question of Time

We've seen how weighting is used to determine the mean (first moment), the variance (second moment), and skewness (third moment), applied to distributions as different as polls and the stock market. Weighting also applies in the time domain—for example, for loans and mortgages—albeit a bit more complicated because of a hidden *exponential* weighting.

Pay Me Now or Pay Me More Later

Interest rates are set by the Federal Reserve in the United States and by the European Central Bank in Europe, both state organizations independent of political affiliation. The rate is set as a matter of monetary policy to macroengineer the economy—increase the rate when spending is high to reduce spending and decrease the rate when spending is low to stimulate investment—and is typically passed on to private lending institutions. When a new rate is announced, the economic dominoes fall as central banks pass on the change to private banks, who in turn pass it on to their customers.

But how are loans and mortgages calculated from these interest rates? It seems reasonable to get a loan for which you have sufficient collateral or the promise of future earnings, determine when you can pay it back, and divide the amount (A) by the number of months (n), plus some money to pay the bank for services rendered (m): Payment $P = (A + m) / n$.

Or, as is becoming popular in poorer countries, one can obtain "microfinancing" (or people's banking), where loans are advanced for as little as $100 without collateral. Two such lending institutions, Bangladesh's Grameen Bank[24] and Bolivia's BancoSol, reported that "less than 3 percent of loan repayments are late and that default rates are still lower—a record that is superior to that of corporate customers in many developing nations" (Stix, 1997).

In a way, that's how friends make loans. If I borrow from a friend, I return the money at an agreed future date along with my thanks (and perhaps a potted plant or a meal out). If I pay back the loan a little at a time, I still do the same when the debt is cleared to show my appreciation. One wonders if that was how borrowing began, before the big banks came along with their profit-motivated interest charges.

In the case of banks, of course, we borrow and pay back in money. That's what a bank is, a lender of money to whom interest must be paid along with the loan. The amount can be a flat rate that decreases in proportion as the amount increases, for example, at an ATM where a pensioner who withdraws $20 and a businessman who withdraws $200 both pay a flat amount (e.g., $2). Or it can be a percentage of the borrowed amount, for example, with traveler's checks or prepaid debit cards, where I pay $14 to buy $200 or $140 to buy $2,000 (e.g., at 7%).

In the case of loans and mortgages, however, the charges are compounded in time according to the equation $FV = PV(1+i)^n$, an exponentially weighted equation that underpins the whole of our consumer world. Not to worry, PV (or present value) is just fancy talk for how much you borrowed and FV (or future value) is how much you pay back (i is the interest rate, and n is the number of periods of the loan). But here's where things gets funny—the derivation of this equation requires a PhD in mathematics.

Suffice it to say, you pay a minimum amount of principal and a corresponding maximum amount of interest at the start of the loan, where with each repayment the principal increases and the interest decreases throughout the term of the loan, such that by the last payment you will have paid back considerably more than you borrowed (i.e., interest is charged on the interest). For example, for a home loan of $100,000 at 5% interest monthly over 15 years, you will have paid back $211,370.40 or more than twice what you borrowed ($211,370.40 = $100,000 $(1+.05/12)^{180}$). On a $250,000 house mortgage over 30 years at the same terms, you will have paid back more than 4 times the purchase price ($1,116,936.08 = $250,000$(1+.05/12)^{360}$). It's not hard to figure out how the rich get richer or how consumer debt has reached record levels.

Some will say if you want to buy a house and don't have the money, there's nothing else you can do. We could argue the value of banking practices that make house ownership more expensive, especially for lower-wage earners. We could also argue that a simpler (flat) loan repayment plan would make lending simpler and easier to understand and, thus, more accessible: $P = (A + m) / n$. Instead, I will suggest a strategy for not paying any interest, although only for the less expensive items that go into all the overpriced homes. The banks aren't going to give up their lucrative usurious practices for nothing, but we can at least not play them at their game.

We've all heard the spiel, "Immediate delivery, no money down, don't pay a cent until next century, no interest for a year, special offer," all from some guy yelling on television that the world is about to end. "Come on down, don't delay, limited supply, we're practically giving it away." However, they know you can't afford it—that's how they get you. Why else would they be "practically giving it away"?

As such, when the payments begin, you dutifully send the checks or have the money direct debited from your bank account each month for no-fuss maintenance. And thanks to exponential interest (and that PhD equation), by the time you pay back the $2,000 reduced to $1,000 on the must-sell-today, pullout sofa, you will have paid back much more than the supposed limited-supply, reduced price—pretty clued-in behavior for a must-sell seller, all thanks to the financial edge of exponential weighting.

So what can one do? Well, the slow and sure way to pay less is to make payments into a dedicated account immediately after you buy a don't-pay-for-a-year, future-pay product, such that when the first, all-interest, no-principal payment comes around, you already have the purchase amount. I know that sounds like saving—anathema in our heady, free-spending world—and it is, but with a trick. You have to make the payments *anyway*, so why not make them *immediately* into a dedicated account to take advantage of the no-interest-for-a-year period, rather than waiting and ultimately paying more. Saving is saving *before* you buy a product, but "no-money-down saving" is saving *after* you acquire a product but *before* you make the payments. You still take advantage of their offer; you just don't pay their ridiculous compounded interest charges.

Should Credit Cards Be Called Debt Cards?

In the same way, it's crazy not to pay off the full balance on a credit card every month, especially since credit buying has become an integral part of everyday living. Started by an enterprising American automobile industry in the 1920s, credit seemed the obvious solution to ease the burden of a purchase price beyond the reach of most buyers—up to 20% of family income (Boorstin, 2000, p. 426). Facilitated by GM, Ford, and other niche lending cards (p. 426), credit quickly became entrenched as the currency of our times and is now worth about $2.4 trillion or $8,000 per person, a third of which is "revolving," i.e., the ongoing credit card type ("G.19 Consumer Credit," 2011). According to the U.S. Census Bureau (2011a), almost 1.5 billion credit cards were in circulation in 2008. What's more, some young-adult households spend about 25% of their income on debt repayments (Barber, 2007, p. 137), not to mention so-called "rent-to-own" scams (which can end up costing more than 3 times the purchase price).

Of course, casually maxing out a credit card and letting the maximum sit while paying double-digit interest rates is just plain stupid (store cards are the worst for rip-off rates, which can be up to 30%) and amounts to billions of dollars in penalty charges each year. Or, worse, using one credit card to pay off another. Or, worse still, paying a mortgage with a credit card, a practice that has increased significantly since the 2008 downturn. You know you are in serious trouble when you are borrowing from Peter to pay Paul.

To be sure, credit card companies are in the business of making money and will do what they can to enhance their bottom line. As noted by the editor of *Which?*, a magazine dedicated to helping consumers make informed choices,

> The credit industry has an alarming number of tricks up its sleeves to wring every last penny out of its customers. Lenders seem to have no qualms about persuading people to take on more debt than they can afford and they'll carry on doing it as long as they can get away with it. (Bennetto, 2005)

And the credit worthiness of the customer doesn't seem to matter. As Boorstin (2000) noted regarding the origins of credit debt in early car ownership, "As installment buying became the normal way, the personal qualifications for securing installment credit became lower, or virtually disappeared. Almost anyone could buy a car on time" (p. 426). At the same time, "the lenders' risks were being justified by hidden high interest rates, by the accumulation of voluminous statistics on the resale value of cars of all makes and ages, and by information available from consumers-credit-rating agencies" (p. 426). As Galbraith (1958/1999) noted, the making of goods is intimately connected to "the process of persuading people to incur debt, and the arrangements for them to do so" (p. 145). To be sure, the moment credit became available, it fast became a fact of life in a revamped American consumerist society.

It has also been argued that excessive personal debt, primarily in the form of mortgages (Krugman, 2009, p. 247), played a major part in the 2009 financial meltdown, when U.S. home sales slumped because inflated prices bought by cheap credit could not be sustained. As Jackson (2009) noted, "[maintaining] debt-driven materialistic consumption . . . destabilized the macro-economy and contributed to the global economic crisis" (p. 178). He added, "The danger is that many advanced economies are already at the limits of consumer indebtedness and face a sharply rising public sector debt as well. Pushing these any further stretches the boundaries of financial prudence" (p. 105).

Quite simply, we owe too much, both personally and collectively (as in the national debt, which we will look at more in Chapter 9), although one has to wonder to whom all this debt is owed, given that the economic system is a closed one. To be sure, not all of us are in debt; only most of us.

What's more, credit cards are not given to customers because of their *ability* to pay but because of their *inability* to pay—why do you think they keep raising your limit? But for those who are at the end of their tether, one can refinance credit card debt in a bank loan at a considerably lower rate, although the credit card companies—mostly fronted by banks—don't tell you that (the best place to start is with a nonprofit credit counseling agency). The important message is, don't be fooled by today's institutional sellers, who know their clients' weakness and encourage them to stay indebted, as though pretending to throw a lifeline to a sinking swimmer.

Of course, the real question is, why are we so casual about buying on credit? Continually putting off paying is insane and suggests a deeper psychological problem with how we manage money, perhaps in line with a constantly stressed-out, carrot-and-stick, paycheck-to-paycheck mentality, where life is held together by elastic bands and sticky tape. Sadly, we're all in it now, but one would think we could at least legislate against rip-off credit card companies.[25]

Perhaps if credit cards were called *loan* cards, the penny might drop. In fact, credit cards should be called loan cards, and bad loan cards at that. Living beyond one's means may be the modus operandi of our modern consumerist society, but it shouldn't be as a matter of course. Removing personal (and national) debt should be everyone's priority.

Better yet, a return to more modest living within one's means is called for. The simple premise of a popular book in the '80s called *The Wealthy Barber* was to put 10% of each paycheck aside immediately, no matter what your occupation, and in 20 years you would be a millionaire. *The Wealthy Barber* was a prescription for keeping wants in line with means, as in the corollary to Parkinson's law stating that needs expand or con-tract to meet income. Essentially, have more, spend more; have less, spend less. An earlier article, "Everyone Ought to Be Rich,"[26] published in a 1929 *Ladies' Home Journal*, said more or less the same. The simple message of all such self-help guides is, "Don't spend beyond your means." In fact, saving is rewarded. But whether one becomes a million-aire depends on staying the course, and, indeed, the first step is the hardest, as in any 12-step program.

Unfortunately, many of us can't save a dime and are exhausted like Aesop's hare after maxing out our credit cards—unlike the smart tortoise, who saves bit by bit and comfortably lounges after the race in her new pullout sofa, not having paid one red cent of interest. Whether the pull-out sofa is worth the supposed knockdown price, well, that's another story. And why "must-end-soon" sales NEVER end, well that's a whole other story altogether (see Chapter 10).

As for the extra money we end up spending because of these cooked-up equations, the mathematics is reasonably straightforward. As we saw above, $FV = PV(1+i)^n$, where FV is what you owe (future value), PV is the loan amount (present value), i is the interest rate per month, and n is the number of periods. And thus, for a $1,000 loan over 36 months at 23% interest, you pay $1,980.74—almost twice the amount of the loan, as shown below.

23.00%: $FV = \$1,000(1 + .2300/12)^{36} = \$1,980.74$ or $55.02 per month

Of course, the numbers are never given in even amounts, and, thus, we get rates such as 22.99% on a $999.99 purchase, where we're meant to think that what we're buying costs less. The difference in using such fractured numbers?—a whopping 2 cents a month (less than $1 over the term of the loan).

22.99%: $FV = \$999.99(1 + .2299/12)^{36} = \$1,980.14$ or $55.00 per month

Obviously, the $0.01, 0.01% difference isn't that much, but when mul-tiplied by millions of customers (say, the adult population of the United States) it does add up—to the tune of more than $100 million. Again, lots of small adds up to one big, but imagine how much they're making on the *overall* loan if they can afford to throw away $100 million on a $0.01, 0.01% *difference*.

To be sure, the devil is in the detail, the decimal and exponential detail—not to mention larger loans and mortgages, which when tied to impossible repayments by owners who have no associated risk can have disastrous effects on the economy (as we will see in Chapter 12). But any way you slice it, the practice is usurious—that is, charging interest on loans or making money from money.

Once considered a mortal sin by the Church, usury has become an established way of life. Galbraith (1991) cited two of our great moralists on the subject of usury and interest: Aristotle and Saint Thomas of Aquinas. Aristotle (384–322 BC) wrote, "The most hated sort [of money-making], and with the greatest reason, is usury. . . . For money was intended to be used in exchange, but not to increase in interest" (p. 12). More than 1,500 years later, Saint Thomas of Aquinas (1225–1274) had this to say:

> There are two kinds of exchange. One may be called natural and necessary, by means of which one thing is exchanged for another, or things for money to meet the needs of life. . . . The other kind of exchange is that of money for money or of things for money, not to meet the needs of life, but to acquire gain. . . . The first kind of exchange is praiseworthy, because it serves natural needs, but the second is justly condemned. (p. 27)

Today, one has only to turn on the television to hear the great moralists of our time shouting at the top of their lungs for us to buy, where lending with interest is now a free-for-all. Go forth, son, and forever be indebted.

One wonders if a loan shark might be a better option. The rates are higher, but then a loan shark doesn't insult one's intelligence with 0.01%, $0.01 reductions, or must-end sales that NEVER end. Nor does he bother with a PhD equation. Future value, 22.99%, $999.99—there ought to be a law.

5.10. The Push and Pull of Change: Prices and Inflation

Historical or "longitudinal" data, as economists call data over time, is essential to predicting trends and is another example of weighted data. Stock prices are given over time to calculate the changing equity of a company, as are the Dow Jones Industrial Average, FTSE 100, and other indices that attempt to represent the market. Inflation is also recorded versus time to measure changing prices or, perhaps more important, for the financial world to determine the worth of future investments—there is no point in buying a government bond at 4% if inflation is running at 5%. Similarly, social security or pensions are of little use if not indexed to inflation.

And yet, calculating inflation or comparing economic data from one era to another is hard, just as comparing an athlete from the past to one today is hard, because of changing standards (e.g., competition, technology, fitness). For example, does a seventyfold increase in the price of a house from $10,000 in 1970 to $700,000 in 2010 correctly reflect the increase in price when inflation has been running at 5% over the same 40 years? The effect of inflation must be taken into account.

Traditionally, a basket of items is used to gauge inflation, which depends on the items in the basket. Basket 1 (milk, bread, eggs), basket 2 (color TV, computer, mobile phone), basket 3 (Ford Pinto, Concorde flight, gas), and basket 4 (actor salaries, athlete salaries, teacher salaries) will produce different sample inflations. What's more, a sample basket can introduce significant errors if it doesn't take account of constantly changing products or excludes new products. One must compare like with like. And apples aren't oranges.[27]

The mathematics, however, is straightforward to calculate the effect of inflation, as in the house example above. Assuming a flat inflation rate of 5%, the increase of $10,000 over 40 years is $1.05^{40} = \$70,400$;[28] thus, if the house sells for $700,000, only about 1/10 of the increase is attributed to inflation and the rest is due to an inflation that hasn't been included in the gauge. To achieve a seventyfold increase over 40 years, the real inflation rate is more like 11% (i.e., $1.1121^{40} = 70.11$).

Economics 101 says that an increase in price is due to an increase in demand or a decrease in supply, i.e., more people *want* houses than there *are* houses. The big question then is, how can the price of a house increase almost 10 times more than that due to inflation but not be reflected in the given rate? The real concern is that house prices are not properly weighted in the rate of inflation and that by quoting a phony inflation one's purchasing power is devalued and comparisons to increased wages, taxes, and living can't fairly be made.

Fake house prices generated by government-sponsored housing booms are also not the result of a real interplay between supply and demand and serve only to increase government coffers through increased taxes. Couple that with a banking sector only too happy to increase leverage to extend lending and we have a recipe for disaster.

Nonetheless, real inflation is not easy to gauge. The Department of Labor provides the Consumer Price Index (CPI), the percentage change of which is the government-recommended measure of inflation—although somewhat skewed because people don't buy the same things from month to month and it's weighted according to category. To explain what goes into the measure, the U.S. Census Bureau (2009) gave this summary and described the general contents of the standard inflation basket:

> The CPI is a measure of the average change in prices over time in a "market basket" of goods and services purchased either by urban wage earners and clerical workers or by all urban consumers. . . . BLS [Bureau of Labor Statistics] publishes CPIs for two population groups: (1) a CPI for all urban consumers (CPI-U), which covers approximately 80 percent of the total population; and (2) a CPI for urban wage earners and clerical workers (CPI-W), which covers 32 percent of the total population.
>
> The current CPI is based on prices of food, clothing, shelter, fuels, transportation fares, charges for doctors' and dentists' services, drugs, etc. purchased for day-to-day living. Prices are collected in 87 areas across the country from over 50,000 housing units and approximately 23,000 establishments. . . . All taxes directly associated with the purchase and use of items are included in the index.

Using the CPI as described above, Figure 5.16 (top) shows the changing price of goods in the United States from 1946 to the present ("Historical CPI-U Data From 1913 to the Present," 2011), showing how a CPI-averaged item that cost 18 cents in 1946 now costs more than $2, a more than tenfold increase in almost 60 years. In the bottom graph in Figure 5.16, the inflation rate is constructed from the CPI data (as we did in Chapter 3 with regard to the changing Dow Jones).

From the figures, one can easily see the particularly high inflation rate of the 1970s, which brought about tough anti-inflation monetary practices by the Federal Reserve and have arguably contributed to the more freewheeling investment times of the past 30 years. Note also the sharp drop in 2009 as a result of the credit crisis, when decreased consumer confidence brought about greatly reduced prices.

But can we be certain of the numbers? In China, where inflation is starting to increase regularly, the debate about how the CPI should be weighted is also heating up. As noted in *China Economic Review*, China's CPI was recently reweighted, with food, entertainment, tobacco, alcohol, and other goods decreasing while housing increased ("Let the Deed Show," 2011). Furthermore, as noted, "Statistics in China have long been suspect. At one time, discrepancies were intentional; today most problems arise from the inherent difficulties of surveying such a large country." Here, the measure is made all the more difficult by sheer size, although the limitations in the counting method still apply. As for real inflation, the calculation is anyone's guess.

Figure 5.16	Consumer Price Index (top) and corresponding inflation rate (bottom) from 1946 to present

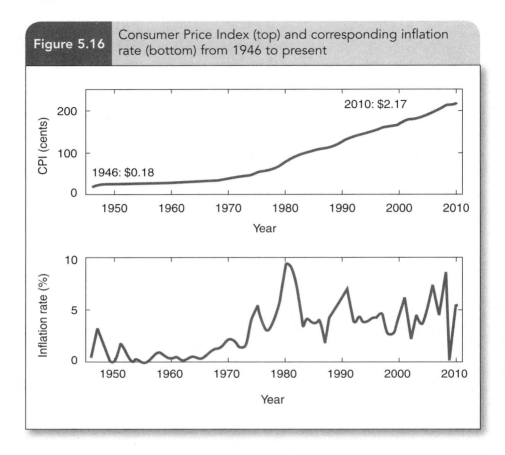

Oil Prices and Phil Mickelson's World Ranking

To illustrate how inflation confuses the real price of goods, Figure 5.17 plots oil prices from 1946 to the present ("Historical Crude Oil Prices," 2011) in actual dollars and adjusted for inflation. Here, we see that one cannot realistically use oil prices as an economic measure unless inflation is accounted for, since the adjusted values can be quite different from the actual values—historically, much less.

A particularly good example of such weighting can also be found in the PGA World Golf Rankings, which reflect a greater importance for recent tournaments. In an attempt to gauge current form as well as past performance, the PGA updates its rankings weekly, weighting all tournaments in the previous 3 months by a time factor of 1, while adjusting tournaments from the previous 21 months by a decreasing factor (a type of "exponential smoothing" that can aid statistical forecasting).

For example, when Phil Mickelson won the 2010 Masters, he received 100 ranking points, giving him 100 points at the time, but when the 2010 rankings were calculated 39 weeks later, his Masters win was adjusted to 86.96 points (100 × 0.8696). As such, a weighted rolling average attempts to convey meaning to current and past performance. To illustrate the overall effect of the PGA weighting, Figure 5.18 shows Mickelson's weighted and unweighted ranking (a result not unlike the effect of inflation on oil prices).

The problem with weighted data, however, comes when measures are not quoted with sufficient frequency and are too slow to represent the change effectively. For example, there is no point reporting a quarterly inflation rate or GNP, especially during times of high volatility (as we addressed briefly in Chapter 3).

One example that highlights the absurdity of quoting economic measures over a 3-month period occurred in the last days of the Gordon Brown-led British government, when he announced that Britain was finally out of recession, stating that the economy had grown by 0.1% over the previous 3 months. But 0.1% is so minimal (about £2 billion or £25 per capita) that it is less than its associated error and may well be meaningless, not to mention that such a perceived short-term economic improvement may not have been the result of government policy.

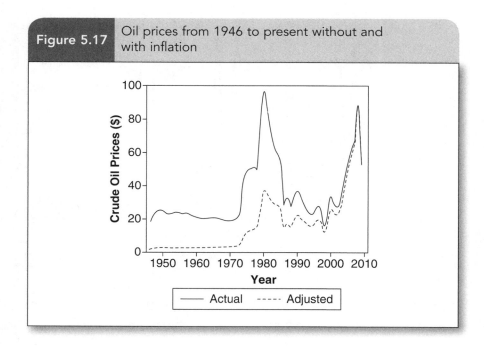

| Figure 5.17 | Oil prices from 1946 to present without and with inflation |

What's more, the stated 0.1% growth likely wasn't evenly distributed such that the measure even applied at the time of the quote (not dissimilar to a golfer going through a rough patch but still being ranked highly, as seen in Phil Mickelson's ranking). More likely, the change was stepped and the measure different from that quoted. It is even possible that Britain's economy was shrinking at the time the measure was announced.

Of course, such figures are politically motivated, in this case allowing the troubled Labour party to claim that Britain was out of recession—a claim not backed up, however, by the quoted quarterly growth figures. Furthermore, as noted by Mason (2009), growth statistics take up to 6 months to solidify, which during the 2008 meltdown gave "the world's finance ministers a whole half-year to avoid using the word *recession*" (p. 115).

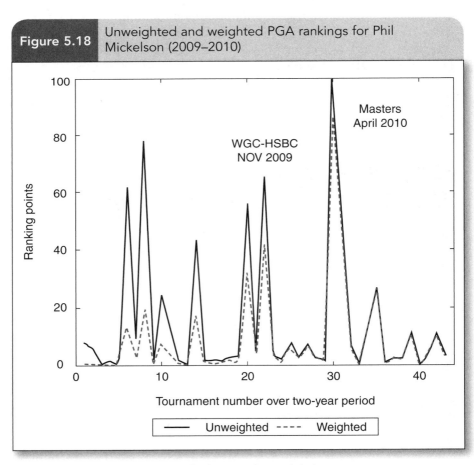

Figure 5.18 Unweighted and weighted PGA rankings for Phil Mickelson (2009–2010)

SOURCE: http://www.officialworldgolfranking.com/home/default.sps

Wind power comes under the same scrutiny—for example, when the world's largest offshore wind farm was credited with powering 200,000 homes. Wind power is intermittent and the deliverable power highly varied (as well as requiring a constant alternative backup). At any time, interval measures can differ greatly from instantaneous measures; so it does no good to rate a turbine at maximum efficiency if the average wind speed is a quarter of maximum load (say, 10 knots out of a typical maximum 40 knots), or indeed when there is no wind. And it does no good to quote a static measure in rapidly changing times.

Changing With the Times

Governments also regulate prices beyond the supposed free-market interplay of supply and demand, such as prices for public transit, health, and utilities, not to mention taxes, regulations, and bailouts (which we will look at more in Chapter 9). Prices can also be manipulated from their so-called true point, as in oil cartels jacking up prices for profit, excess farm produce being destroyed to maintain prices, and even Christmas "sales" cynically playing on emotional giving and, thus, not being correctly reflected in the index. But if the true cost of goods and services are so manipulated, how can inflation be properly gauged?

In many countries, the RPI (Retail Prices Index) is used, but this index doesn't accurately reflect economic trends either. In early 2009, Britain's RPI dropped to 0%—a 50-year low—while Spain reported a decrease of 1%—the first euro-currency country to do so, although similar patterns soon emerged in Portugal and Ireland. In fact, during the height of the downturn it was possible to get 20% off a meal almost anywhere in Dublin—one such popular restaurant cut prices by 40%—yet price indices were little changed. Suffice it to say, changing prices confuse the economic landscape, such that one can't gauge inflation or deflation with any degree of confidence. Moreover, in today's fast-changing world, one can't gauge change even from one day to the next.

Of course, some believe that inflation is kept low primarily to keep bondholders happy—no point in investing in the future if inflation is going to wipe out the gains at low bond yields—while others believe that some countries, including the United States, may soon choose to induce high inflation as a strategy to devalue their debts, despite the hit to bonds and the resulting disincentives to save.

However, without stable prices, we all suffer, especially those on fixed wages not indexed to inflation (e.g., minimum wage, welfare, and social security). Furthermore, constantly changing economic times produce spirals—increased prices that increase wages that increase prices and so on as politicians tinker with their spreadsheet economies, sometimes with alarming consequences, as in the early 1970s when wage and price controls were introduced in response to excessive double-digit inflation. Or in 2009, when prices tumbled, ultimately bringing wages with them and the threat of a deflationary spiral. In such deflationary times, debt becomes doubly worrying as prices and wages slump but debt stays the same. As Mason (2009) noted, "This 'debt-deflation spiral' was a real possibility in the first three months of 2009" (p. 175).

Such imbalances can be compared to pulling a load, where the rope becomes taut and then slack as the pulling force changes. Here, the rope is the measure, but to understand the changing reality, one needs to look at the load, not the rope. Reducing lag time in an economy or business is paramount to success, as can be seen in the failure to recognize changing trends and emerging needs.

What happens when needs change but an industry doesn't is especially seen in the American auto industry, which failed to introduce smaller fuel-efficient models in a changing market, losing out to foreign competition with catastrophic consequences—alarmingly evident in Detroit, where the population has halved since the 1970s (Mason, 2009, p. 83) and unemployment is now higher than the worst of the Great Depression (Reid, 2009). In this case, mismanagement was directly related to a poor understanding of rapidly changing times.

Competition from China, which plans to add almost 100 million cars to its roads by 2020 (Zakaria, 2009b, p. 31), and Ontario, which provides workers with subsidized health care amounting to $1,500 per car (Zakaria, 2009b, p. 208), means that GM, Chrysler, and Ford must produce more efficiently to keep up. The opportunities are there, but American car companies can no longer expect to be the biggest and best in the world (GM was global sales leader for 77 straight years until replaced by Fiat in 2008; Moyo, 2011, p. 78). They must compete in a world where "everyone is playing America's game, and playing to win" (Zakaria, 2009b, p. 206). Simply put, institutional inertia is not good for the bottom line.

Lowering prices could also help Detroit compete more, but wage restraints make this difficult. If automakers could function as part of an independent economy, they could also adjust exchange rates when sales decline—the ultimate in monetary feedback control, a policy that is nonetheless unavailable to a regional spoke in a national wheel. Similar problems exist within the European Union, where misfiring smaller member states are stuck with a single exchange mechanism and cannot devalue their currency to make up for lower economic activity.

Whether such unbridled competition is good for the overall economy is another question, one we'll tackle later with regard to interconnected systems and zero-sum and non-zero-sum games. For now, we recognize only that the change in a measure is as important as the measure itself and is essential if we want to know how or whether things are improving. In today's fast-changing world, no one can afford not to recognize change—in the landscape or in the data. And no one can afford not to recognize the effect of weighted data.

Notes

1 **American advertising expenditure**: Robert J. Coen, forecasting director at McCann-Erickson for more than 60 years, reported that revenus were about $5 billion in 1948 (Elliott, 2009a) and are now almost $300 billion (Coen, 2007). American advertising also accounts for almost half of global advertising (Barber, 2007, p. 11; Coen, 2007). Internet advertising, which is growing at the expense of print and television advertising, is expected to account for about $25 billion or more than 10% of advertising revenues in 2011.

2 **Myers-Briggs modified Jungian archetypes**: There are 16 sets (2^4) of either/or opposite pairs used to determine personality type (extrovert/introvert, intuitive/sensing, feeling/thinking, perceptive/judging).

3 **The mean, median, and mode**: The term *mean* is used interchangeably with *average*. *Median* (the middle value in a distribution) and *mode* (the most common value in a distribution) are not used to discuss trends.

4 **Hollerith constants**: In honor of Herman Hollerith, early versions of FORTRAN used "Hollerith" constants to manipulate characters.

5 **2000 United States presidential election vote count**: Al Gore, 50,999,897 votes (48.4%); George W. Bush, 50,456,002 votes (47.9%), a difference of fewer than 500,000 votes in a total of more than 100 million cast (State Elections Offices, 2001).

6 **2010 British general election seat and vote count**: The Liberal Democrat party received 55 out of a total 649 seats (8.5%) on 6.5 million out of a total 28.5 million votes cast (23%).

7 **Alternative voting (AV) referendum**: The resultant AV referendum was defeated 69% to 31%.

8 **Scrabble probabilities when *M* is drawn first**: With 99 tiles in the bag after an *M* is drawn, the probability of drawing less than an *M* is 54% (53/99) and the probability of drawing greater than an *M* is 45% (45/99). Since there is also one *M* left in the bag, there is a 1% chance (1/99) of drawing another *M* and thus necessitating a redraw.

9 **Summation becomes integration**: $\Sigma \ f(x) \cdot x \rightarrow \int f(x) \cdot dx$.

10 **Longitude and latitude**: One degree latitude (111 km or 69 miles) is about two degrees longitude (56 km or 35 miles) at the equator. To calculate the distance between two degrees of longitude at other latitudes, multiply by the cosine of the latitude.

11 **The flat earth**: Note that the Internet is less susceptible to the same geographical constraints since fiber-optic cables and routers have achieved sufficient global saturation to

"flatten" the earth. Photons and electrons travel much more efficiently than do people and goods.

12 **Population scale and aspect ratio**: For all population figures, the population has been normalized to the largest city and the aspect ratio maintained such that 1 degree latitude = 2 degrees longitude times the cosine of the midlatitude. There will be some distortion because of the projection of spherical coordinates onto a rectangular grid, which exists in all map projections (see the Gall-Peters and Mercator projection maps in Chapter 3).

13 **Ireland and the United Kingdom**: The 10 largest cities in Ireland, including Northern Ireland, and the 10 largest cities in the United Kingdom, excluding Northern Ireland, were used.

14 **Root mean square**: Summing deviations from the mean and then averaging will always add to zero. For example, in our test scores, the deviations are −6, −1, 3, and 4, which sum to 0 and are thus of no use to calculate error. In the same way, root-mean-square (rms) values are used for electrical voltage, since the average of a sinusoid is always zero (equal amounts above and below zero sum to zero). Grid voltage in North America is rated at 120 V (rms), but the peak voltage is actually 170 V (340 V peak-to-peak).

15 **Number of data points**: The number of data points will affect the variance by lowering the importance of a single outlier, as one would expect when the size of a data set is increased.

16 **Standard deviation formulae**: The population standard deviation uses n in the denominator, whereas the sample standard deviation uses $n − 1$. In the examples given here, the population standard deviation is calculated.

17 **Standard deviation of a normally distributed data set**: If the data is normally distributed (e.g., coin flips, chest widths, bakery bun sizes, etc.), the mean ± 1 standard deviation represents 68% of the data, the mean ± 2 standard deviations 95%, and the mean ± 3 standard deviations 99%. Standardized z scores are used to calculate a probability from a width or conversely a width from a probability, where the area under the normal curve equals 1. If the distribution is flatter than the normal curve it is "platykurtic" (from the Greek *platus* for fat) and if thinner "leptokurtic" (from the Greek *leptos* for thin).

18 **Something for nothing**: A casual survey will show up the likes of After Eight ("450g for the normal price of 300g"), Roma spaghetti ("750g for the price of 500g, giving you more wholesome ingredients"), or Dubliner cheese ("300g for the price of 200g"). Many products spout the dubious claim, "buy one, get one free."

19 **Bernoulli's Theorem**: As Bernoulli elegantly wrote in his *Ars Conjectandi* of 1713, "The more observations that are taken, the less the danger will be of deviating from the truth" (Hald, 2003, p. 257).

20 **A power-law stock market mechanism**: Preis and Stanley (2011) noted that stock markets may, in fact, follow an inverse quartic power law rather than a binomial random walk (for which large standard deviations rarely occur); for example, if there is a probability, p, of a $5 price change, then there is a probability, $p/16$, of a $10 price change. Furthermore, they noted that bubbles and crashes are scale free, operating similarly on the millisecond and quarterly scale, knowledge of which is essential for spotting upturns and downturns.

21 **Moments of a distribution**: $\mu_n = \Sigma f(x) \cdot x^n / \Sigma f(x)$, where $n = 1$ for the mean, $n = 2$ for the variance, $n = 3$ for skewness, and $n = 4$ for kurtosis (a measure of the shape of the distribution). Note that for $n > 1$, moments must be centered about the mean.

22 **Beat the house and the stock market**: Blair Hull, one of the original MIT team members, read Edward Thorp's seminal books *Beat the Dealer* and *Beat the Market* and

parlayed his $25,000 MIT stake into a high-frequency options trading firm that was sold to Goldman Sachs for $531 million (Patterson, 2010, p. 60).

23 **Credit default swap (CDS)**: A type of derivative whose value is related to an underlying security. Patterson (2010) noted that CDSs are not about buying and selling but the perception of default (p. 94).

24 **Grameen Bank**: Muhammad Yunus, the 2006 Nobel Peace Prize winner, set up Grameen Bank in the 1970s in his native Bangladesh to help reduce poverty in the poorest rural areas, particularly for women who could not obtain credit. The prize was awarded to Yunus and Grameen (which means "village") "for their efforts to create economic and social development from below." In the Grameen setup, borrowers create a group of five women, two of whom get 1-year loans. After the repayments are made, the other women receive their loans. There are more than 7 million customers with $7 billion in loans in Bangladesh. The repayment record is 98.5%.

25 **Credit card legislation**: In the United States, the Credit Card Accountability, Responsibility, and Disclosure (CARD) Act was passed on February 22, 2010, and called for credit card companies to stop charging hidden fees, allow customers to opt out of new limits, and give 45 days' notice of rate hikes. In Britain, the government has introduced some restrictions on credit card companies, such as not changing the interest rate midpayment and requiring customer permission to increases limits.

26 **"Everyone Ought to Be Rich"**: John J. Raskob's article was published 2 months prior to the great 1929 crash, which may have thrown a small wrench in his planned rate of returns. Nonetheless, all such prescriptive wealth guides offer sound saving advice and will work for those who can stay the course, provided, of course, that not everyone does the same. Monetary wealth is relative, and, thus, it is impossible for everyone to be rich (see the zero-sum game in Chapters 9 through 12).

27 **Irish inflation**: A former Irish talk show host quoted the price of a basket of goods on his show, commenting how it didn't change much, until it was discovered that the supermarkets knew which products were in the basket and had rigged the prices (Casey, 2006).

28 **CPI inflation data**: If one uses the CPI-U inflation index data instead of a flat 5%, the increase in house prices due to inflation over 40 years is even less (5.7 times less, or $57,000 instead of $74,000) and, thus, is even less accounted for by inflation.

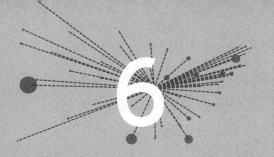

A CASE FOR CORRELATION

Sampling and Inference, From Shakespeare to Death Row to Sports Competitiveness

With the basics of statistics under our belt, we can now move on to correlation and attempt to infer cause and effect from a data set of independent and dependent variables. Any distribution of data can be used to help understand statistical inference, from Shakespeare's plays to death row data to sports league competitiveness. Does the length of Shakespeare's plays tell us anything about his style as he aged; do death row numbers decrease with religious affiliation within a state; does the spead of winning percentages in a sports league change from year to year? The correlation should tell us as much. Fortunately, the same analysis applies in each case.

6.1. A Simple Comparison: Shakespeare's Plays by the Numbers

In statistics, if two variables are correlated (e.g., x and y), then one is a function of the other. The data is referred to as independent (x) and dependent (y) because one variable *depends* on the other; by convention, $y = f(x)$ or y is a *function* of x. Determining how much a variable depends on another and whether one is the cause of the other is the main goal of statistical inference and regression analysis.

To highlight the basics of correlation, we first look at three of Shakespeare's plays—*Romeo and Juliet*, *Hamlet*, and *Macbeth*—from which we can calculate the correlation (if any) between two variables, such as whether a playwright has a particular style determined by letter or word frequency. To get us started, the letter frequency is calculated for each of the three plays, as shown in Figure 6.1 (number of each letter divided by the total number of letters).

From this simple analysis, one can see that e is the most frequent letter, typical of English, followed by t and o, and that the distribution is similar regardless of play, suggesting that the language and not the play is a greater indicator of letter frequency. For example, *Macbeth* in German, *Hamlet* in French, and *Romeo and Juliet* in Arabic will produce different letter distributions, indicative of the difference in the languages and not the plays (also of importance in cryptography,[1] as we will see in Chapter 8).

Furthermore, we can see that the letters r and j are more frequent in *Romeo and Juliet* and the letter m equally as frequent in *Macbeth* and *Hamlet*, as might be expected given the stage time of the leads. We could calculate the frequency of letters with each character's name removed from the script to factor out their occurrence, although, interestingly, h is less frequent in *Hamlet* than in the other two plays, perhaps suggesting that Hamlet doesn't speak as often as Shakespeare's other leads, Romeo, Juliet, and Macbeth.

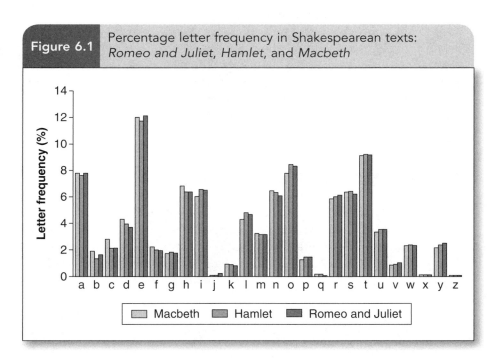

Figure 6.1 Percentage letter frequency in Shakespearean texts: *Romeo and Juliet*, *Hamlet*, and *Macbeth*

Further statistics such as average word and sentence length and the variance in average word and sentence length could also be used to analyze the changing style of an author over a career, to compare different authors' styles, or to determine the authenticity of an unknown work that displays a telltale statistical signature. At the very least, we can begin to see how a computer extracts information from a data file (in this case, words), say, to compile a précis for a large text; check a paper for plagiarism; or query whether a text contains competitive, illegal, or coded information without having to be read by a human, an extremely time-consuming process, as depicted in Robert Redford's CIA job in the movie *Three Days of the Condor.*

Whether a book or article is balanced between opposing views can also be gauged by its word count, and current trends analyzed by determining the increased (or decreased) use of words in various media today. For example, Veseth (2005) noted that "in the early 1990s . . . 'globalization' appeared in the *Times* about once a week or less on average. By the end of the decade, however, it was being used once a day or more" (p. 22).

To determine whether Shakespeare was more economical in length or style as his career progressed, however, we must turn to a more formal means of correlation or regression analysis, i.e., calculating dependency, if any, in the data. Figure 6.2 shows play length versus completion date for 37 of Shakespeare's plays (file size is used for play length normalized to the longest play, i.e., Hamlet = 100), where we can see that although his play lengths oscillated in time, a line can nonetheless be drawn through the data points, showing that his plays lengthened as he aged (the line is drawn such that the difference between the line and each point is minimized, a so-called "least-squares fit"). How much?—well, that depends on the numbers, as we will see.

The harder analysis, however, is determining cause and effect between the two variables. For example, is the oscillation of long and short plays because of alternating market demand and fatigue—a plausible explanation—or are other factors at work? Did Shakespeare's plays lengthen because he was more successful? Letter frequency and language may correlate well, but does play length versus time suggest anything about an author's style or market vagaries?

6.2. Cause and Effect? Executions by the Numbers

We can apply the same analysis to any paired set of data, no matter how seemingly unrelated—for example, state executions and religious

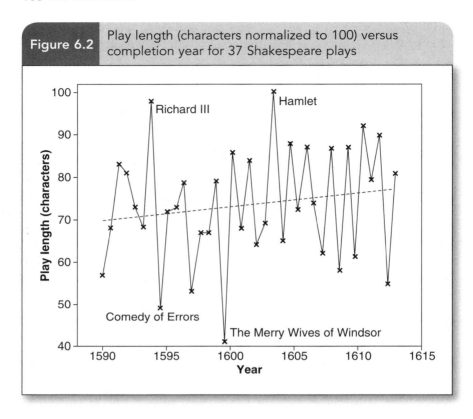

Figure 6.2 Play length (characters normalized to 100) versus completion year for 37 Shakespeare plays

affiliation in the United States. One might think the two variables shouldn't correlate, but in a country where 77%[2] identify themselves as Christian (Kosmin, Mayer, & Keysar, 2001), the death penalty—a seemingly un-Christian-like means of punishment—is still used. But what do the numbers say?

Table 6.1 shows the number of executions since 1976, along with the population and percentage of those who consider themselves religious in the 10 highest and 10 lowest execution states. As one might expect, Texas leads the way with as many executions as the next six states combined, but, as presented, it is difficult to make comparisons between the number of executions and religiosity. When one prorates the data (number of executions per 1 million), however, an underlying trend begins to appear, i.e., the more religious a state, the more likely it is to exact the maximum penalty.

When prorated by population, Texas slips to No. 2, with Oklahoma becoming the death penalty capital of the United States. Delaware rises to No. 3 with 15.6 executions per million (not shown in the table), while

Table 6.1		Most and least U.S. state executions since 1976			
Rank	State	Number of Executions Since 1976[1]	Population[2]	Percentage Religious[3]	Number of Executions Since 1976 per 1 Million
1	Texas	475	25,145,561	89 (8)	18.9 (2)
2	Virginia	109	8,001,024	88 (9)	13.6 (4)
3	Oklahoma	96	3,751,351	86 (17)	25.6 (1)
4	Florida	70	18,801,310	88 (9)	3.7 (16)
5	Missouri	68	5,988,927	85 (18)	11.4 (6)
6	Alabama	55	4,779,736	94 (1)	11.5 (5)
7	Georgia	52	9,687,653	88 (9)	5.4 (10)
8	Ohio	45	11,536,504	85 (18)	3.9 (14)
9	North Carolina	43	9,535,483	90 (7)	4.5 (11)
10	South Carolina	43	4,625,364	93 (2)	9.3 (7)
24	Nebraska	3	1,826,341	91 (4)	1.6 (21)
24	Kentucky	3	4,339,367	87 (14)	0.7 (26)
24	Pennsylvania	3	12,702,379	88 (9)	0.2 (32)
27	Oregon	2	3,831,074	79 (32)	0.5 (28)
28	Wyoming	1	563,626	80 (30)	1.8 (20)
28	South Dakota	1	814,180	92 (3)	1.2 (22)
28	Idaho	1	1,567,582	81 (28)	0.6 (27)
28	New Mexico	1	2,059,179	82 (27)	0.5 (29)
28	Connecticut	1	3,574,097	88 (9)	0.3 (31)
28	Colorado	1	5,029,196	79 (32)	0.2 (33)

1. Death Penalty Information Center (2011).

2. U.S. Census Bureau (2011b).

3. Kosmin et al. (2001).

Florida falls from 4th to 16th because of its relatively larger population. Although North Carolina executed more people than South Carolina did, South Carolina executed almost twice as many per capita as its northern cousin did. At the same time, the top execution states per capita are more religious than the bottom states, with a lower summed ordinal ranking of 98 to 188, including the two most religious states in the top 10, Alabama

(94%) and South Carolina (93%), and the two least religious states in the bottom 10, Oregon (79%) and Colorado (79%).

Any number of such comparisons can be made, but until we compare all the states in a measured way, we can't calculate an overall correlation.[3] We must also be mindful of any inferences we make from the data. In this case, however, it appears there is something counterintuitive about executions and the religiosity of a state. If the death penalty was enacted less in more religious states, one could make a case for leniency based on religious compassion, citing forgiveness, future repentance, and turning the other cheek, typical values associated with Christian charity and tolerance. Given the apparent opposite trend, however, there may be something else at play, perhaps based more on retribution and vengeance.

Perhaps the death penalty is being used as an act of revenge, rather than as punishment. Of course, the numbers help us draw such conclusions and can be used to quantify any correlation, as we will do now.

6.3. Correlation: From Fingerprint Matches to *Roe* v. *Wade*

Regression (or comparison with a trend) is easy; inference is not. It may be true that letter frequency is a function of language but not that an author's word count or average word length increases as he gets older because of experience—he may be lazier, may be after a quick buck, or may not need to prove himself as much. Execution and religious affiliation data should also be scrutinized to ensure that valid inferences are made. It may be true that poverty played a greater part than religious affiliation in the percentage of executions per state, and further multivariate analysis is needed, as in a more involved ANOVA (analysis of variance) rather than a seemingly simple two-variable linear correlation.[4] The data may also not be linear, and other nonlinear regressions may explain the data better.

One of the best examples of whether cause can be inferred from effect is the famous ice cream and heart attack comparison. It is true that as the number of ice creams sold in a day increases, so does the number of heart attacks. But does eating ice cream cause a heart attack or increase the likelihood of one? More likely, ice cream sales increase because temperature increases, and people are more susceptible to heart attacks in hot weather. It would seem that high temperatures cause heart attacks or play a part, not ice cream, although the data nonetheless shows a high positive correlation.

Many facetious inferences can be made from incorrectly correlating data as a trend. One American study suggested that those who quit smoking are twice as likely to develop type 2 diabetes as those who con-tinue smoking (Rose, 2010), but the associated weight gain is what increases the likelihood of diabetes, *not* quitting smoking. Quitting smok-ing (which is a very healthy thing to do) *as well as* watching your weight when you quit will not increase your chances of diabetes but, rather, will significantly enhance your future health and well-being.[5] All such health correlations need to be scrutinized, as in the claim that the incidences of cancer are increasing because of environmental reasons, when in fact more people may be dying of cancer because they are living longer.

Prince Edward believed that the number of people enlisting in the Duke of Edinburgh's Award program increased in its early years after the death of a young boy was reported, stating that students were more keen to get involved in life-threatening adventures, but the increase could be the result of more exposure because few knew of the program prior to its sensationalist reporting. Indeed, I would have thought that *fewer* people would get involved having learned of a death.

Nonetheless, given a good, representative data set, regression analysis can tell us how two variables are correlated; i.e., a good regression can predict a trend, where one variable depends on another. Does weight depend on height? How do jobless numbers affect GDP? Does inflation depend on housing starts, oil prices, consumer confidence? Mathematically, $y = f(x)$—that is, y is a function of x. Even fingerprints can be correlated accordingly, as depicted in any number of police television shows, where a computer program compares the swirls and lines in a suspect's finger-print, encoded by a set of numbers in a feature-identification algorithm, to a fingerprint database, such as the FBI's Automated Fingerprint Identification System.

But how does correlation work? How do we compare one set of num-bers or swirls and lines to another? It all has to do with error, as shown in a simple graph of y versus x with three possible trend lines superimposed over 16 individual data points (see Figure 6.3).

Figure 6.3 Pictorial correlation: "Best fit" error lines

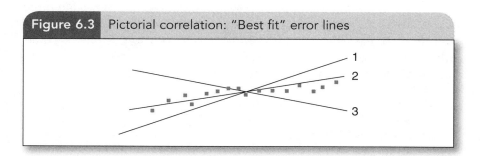

Here, even the eye can tell that the middle line "best" represents the data, visually minimizing the difference above and below the line. Not unlike the eye, one mathematically constructs a line with the "best" fit or "least" error to the data, using a "least-squares fit," where a line is chosen such that the summed squared errors between the line and each point are minimized. If the variables are highly correlated, the sum of the errors between the best line and each data point will be small, where mathematically a higher value of r (the correlation coefficient) gives a higher correlation ($r = 1$ is a perfectly linear fit, $r = -1$ a perfectly inverse linear fit, and $r = 0$ a perfectly random distribution[6]). Note that the data is squared and square-rooted to avoid a zero sum (as we did with the standard deviation), since it is the distance from the best-fit line that matters, not the direction.

As in the examples above, for which we showed only a loose correlation, a measure of the correlation can now be calculated in this way, resulting in $r = 0.16$ for Shakespeare's play lengths versus completion dates and $r = 0.25$ for the number of executions versus a state's religious affiliation. Any two variables can be correlated as such, by calculating the r value, and we can more confidently infer correlation from this calculation. Has the introduction of 24/7, online casinos increased gambling addiction? Is poverty an indicator of mortality rate in a particular area? Does the number of Olympic medals depend on population, or oil consumption on GDP (two examples we will look at below)? In each case, the regression coefficient measures the correlation.

Finding cause and effect in some correlations, however, is much harder than in others. Have homicides been reduced because a particular governor took office and enacted new anticrime laws with more policing or because of a new job creation program, as he or she might like to claim? Or are homicides down because of mobile phone technology that can call for help faster and improved paramedic methods that keep attack victims alive on their way to the hospital?

Levitt and Dubner (2006) stated that crime decreased in the United States in the 1990s as a result of the landmark 1973 *Roe v. Wade* Supreme Court decision, which overturned state laws outlawing abortion. Basically, they argued that unwanted children are more likely to become criminals when they grow up, but because of better access to abortions unwanted children will instead be aborted, resulting in reduced crime:

> When a woman does not want to have a child, she usually has good reason. She may be unmarried or in a bad marriage. She may consider herself too poor to raise a child. She may think her life is too unstable or unhappy,

or she may think that her drinking or drug use will damage the baby's health. She may believe that she is too young or hasn't yet received enough education. . . . For any of a hundred reasons, she may feel that she cannot provide a home environment that is conducive to raising a healthy and productive child. (p. 138)

They further suggested that having an unwanted child in any of these situations leads to a higher likelihood of a poor upbringing and, hence, a higher likelihood of a life of crime for that child (p. 139).

But crime may have been reduced in the 1990s because of a whole host of reasons, such as better policing—although Levitt and Dubner (2006) dismissed this possibility using data in pre-*Roe v. Wade* New York City (p. 141)—more-integrated schools, low-income subsidies, increased incarceration levels from recent prison building programs or harsher sentencing, or even seemingly unrelated factors such as better health care, warmer weather, or economic policies that put more money into the system to increase welfare or promote business entrepreneurship. Just because one variable decreases does not mean it was caused by another that also decreased.

Similarly, state governors routinely credit job creation programs with reducing crime, since the unemployed are also more likely to become criminals. Others cite a poor background as the cause of crime and say that more education translates to financial success. Krugman (2009) noted that changing demographics significantly affected crime rates when job opportunities severely declined for millions of migrant workers, particularly for Southern blacks moving into increasingly crowded Northern cities (p. 88). But does one cause the other? Correlation does not always follow. Furthermore, can we weight the effects of different crimes, especially the less noticeable and underreported white-collar crimes?

Some believe that harsher penalties reduce crime, although Ball (2005) questioned this, stating that "there is no unambiguous evidence" (p. 396). Interestingly, Ball also noted, "There is no reason, however, to believe that [understanding and compassion] did any better either" (p. 396), but "a culture steeped in social etiquette and peer pressure, this is perhaps a more powerful deterrent" (p. 398).

It would seem that one's socioeconomic standing likely contributes significantly as a motive for crime. Necessity may be the mother of invention, but unfulfilled needs and wants are the mothers of crime, from the need for food and shelter to the desire for information and entertainment. Here, simple numbers show that increased satisfaction with one's place as a consumer in the 1990s and beyond also played its part in reducing crime. We used to join clubs to stay off the streets and out of trouble, but today, countless Internet games, entertainment sites, and home-based Wii, Nintendo, and PlayStation groups do the job and are so great in numbers that crime must necessarily lose out. Much of this analysis is qualitative, but the numbers show a dramatic increase in accessible, at-home entertainment over the past two decades (one estimate puts the number of Facebook users at more than 800 million).

Interestingly, Levitt and Dubner (2010) suggested that more TV watching by young people leads to more crime, a notion they advanced from TV-watching and crime statistics in the 1960s (p. 109). One could just as easily argue that poorer parents, who didn't care how much TV their kids watched, were the likelier cause—poorer, that is, in both their ability to parent and in their income. Does ice cream cause heart attacks?—only if there is a causal link. To be sure, not all horses pull carts.

Education is also routinely cited as a key to reducing crime as well as the best path to financial success, but better education does not necessarily mean more money or reduced crime. Huff (1954/1993) noted that the bright and rich would likely make more money regardless of whether they went to college, and in fact may make *less* money by spending their prime earning years in college (pp. 93–94). One could also argue that better education increases crime (successful crimes anyway).

Nonetheless, Huff (1954/1993) noted how easily specious inferences are made:

> It is easy to show a positive correlation between any pair of things like these: number of students in college, number of inmates in mental institutions, consumption of cigarettes, incidence of heart disease, use of X-ray machines, production of false teeth, salaries of California school teachers, profits of Nevada gambling halls. To call some one of these the cause of some other is manifestly silly. But it is done every day. (p. 97)

To be sure, we must be ever mindful of how statistics and numbers are shaped to support specious reasoning from which incorrect inferences are made. Furthermore, what is most important to remember in any statistical analysis is that correlation does not mean causation.

6.4. Correlation and Trend Lines

If sufficient data does exist, however, and the independent and dependent variables are clear, reliable inference can be made. Figure 6.4 shows

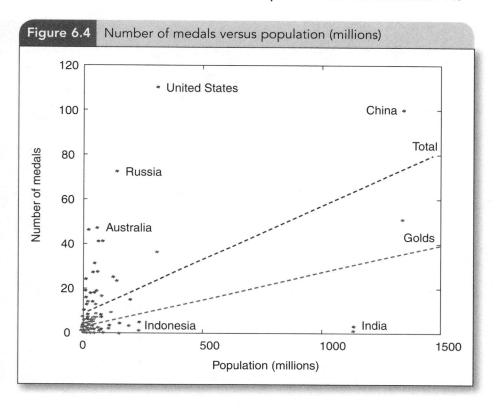

Figure 6.4 Number of medals versus population (millions)

the correlation between number of Olympic medals and population for 86 countries competing in the 2008 Beijing Olympics, where both total medals ($r = 0.49$) and golds ($r = 0.60$) indicate a high correlation with population. One can also see a bias toward larger countries to win gold rather than total medals (the golds slope is lower than the total medals slope), which highlights the value of giving multiple medals since fewer countries can win gold than can win any medal.[7]

Here, the data is highly dependent on outliers (not to mention the more than 100 countries that didn't win any medals or send many athletes). Note that outliers are the data points with the greatest and lowest slopes (not necessarily the farthest points), as Table 6.2 highlights, and suggest that other inferences can be made from the data (i.e., in a multivariate relationship). Thus, if one wants to infer that populous countries are successful in the Olympics, there are many exceptions, and one should also look at medals per population or the strengths of national athletic programs (e.g., Bahamas, Jamaica, and Iceland excel, whereas Egypt, Vietnam, and India do not).

Table 6.2	Largest apparent outliers above and below the regression and best performing (medals per millions)		
Country	Number of Medals	Population (Millions)	Medals per Thousand
Above			
Australia	46	22.4	2,054
Russia	72	142	507
United States	110	310	355
China	100	1,339	75
Below			
Indonesia	5	234	21
India	3	1,133	3
Outliers			
Bahamas	2	0.34	5,882
Jamaica	11	3.0	4,074
Iceland	1	0.32	3,125
Egypt	1	79	13
Vietnam	1	86	12
India	3	1,133	3

It would seem that richer countries with strong sporting associations fare better than average (e.g., Australia, the United States, Russia, and China), as do a few smaller countries with long-standing athletic traditions (e.g., Jamaica), whose many wins belie their smaller numbers. Nonetheless, the poorer countries aren't well represented at the Olympics, as can be seen in the correlation of medal count with GDP, where $r = 0.82$ shows strong agreement. Note again that the data does not include countries that won no medals, which if included would produce an even greater correlation between wealth and performance.

With good correlation and proper inference, we can also construct an equation to represent dependence, such as oil consumption (OC) versus GDP, which has an almost perfect correlation ($r = 0.97$). Here, a linear relationship exists, from the smallest to the largest country, and because the correlation is almost perfect, a straight-line equation can be constructed, as shown in Figure 6.5. This equation gives OC if GDP is known—i.e., $y = 0.44x$, where x is GDP (in trillions of dollars) and y is OC (in billions of barrels per year).

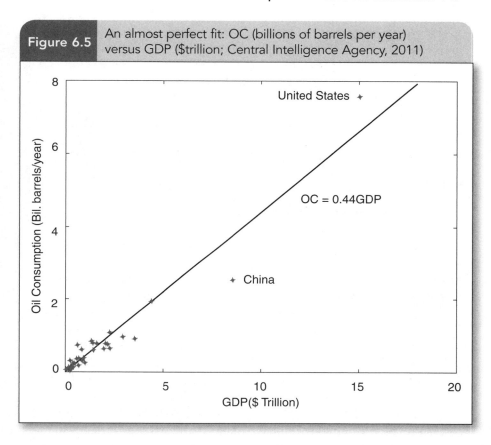

| Figure 6.5 | An almost perfect fit: OC (billions of barrels per year) versus GDP ($trillion; Central Intelligence Agency, 2011) |

All one has to do is plug in the GDP for a country and out pops the corresponding oil use, as shown for a few countries in Table 6.3. Note that the error for any individual country varies depending on how far it is from the trend line, but the error is minimized over the whole data set. Note also two important outliers that especially highlight how the world is changing: The United States is above the trend line, indicating a higher-than-average consumption (and an overreliance on oil), and China is below the trend line, indicating a lower-than-average consumption (and an expected rise in OC as GDP grows).

6.5. Analyzing Statistical Data: The Qualitative Becomes Quantitative

With a better grasp of the mean, standard deviation, and regression, we can use such statistics to look at any distribution of data that strikes our

Table 6.3		OC (billions of barrels per year) versus GDP ($trillion) for top and bottom three countries		
Country	Rank	GDP ($Trillion)	OC (Bbpy)	OC (0.44GDP)
United States	1	15.1	7.59	6.64
China	2	8.56	2.53	3.77
Japan	3	4.40	1.95	1.94
Sao Tome	187	3.1×10^{-4}	2.4×10^{-4}	1.4×10^{-4}
Falklands	188	1.1×10^{-4}	0.9×10^{-4}	0.5×10^{-4}
Nauru	189	0.5×10^{-4}	3.8×10^{-4}	0.2×10^{-4}

fancy, from sports scores to the stock market. The goal is to uncover anomalies and quantify trends to see if the past tells us anything about the future (the statistical oracle), important for financial analysts and bettors alike, not to mention regulators and oddsmakers.

Here, we look at North American sports leagues—Major League Baseball (MLB), the National Football League (NFL), the National Basketball Association (NBA), and the National Hockey League (NHL)—for which a wealth of longitudinal data exists. Our goal is, first, to see if we can divine a winner and, second, to determine if they have become more (or less) competitive over the years and, if so, by how much (something we looked at qualitatively in Chapter 4). We may even be able to tell if a sport is fixed.

Determining whether a league winner (team with the highest winning percentage[8]) will win its corresponding league championship (World Series, Super Bowl, O'Brien Trophy, or Stanley Cup) is straightforward. In the NFL, the league winner won the championship 26 times from 1960 to 2010, or slightly more than half the time.[9] The other sports show similar numbers, from a high of 48% in the NBA to 34% in the MLB (calculated after the introduction of a four-team playoff in 1994[10]). From this simple analysis, it seems that winning the regular season is not much of a metric for predicting the eventual champion.

A better, or more predictive, statistic might be how often the higher-ranked team wins the championship, i.e., how often the favorite beats the underdog—always an important talking point in any final. Here, we see a striking result in one sport that bucks a previously long-standing trend, whereas in another, the result is no better than random.

In the NHL and NBA, the higher-ranked team won the final 80% and 72% of the time, respectively, percentages that compare well with the

NFL's 74%. In fact, prior to Super Bowl XV, the lower-ranked team didn't win a single Super Bowl, and only rarely afterward. That is, until 2006, when the lower-ranked team won four of the next six Super Bowls, including the lowest-ranked winner ever, the seventh-ranked New York Giants, who upset the top-ranked New England Patriots in 2008. Whether the trend will hold is anybody's guess, but things seem to be changing in the NFL.

On the other hand, in MLB, the higher-ranked team faired quite poorly in 40 years of pennant races, winning the league championship only 58% of the time (a high of 65% in the American League compared with a middling 51% in the National League, essentially a coin flip). It seems that how teams play during a 162-game regular season is not indicative of performance in a shorter playoff.

We can also gauge competitiveness, a hallmark of any sport, from the average rank of the two finalists. The Super Bowl champion is higher ranked than champions in the other sports (2.1), likely because of the reduced number of playoff games and shorter season that restricts the number of teams, which may also account for the recent underdog trend because of the introduction of more playoff games. Hockey's average winner (2.7) comes from farther down the ranks, likely because of its more accessible playoff format (more than half the teams qualify). Not surprisingly, the losing finalists are lower ranked and the spread (± error) is also greater (see Table 6.4).

From the numbers, the Super Bowl winner most represents its league champion, although the NHL playoffs are so long (as much as 25% of the total season), it is not surprising that the Stanley Cup winner is not as highly ranked. Many more teams qualify, and any team with a good playoff run can make it to the final, e.g., the 18th-ranked Edmonton Oilers, the lowest-ever finalist (0.50), who lost to the Carolina Hurricanes in 2006. Even sub-500 teams can occasionally make the finals, as did the

Table 6.4	Against the grain: Winners and losers by league rank			
League	Average Winner Rank ± σ	Average Loser Rank ± σ	Lowest Winner	Lowest Finalist
NFL	2.1 ± 1.5	3.1 ± 2.5	7th	11th
NBA	2.2 ± 1.9	3.7 ± 2.7	10th	11th
MLB	2.3 ± 1.3	2.7 ± 1.6	5th	8th
NHL	2.7 ± 2.2	5.6 ± 4.5	9th	18th

1982 Vancouver Canucks (0.48) and the 1991 Minnesota North Stars (0.43)—a feat matched only once in any other league (the NBA)—although none won.

In the NFL, the lowest-ever finalist was the 11th-ranked Los Angeles Rams (0.56), who in 1979 lost to the league-leading Pittsburgh Steelers. In the NBA, the 11th-place Houston Rockets (0.49) lost to the top-ranked Boston Celtics in 1981, whereas MLB saw the 8th-ranked Los Angeles Dodgers (0.52) lose to the 2nd-place Philadelphia Phillies in the 2008 National League Championship. To be sure, as the leagues add more playoff rounds, more teams have a chance, creating a more competitive league or, depending on one's perspective, a winner less representative of the league champion.

Gauging Competiveness: Absolute and Standard Deviation

Another measure of competitiveness is whether winners are winning less and losers winning more from one season to the next, important to owners hoping to fill expensive seats during the regular season. Here, the more formal statistical measures that we have already seen (e.g., the standard deviation and correlation) are especially useful.

The graphs in Figure 6.6 show the highest and lowest team winning percentages over 50 years for each league, revealing a fascinating telltale signature of competitiveness versus time. The absolute deviation (highest–lowest) is also included with an associated correlative trend line that especially highlights longitudinal changes.

With these statistical figures at hand, some obvious trends jump out. First, we can begin to see marked differences between the leagues, where the NFL is the least competitive (greatest absolute deviation between top and bottom teams), a disparity that continues reasonably unchanged over time (flat trend line), and MLB is the most competitive (so much so that the deviation is less than the minimum).[11] Second, we can see that the NBA has become less competitive over time (a startling effect evident in its increasing trend line). Furthermore, after a steady decrease in competitiveness in the '60s and '70s, the NHL is more competitive today.

Of course, many things have changed over the years (expansion, number of games, salary restructuring, free agency), but the trends are there. Why they exist may be harder to determine, but the length of a season is certainly a major factor, from a low of 16 games in the NFL to 82 in the NBA and NHL and 162 in MLB. It seems that randomness has more of an effect in a longer season, pushing winning percentages

Figure 6.6 | NFL, NBA, NHL, and MLB (American League only) (1961–2010): Highest winning percentage (top curve), lowest winning percentage (bottom curve), and absolute deviation with trend line (middle, dashed curve; bottom curve for MLB).

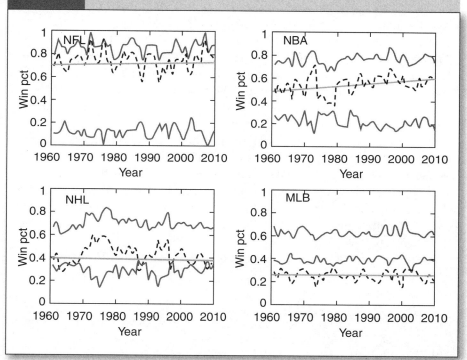

together. This may be counterintuitive, since good teams should stay good throughout a season, but perhaps the rigors of a longer schedule equalize the teams. In MLB, one can also argue that the game itself produces closer games and, thus, a general seasonal narrowing (only three teams since 1960 have won more than two of every three games played over a whole season).[12]

However, we can be more certain about the causes of a changing trend. From the figures, one can easily see that the NBA has become less competitive—likely because of its smaller teams, which makes it easier for one player to dominate (e.g., Michael Jordan and the Chicago Bulls 1991–1993 and again 1996–1998), and the effect of expansion, which greatly watered down the player pool—whereas in the NHL, perhaps because of the increased tightness of the game and less reliance on individual stars, no one team can reign for long. The effect of free agency can also be seen in MLB, which saw an extraordinary flip-flopping of winning percentages in the American League from 1995 to 2001. Many factors contribute, but backed up by good statistical data, such conclusions are given more credence.

One of the problems with absolute deviations, however, is that they don't account for the spread of the data. For example, 15 teams with perfect winning records (1.0) and 15 teams with perfect losing records (0.0) is much different than 1 team with a perfect winning season, 1 team with a perfect losing season, and 28 teams with 50/50 seasons. For example, in the NFL, the absolute deviations were identical in 1974 and 1975 (0.74), but 1974 was much less spread out, primarily because six teams had even records.

To correct for the limitations of the absolute deviation statistic and better account for the "tightness" in a distribution, the standard deviation (σ) is used.[13] The number of data points is also included in the calculation, making it easier to compare different-sized distributions (e.g., the 12-team NFL of the 1960s and the 30-team league of today). Figure 6.7 shows the absolute deviation (left) and standard deviation (right) of the four leagues, revealing subtle but important differences in the two measures, although some general trends remain (note the scale change).

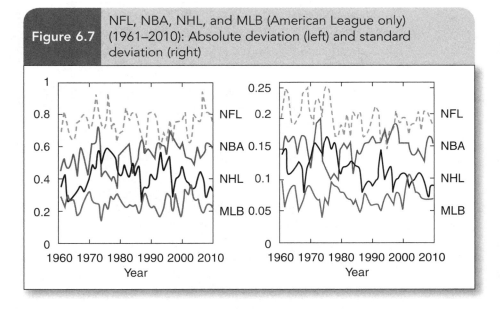

Figure 6.7 NFL, NBA, NHL, and MLB (American League only) (1961–2010): Absolute deviation (left) and standard deviation (right)

Outliers in the Data

With such numbers at hand, we can now easily spot anomalies and attempt to explain them. The least competitive years (highest σ) for each league were 1976 (NFL), 1973 (NBA), 1977 (NHL), and 2002 (MLB). In 1976, the NFL expanded and the newly formed Tampa Bay Buccaneers didn't win a game, a first for the NFL, producing the largest-ever spread in the standings. In the NBA in 1973, the Boston Celtics' (0.83) domination coupled with an abysmal season by the Philadelphia 76ers (0.11) produced the most varied NBA season. In the NHL in 1977, the Montreal Canadians (0.83) had the highest winning percentage ever, losing only eight games, a single-season league record. Baseball's most lopsided season was 2001, when the Seattle Mariners (0.76) ran away with the title with a league record 116 wins, although the greatest spread occurred the next year (highlighting the difference between absolute and standard deviation), likely because of the topsy-turvy effect of free agency (e.g., the Florida Marlins won the 1997 World Series in their fifth season but finished dead last the next year, an astonishing affront on traditional league competitiveness).

Changing Statistics Over Time

We can also quantify how competitiveness changes over time, using our newly constructed average standard deviation of winning percentages statistic. Without such a statistic, we could make only qualitative remarks about isolated seasons or show numerous graphs with numerous changing trend lines. Here, our statistical metric works perfectly and shows that despite being the least competitive league, the NFL has nonetheless become more competitive over time. We can also see how competitiveness has increased in the NHL and MLB (the National League more so than the American League), yet decreased in the NBA (see Table 6.5).

The number of same-team winners and different-team winners is also an indicator of competitiveness, showing its own telltale signature. Interestingly, despite the lack of competitiveness during the regular season, the NFL shows a remarkable mix of championship winners, likely because of its egalitarian salary caps (calculated as a percentage of league revenues), which limit trading and financial manoeuvring prevalent in the other sports. Within any one year, competition is reduced because of the shorter schedule, although year-to-year dynasties are less likely. If fair-mindedness (i.e., the likeliness of different champions) is a goal in sports, football gives the most equal reward to its members,

Table 6.5	League competitiveness: Average standard deviation of winning percentages (ASDWP) and dynastic titles					
League	ASDWP 1960–1985	ASDWP 1986–2010	Most Times Champion 1961–1985	No. of Different Champs	Most Times Champion 1986–2010	No. of Different Champs
NFL	0.21	0.19	Green Bay, 5 Pittsburgh, 4 6 tied with 2	12	Dallas, 3 San Fran, 3 NY Giants, 3	13
NBA	0.15	0.16	Boston, 11 Los Angeles, 4 2 tied with 2	9	Los Angeles, 6 Chicago, 5 San Antonio, 4	7
NHL	0.13	0.10	Montreal, 10 Toronto, 4 2 tied with 3	6	Detroit, 4 3 tied with 3	13
MLB National League	0.08	0.06	Los Angeles, 7 Cincinnati, 5 St. Louis, 5	8	Atlanta, 5 Philadelphia, 3 St. Louis, 3	13
MLB American League	0.08	0.07	New York, 7 Baltimore, 6 Oakland, 3	8	New York, 7 Boston, 3 Oakland, 3	10

whereas baseball has become more swayed by market forces (e.g., the New York Yankees routinely fail to adhere to the salary cap).

Furthermore, basketball, which has always been the most susceptible to dominance by one player and one team, is becoming less competitive, while hockey, because of its ever-longer playoff season, produces a champion least indicative of its regular season. Aside from the NBA and MLB's American League, the dynasty metric shows how hard it has become for one team to dominate.

Standard Metric to Analyze Sports Leagues

Most important from our method of statistical analysis, however, is that we now have a simple metric—the average standard deviation of winning percentages (ASDWP)—which can be used for any sport, from tiddly-winks to darts (and can apply in many diverse areas, as we will see in Chapter 9 with regard to pop charts and economic disparity). As such, we can now test our assertion in Chapter 4 that North American sports leagues are more competitive than European football leagues because of reverse-order player drafts and salary caps that mix up the league and reduce the chance of one team dominating.

As seen in Figure 6.8, the numbers back up our conclusion, showing a lack of competitiveness particularly in English football, where the imbalance is evident and has been worsening since the introduction in 1992 of the revamped English Premier League (EPL) and the European Champions League. In fact, the change in competitiveness is striking, showing an increase in ASDWP from 0.11 to 0.13, increasing even more in the past 10 years to 0.15. The absolute deviation regression trend line highlights the dramatic change over the past 50 years.

On preliminary examination, the Spanish La Liga (SLL) appears to be unchanged over the past 50 years, but here our ASDWP statistic is a bit deceiving, appearing flattened out despite periods of obvious increasing and decreasing competitiveness. When coupled with the dynasty metric, however, it is clear how uncompetitive the SLL is, where Real Madrid has won half of all league titles while only seven other teams have won one. The same mismatch appears in English football, where two teams (Manchester United and Liverpool) have won more than half the league titles between them (see Table 6.6).

In European football, the rich are indeed becoming richer, primarily because of big-money television deals and Champions League qualification

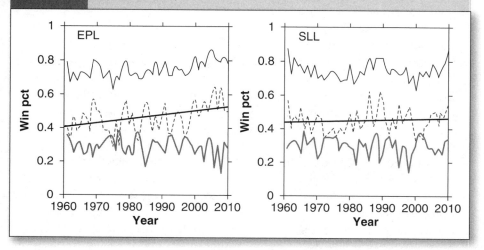

Figure 6.8 English Premier League (EPL) and Spanish La Liga (SLL): Highest winning percentage (top curve), lowest winning percentage (bottom curve), and absolute deviation (middle, dashed curve).

Table 6.6	EPL and SLL competitiveness: Average standard deviation of winning percentages and dynastic titles (league title only)					
League	Ave σ 1960–1985	Ave σ 1986–2010	Dynasties? Champs From 1961–1985	No. of Champs	Dynasties? Champs From 1986–2010	No. of Champs
EPL	.11	.13	Liverpool, 10 Everton, 3 3 with 2	11	Man United, 11 Arsenal, 5 2 with 3	7
SLL	.11	.12	Real Madrid, 14 Atlético de Madrid, 4 3 with 2	6	Real Madrid, 11 Barcelona, 10 Valencia, 2	5

(the top four teams advance to the lucrative European Champions League). Instead of mixing up the talent in reverse-order player drafts and regulating team payrolls via salary caps as in North American sports, the top players continually go to the top teams, whereas the lesser teams have no chance to compete without massive infusions of outside money, the result of a high-stakes winner-take-all market (which we'll look at more in Chapter 9). Maturing homegrown talent doesn't cut it anymore, where to the victors go the ever-increasing spoils. Of course, no one begrudges a good team built on the expertise of a manager who scouts for good talent and a coach who gets the best out of his or her team, but stacking the deck is hardly sporting.

If competitiveness decreases, the very existence of a league can be at stake. When teams lose year after year, franchises suffer and can be wound down or moved to another city. The concern is greater in North America, where most teams are many miles apart and depend primarily on local support. In Europe, the imbalance is maintained as a matter of course, partly because allegiance is not entirely related to geography but also because elitism is promoted (e.g., entry into the Champions League).

Nonetheless, European football will see a decline if lack of competitiveness continues, as perennially losing teams' fan bases diminish and clubs fail to make financial ends meet. Lots of money is at stake, but why European football is so reluctant to endorse fairness is puzzling since a highly competitive league can be just as exciting and rewarding.

North American sports, however, should take note, because a similar lack of competitiveness is appearing in its own leagues as money plays a greater part in determining team makeup and competitiveness. The NFL locked out players over the 2011 off-season in an ongoing dispute over spiraling costs and revenue sharing ($9 billion annually), doing away with the salary cap along with its collective bargaining agreement. MLB is planning to further undermine its tradition of determining a champion over a long season by introducing more playoffs, ultimately lowering the importance of the regular season and likely serving only to expand revenues for select teams,[14] and the NBA is becoming increasingly dominated by big-market teams with limited adherence to salary caps. We should remember, however, that a mix of fairness and competition is what drives sports, not the same financially manufactured teams winning year in and year out.

Plenty of straightforward and involved analysis can be done on any number of data sets. Picking the right measures, ensuring the right samples, and making the right inferences is no easy task, but the analysis can be made with a proper understanding of the basics, particularly with derived statistics and correlation.

Notes

1 **Simple code breaking**: Singh (2000) noted that "the letters *a* and *l* are the most common in Arabic, partly because of the definite article *al-*, whereas the letter *j* appears only a tenth as frequently" (p. 17), a fact which would lead to the first great breakthrough in cryptanalysis.

2 **Religious affiliation in the United States**: At the time of the American Religious Identification Survey (ARIS) in 2001, the percentage of the population classified as Christian was 77%, a decrease from 86% in 1990 (Kosmin et al., 2001). If a similar downward trend continued, the percentage in 2011 would be 68%.

3 **Death penalty states**: Note that 13 states do not have a death penalty, none of which are in the South and 6 of which border Canada, which outlawed the death penalty in 1989 for all crimes (last execution in 1962).

4 **Multivariate covariance**: Covariance (or the lack of covariance) can be extended to any number of variables. For example, the period (T) of an oscillating spring depends on its mass (m) and spring constant (k) but not on its volume (V) or gravity (g). In an experiment on different masses and springs, plotting T versus m and k will show both covariant relationships but not T versus V or g.

5 **Smoking and health**: Hospital admissions for heart and respiratory problems have been cut by one third in Canada since the introduction of no-smoking laws in public places.

6 **Correlation measure**: The coefficient of determination, r^2, is a measure of correlation (from 0 to 1), typically used in many canned software programs to construct a linear relationship in the data (if any), and estimates the fraction of the variance in the dependent

variable (y) explained by the independent variable (x), where the scatter in the data has not been explained by the regression. The correlation coefficient, r (aka Pearson's r after Galton), is the square root of r^2 and better shows the extent of the correlation as well as the direction. Other regression measures are also available if the output depends on more than one variable.

7 **The effect of the most populous country**: If China were removed from the analysis, the original regression slope would be halved, giving a powerful measure of the effect of China (or any dominant variable) on an overall analysis.

8 **Winning percentages**: Two points are given for a win and one for a draw (where applicable), such that the summed winning percentage of all teams is one-half the total number of teams. League data may differ because of the number of teams and number of games (e.g., in the 12-team NFL of the 1960s or the 6-team NHL prior to 1967), although the effect is minimal.

9 **The Super Bowl**: The Super Bowl began in 1967. Data prior to the Super Bowl is for the NFL.

10 **MLB comparison**: MLB is slightly different because of its reduced playoff structure and noninterleague play. Interleague playoffs began in 1969 with two teams and expanded to four in 1995. As such, the American and National leagues are considered separately. Limited interleague play also began in 1997, which will skew the summed winning percentages of each league, although the effect is minor.

11 **MLB results shown for American League**: For presentation purposes, the results are presented only for the American League. Similar trends appear for the National League.

12 **The only three plus 0.67 MLB teams since 1960**: 2001 Seattle Mariners (116–46: 0.72), 1998 New York Yankees (114–48: 0.70), and 1969 Baltimore Orioles (109–53: 0.673).

13 **Minimum and theoretical maximum**: The minimum possible is 0 for a completely equal distribution (all 0.5) and the theoretical maximum is 0.5 for a completely unequal distribution (half at 0 and half at 1).

14 **MLB playoff expansion**: Adding more playoff rounds goes against the whole nature of professional baseball, where a long season has been traditionally used to determine the best teams, who then compete in a short "show" tournament. Adding more playoff rounds will minimize the importance of the regular season and make the World Series a crapshoot, won by the end-of-year in-form team.

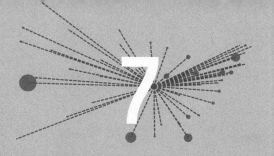

COUNTING AND COMPUTERS

From Boole to Google and Back

More and more of everything today is done by computer, from databases that profile health, wealth, and our general likes and dislikes to loyalty cards that tabulate consumer purchases and correlate buying patterns to the social and gaming networks where we spend more and more of our spare time. With the increased importance of the Internet, the computer now ranks our web searches, suggests our friends, and provides real-time data on anything from stocks to sporting results. And that is just the tip of the computing iceberg.

What's more, probability and chance involving large numbers can be calculated only by computer, where simulations count the possible outcomes in the absence of analytical solutions. Computers and probability are also the guts of today's fast-growing online gaming industry, and a basic understanding of both is needed if we are to keep from becoming sitting ducks for the now-ubiquitous Internet huckster.

To be sure, the power of the computer increases by the day, and in today's wired world, no one can afford to remain technologically ignorant. As such, in this chapter, we briefly look at the history of computing, calculate the numerical odds in a few betting scams designed to make money from our innumeracy, and work out how to win at Monopoly and blackjack using so-called Monte Carlo simulations—all in the hopes of becoming better acquainted with the reality of today's number-crunching world.

7.I. George Boole: The Real Brains Behind the Google Universe

Computing can be said to have begun around 600 BC in China with the abacus, a device with two sections of beads called "heaven" and "hell" used to count in multiples of 5 and 10, likely because of our five-fingered hands. Named from the Greek *abax*, for "counting board," the abacus can be used to add, subtract, multiply, divide, and take square roots, and is still used in modern China with great speed and accuracy.

Before that, the first high priests in our earliest societies calculated when and where the sun, planets, and stars rose and fell each day and the difference from year to year,[1] from which they devised number systems based on 12 (the number of whole moons in a year), still used today (e.g., inches in a foot or items in a dozen); 30 (the nearest whole number to the moon's period); and 60 (a sexagesimal system, which is the smallest number that divides both 12 and 30). They also philosophized about the incommensurate number of days in our 365.254-day year, from which they thought about ratios of numbers and shapes of geometric objects. From these early beginnings, we began to count and think of ever-new ways to reduce the laborious nature of such accounting.

The greatest mathematicians of yore then put their minds to irrational numbers—those that cannot be expressed as the ratio of two whole numbers, such as pi (≈ 3.14) or the square root of 2 (≈ 1.41)—perhaps as they contemplated the incommensurate number of days in the year. What did it mean to have a forever-increasing decimal, and how could such an imperfect number represent their perfect mathematical world? The legacies of these early mathematicians—Pythagoras, Euclid, and Ptolemy—long remain.

We all know Pythagoras for his triangle theorem, where the basics of trigonometry can be worked out from any right-angle triangle,[2] the simplest of which has sides of lengths 3, 4, and 5 ($3^2 + 4^2 = 5^2$). From such simple constructs, right angles on building sites could be fashioned, as in ancient Egypt, and sines and cosines deduced from a projection or shortening of 5 to 3 in one direction and 5 to 4 in a perpendicular direction, essential in modern engineering.

Euclid gave us his myriad axioms, which continue to confound students today but are nothing other than the rules of the mathematical road, how angles give us lengths and vice versa. And it is from Ptolemy's incorrect geocentric planetary system that the days are so named—i.e., Monday, Tuesday, Wednesday, etc., instead of Monday, Saturday, Sunday, etc. (as we will see in Chapter 8). Ptolemy's days can be explained by modulo arithmetic, a method of counting in cycles.

Not much in the way of counting was advanced during the Dark Ages, other than working out the times of eclipses and Easter dates, until John Napier, the laird of Merchiston in Edinburgh, and his calculating bones. Napier was interested in translating the Bible to Scottish and engineering machinery for warfare, but he is mostly remembered today for the logarithm, a calculating method that helped improve our understanding of the world in the 1600s and beyond. His method was based on adding exponents when numbers are multiplied (e.g., $100 \times 1,000 = 10^2 \times 10^3 = 10^{2+3} = 10^5 = 100,000$). It is not easy to multiply two large numbers, e.g., $12,345 \times 67,890$, but with logarithms (or logs), the answer can be worked out via a much simpler addition;[3] thus, $12,345 \times 67,890 = 10^{4.091} \times 10^{4.832} = 10^{4.091+4.832} = 10^{8.923} = 837,529,282$.[4] Instead of a long and tedious multiplication, the answer is worked out by adding numbers found in a log table, which Napier created along with his logarithms.

From the calculation methods devised by Napier and his mechanical "bones," a 16th-century precursor to the slide rule, Johannes Kepler was able to interpret the planetary data as recorded by Tycho Brahe to work out that the planets orbit the sun in ellipses and not circles, "an early example of the scientific method at its best" (Kuhn, 1957/1997, p. 215). From this, Isaac Newton could eventually derive his inverse-square law of gravitation.[5] The resultant scientific analysis wonderfully verified the heliocentric system posited by Nicolas Copernicus 150 years earlier.

Next, there was René Descartes, who designed a coordinate system of ordered pairs of numbers, a system as simple as a children's game of Battleship, where hiding a battleship at A1, A2, A3, and A4 or a submarine at G4, H4, and I4 is as easy as working out the intersection of two lines on a grid (in this case, a column letter and a row number). All of modern analytical mathematics (i.e., lines represent equations) is based on such a coordinate system, e.g., the coordinates of an x–y graph, a spreadsheet, longitude and latitude, even bingo columns and rows.

The Jacquard loom in the early 1800s was the next breakthrough on the calculating scene and used punch cards to store a repeated set of instructions or "program" (the weft or weave motion of a loom corresponds to a hole in the card), much like a player piano or music box encodes sequences of raised metal pegs as musical notes. From there, calculating was well on its way with such inventions as Blaise Pascal's Pascalina and Charles Babbage's unbuilt analytical engine, which had all the conceptual makings of the computer we know today.[6]

The 20th century saw the advent of electromechanical computers, such as Colossus and ENIAC—used for weather calculations, ballistics, and code breaking during World War II—the transistor invented at Bell

Labs in 1947, and the integrated circuit in 1958.[7] The launch of Sputnik, however, changed everything. As Boorstin (2000) wrote, "Never before had so small and so harmless an object created such consternation" (p. 591). With Sputnik—a basketball-sized device that weighed only 183 pounds, took 98 minutes to orbit the earth, and spent 57 days in space before being destroyed during reentry into the earth's atmosphere—the space race and the resultant miniaturized electronic age began. As seen in every computer today, from the fastest supercomputer to the cheapest netbook used only to surf the web, we owe a great debt to Sputnik and the satellites and space programs that followed. And today, Moore's Law continues to take us even further into a world of unimaginable processing power that connects millions of people worldwide, from an estimated 16 million Internet users in 1995 (Ball, 2005, p. 470) to more than 2 billion today.

At the core of every computer and every Google search, however, is the same mathematics—Boolean algebra—as devised in the 19th century by George Boole, a mild-mannered English mathematician who spent his adult life at the University of Cork, Ireland.[8] Boole is the main mathematics man these days and the real brains behind Google searches, with logic no more complicated than a 0 and 1, on and off, true and false. In fact, Boole was attempting to represent human thinking mathematically when he came up with the idea of using sets algebraically, where the seemingly illogical makes sense: $0 + 1 = 0$ (e.g., the set of Californians *and* the set of Americans) or $x = x^2$ (e.g., the set of males and the set of males). Such logical systems form the basis of "truth tables" as well as the electronic workings of computer circuits, as shown in Figure 7.1.

As for Internet search engines, when we type two names into a Google search, we are performing the same mathematical operation, known as the union of two sets (which we also saw in Chapter 4 with regard to Venn diagrams and testing). For example, if A = *Bill* and B = *Clinton*, typing *Bill Clinton* into Google returns all webpages that have either the word *Bill* or the word *Clinton*, called the union of A and B (i.e., A *or* B). One result might contain a story about a dog named Clinton and another about a boy name Bill, but the data can be further restricted by typing *Bill Clinton* in quotes ("Bill Clinton" or *Bill+Clinton*), which then lists only the pages with *Bill* and *Clinton* appearing together, called the intersection of the two sets (i.e., A *and* B).

From these basic operations, including the complement of a set—for example, all Bill Clinton pages that don't have the words *Monica Lewinsky*, i.e., *Bill+Clinton – Monica+Lewinsky* (A *and* B *and* not C *and* D)—we see the basis not only of a Google search but the operation of a

Figure 7.1 George Boole's or/and logic

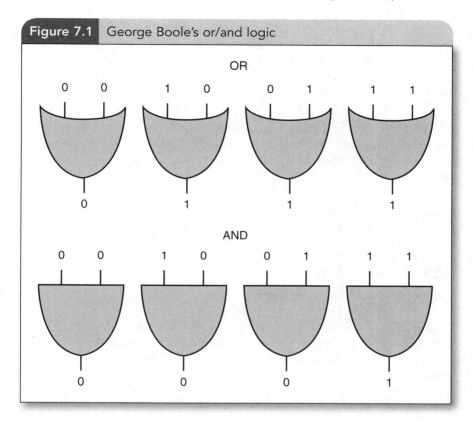

modern computer as it runs billions of logic tests on incoming data each second using nanometer-sized transistors (1 nm = 1 × 10⁻⁹ m). To give you an idea of the number of computer lookups when "googling" the Internet and how search queries restrict data, I got 1,370,000,000 hits for "Bill OR Clinton," 295,000,000 for "Bill AND Clinton," and 89,100,000 for "Bill Clinton" – "Monica Lewinsky" from more than 25 billion pages (calculated from the number of pages retrieved on a search of the letter a).

As for what changed a little-used, university administrative network (ARPANET) into the World Wide Web we know today, that honor goes to Tim Berners-Lee, a British physicist who devised the idea of hypertext and the linking of documents while working at CERN. As he noticed, there were a great many documents on different computer platforms with varying protocols but not a uniform way to access them, showing how interconnectedness rather than loosely connected ideas promotes use. How connectedness arises and, indeed, grows has much to do with the "social physics" of how people interact, whether exponentially doubling or in self-organized, small-world ways (as we saw in Chapters 1 and 2), or by chance and statistical happenstance, as we will see in this chapter.

The ways in which we choose to communicate are at the core of our connectedness, nowhere more prevalent than in computers and computer networks. The subject is fast increasing, and many ideas in information and computational theory are just finding their way into other disciplines, from epidemiology that studies how colds and other viruses propagate to economics that looks at how income distribution and inequality spread or the increasingly significant effects of fluctuating markets. But before we tackle the ideas of connectedness as they apply to social networks (see Chapters 10 and 11), we will first take a look at the power of computing, especially with regard to probability and the Internet.

7.2. How to Make Money in Your Spare Time, From Barbecues to E-mail Orders

Today, as an army of hucksters attempt to exploit our ignorance of numbers, a host of newfangled ways to separate our money from our wallets has been made especially relevant by computers and the Internet. Everything from casinos to lotteries relies on our ignorance, especially regarding chance and probability, where the unlikeliest of events is quantified by the mathematics of probability, a field of study that has intrigued many a huckster and gambler alike. And the field is growing, with computers at the core. Today, at any given moment, millions of people are playing online poker, checking a lottery ticket, or taking a pick on an unknown stock.

Of course, most students would rather give up burgers and fries for life than think about probability, although the subject is not nearly as difficult as it seems. As such, few could work out the basic odds for roulette or the lottery, although, to be fair, the mathematics seems overly abstract and of little use unless one is planning a career as a professional gambler or banker.

To ease our way into mastering probability, we begin with the old favorite about how many people must be in a room before the odds are 50:50 that two will share the same birthday. Surprisingly, the answer is much fewer than you might think. If your interest in odds is purely monetary, you can make a lot of money with this kind of knowledge.

But before I go through the math, let's start with a simpler example to establish the basics: What are the odds in a room of 12 people that no one is born in any given month—say, January? Well, the odds that any *one* person was born in January is 1/12, and so the odds that any one person was *not* born in January is 11/12 (mathematically, $p = 1/12$ and $q = 1 - p = 11/12$, where $p + q = 1$).[9] Since each birthday is assumed to be independent for each of the 12 people (no twins or a roomful of leap-year babies), the odds multiply in the same way as dice (there is a 1 in 6 chance of rolling a 6 on one die and, thus, a 1 in 36 chance of rolling *two* 6s on *two* dice: $1/6 \times 1/6 = 1/36$). Thus, the odds of no January birthdays in a room of 12 people is 11/12 times itself 12 times or $(11/12)^{12} = 0.352$, or a 35.2% chance that in a room of 12 people no one was born in January (or any other stated month).

But what good is that, other than as a party trick or on a high school exam? Well, if you want to turn such numerical acumen into money, you bet. At 2-to-1 odds, you will get twice your money back 35.2% of the time and nothing back 64.8% of the time. So for a $1 bet, you would win $2 \times 35.2\%$ and lose $1 \times 64.8\%$, with a total winning of 5 cents on average ($\$0.70 - \$0.65 = \$0.05$).

You shouldn't get discouraged if you lose the first few bets, because you will—over the long haul—win 5 cents per bet. And, given that a typically innumerate public thinks such a result is highly unlikely, as an astute bettor, you could try better odds—say, 3 to 1, which would win 41 cents for every dollar bet. Similarly, 4 to 1 earns a 76% payout and 5 to 1 more than doubles your money, as shown below.

2 to 1: $\$2 \times 35.2\% - \$1 \times 64.8\% = \$0.05$
3 to 1: $\$3 \times 35.2\% - \$1 \times 64.8\% = \$0.41$
4 to 1: $\$4 \times 35.2\% - \$1 \times 64.8\% = \$0.76$
5 to 1: $\$5 \times 35.2\% - \$1 \times 64.8\% = \$1.11$

For matters more important than winning money, however, the point of including a bet is to show how an expected result enters into a decision and is the basis of profit–loss thinking that measures the value of one event versus another (as we saw qualitatively for the airplane and car industry in Chapter 4). With the numbers at hand, an industry accountant might advise against 2-to-1 odds and a 5% profit for a given plan, especially when fixed costs and such are factored in, but might give the go-ahead at 3 to 1 and a 41% profit. Here, the mathematics helps us understand the implications of our choices. As for the equal odds of two people in a room having the same birthday, the mathematics[10] is a little harder but gives a seemingly unlikely answer of only 23.

$$.5 = 364/365 \times 363/365 \times 362/365 \times \ldots \times (365 - n)/365$$

But is it possible to check the answer without having to work through the math? Yes, empirically, from rosters of professional sports teams, which are easily obtained on the Internet today—a method that would have pleased Gerolama Cardano, the 16th-century Italian mathematician, physician, and gambler credited with devising the idea of a sample space (Mlodinow, 2008, p. 50). Such a method also launches us on our odyssey of using sample sets to calculate probabilities (see Table 7.1). As shown, the odds increase with roster size, where in the NBA, a roster size of 15 gives only a 25% chance that two players are born on the same day, whereas in the NFL, the larger roster size of 64 gives a 99.6% chance that two players share a birthday.

From the table, we see how well the predictions match the real numbers, despite the contrary perception. Only six teams in the NBA (21.4%) have players with the same birthday, whereas in the NFL all teams (100%) have players with a shared birthday, a seemingly unlikely occurrence, yet completely in accordance with the theoretical predictions (99.6% for a 64-player roster). The numbers also accurately match the predictions for MLB, the NHL, and the soccer World Cup.

It is interesting to note, as well, that in the NFL all teams had *multiple* double birthdays, from a low of 3 (for the Oakland Raiders, Minnesota Vikings, and San Francisco 49ers) to a high of 13 (!) for the Denver Broncos, more than one third of the team. Furthermore, of the 32 NFL teams, 12 had triple birthdays, and the Jacksonville Jaguars and Tennessee Titans had a double triple. Again, we see that it doesn't take

| | | | | | Number |
League	Number of Teams	Average Roster Size	Percentage Same Expected	Percentage Same Actual	Same Actual
NBA[a]	28	14.6	23.8	21.4	6/28
Soccer[b]	32	23	50.1	50.0	16/32
NHL[a]	30	24.8	55.5	56.7	17/30
MLB[c]	30	25	56.1	50.0	15/30
NFL[d]	32	64	99.6	100.0	32/32

Table 7.1 Expected and actual shared birthdays for sports teams

NOTES: a. End of 2010–2011; b. 2010; c. Start of 2011; d. 2006

many people in a room to ensure a better than 50:50 chance of a double birthday, multiple double birthdays, or even triple birthdays—we could even calculate the probability. Imagine how much you could win betting that a particular NFL team had a *triple* birthday (12/32 = 37.5%). Who wouldn't give you 100-to-1 odds?

Note that the number of people needed in a room to share a *particular* birthday equal to 0.5 is 251 (i.e., $0.5 = (364/365)^{251}$), which is about 10 times as many as *any* birthday and better matches our expectation of the rarity of such an event. Whether one birthday or birth month *should* be more probable is another question, which we'll look at later.

For now, the moral of the story is to be wary of numbers, especially those quoted by people who have something to prove, whether the author of a book, someone on television, a door-to-door salesman, or, as we will see next, the increasingly common Internet huckster.

A 50/50 Binomial Internet Numbers Scam

The need to understand mathematical basics has never been more present than in the age of Internet computing. Every day, it seems, we have to be more wary about the Internet, and not just the usual e-mails with promises of million-dollar windfalls if you send your bank details to so-and-so, but those of the ever more mathematically adroit. The rise of computational spam has made it even easier to trick, as in the following particularly loathsome example of mathematical Internet abuse.

In this example—previously cited by Paulos (2001) with regard to a pre-Internet mail scam—a stock market trader sends out 32,000 blind letters, half of which state that a particular stock index will rise and half that it will fall (p. 43). The following week, he sends letters to the half that were correct (16,000), again stating that the index will rise on half and fall on the other half. He continues the deception for 4 more weeks, and after 6 weeks, 500 people (32,000/64) have received six correct predictions. The trickster then sends a letter to the 500 who are in possession of six "perfect" predictions, asking them to pay for the next one.

Not a bad little con as cons go—who wouldn't pay an economic guru who can predict the future? Mind you, when you consider the price of stamps and stationery, the possible takings aren't quite up to snuff (32,000 + 16,000 + 8,000 + 4,000 + 2,000, + 1,000 + 500 = 63,500 stamps, letters, and envelopes). At 50 cents a mailing, that's more than $30,000—proof, perhaps, that crime doesn't pay after all.

However, I think our trickster can do a little better, since each week's prediction is independent of the previous week's (as originally assumed).

A smart trickster could send out all six predictions *in one letter*, saving the cost of 31,500 envelopes, letters, and stamps. Let's say the index went up, up, down, up, down, and down (UUDUDD) over a 6-week period. All who received a letter saying the stock would go down the first week would just junk the letter, leaving 16,000 intrigued potential investors with correct first-week letters (U). The next week, 8,000 would be in possession of correct second-week letters (UU), and 4,000 the next (UUD), followed by 2,000 (UUDU) and 1,000 (UUDUD), until, as before, 500 possessed completely correct letters (UUDUDD) after 6 weeks. More important, all the predictions were made *6 weeks previously*—an astounding feat of stock market wizardry sure to get the phone ringing off the hook. Our trickster might even get a few calls from those with five right predictions and one wrong, i.e., 500 from the last week and 2,500 from the 5 previous weeks. All for a savings of almost half in stamps, letters, and envelopes (not to mention the stamp and envelope licking).

Although new, the con still seems a tad expensive for most would-be hucksters, given the original outlay of 32,000 letters, but of course is made all the more relevant today by e-mail, which allows the conning of millions with little effort or cost. The mail-order scam of yesterday has been blown wide open by the Internet, where all one needs is 32,000 addresses, easily found in today's information world. Amazing how the snake is ever inventing new ways of playing old cons.

The odds, however, are not new and are elegantly summarized by Pascal's famous triangle, which shows the odds of any binomial distribution of events (up/down, heads/tails, true/false). Starting with a 1 for the top row and 1, 2, 1 for the second row, one adds the two numbers above to get the next line of coefficients—i.e., 1, 3, 3, 1 for the third row—as marked for the sixth row in Figure 7.2.[11]

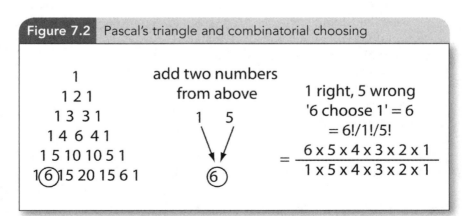

Figure 7.2 Pascal's triangle and combinatorial choosing

Pascal's triangle is a simple way to get the odds for any binomial event, including our trickster's con—a doubling over 6 weeks—where row 6 thus gives us the coefficients 1, 6, 15, 20, 15, 6, 1 for a total number of outcomes ($64 = 2^6$ or $1 + 6 + 15 + 20 + 6 + 1$), and corresponds to 1 all right, 6 five right, 15 four right, 20 three right, 15 two right, 6 one right, and 1 all wrong possibilities. The odds are the coefficients divided by the total, giving 1/64, 6/64, 15/64, 20/64, 15/64, 6/64, and 1/64, which shows that our trickster has a 1 in 64 chance of "predicting" that a stock index will rise for 6 straight weeks and a 6/64 or almost 10% chance of getting five right. Of course, 1 in 64 is not great odds, but remember that our trickster has mailed out 32,000 letters, 500 of which are correct ($32,000 \times 1/64$), or possibly hundreds of thousands by e-mail. Note also that the number of outcomes is 2^6 or 64, as we saw with the geometric series, and the same 1 in 64 can be calculated as $1/2^6$ (although we can't get the coefficients for the other odds this way).

Any binomial probability can be calculated in the same way—for example, a coin toss repeated three times, where one gets 1, 3, 3, 1 for one outcome with all three heads (HHH), three outcomes with two heads and one tail (HHT, HTH, THH), three outcomes with one head and two tails (HTT, THT, TTH), and one outcome with all three tails (TTT) for a total of 2^3 or 8 possible outcomes. Here, the probabilities are again the coefficients divided by the number of outcomes, giving 1/8, 3/8, 3/8, and 1/8. Pascal's triangle is that simple.

To determine the coefficients for higher odds, however, such as multiball lotteries with many numbers to choose, we must use more advanced methods (which we'll look at in Chapter 8). But for small numbers, Pascal's triangle is perfectly capable, where a simple construct gives the same numbers as the more involved mathematics does.[12] Indeed, Pascal's triangle works amazingly well for any small number of binomially distributed events (e.g., 6 weeks of up/down stock indicators or three flips of a coin).

Interestingly, Pascal devised all this when he was working out for a gambling friend how to calculate fair payouts in an interrupted game of chance (Mlodinow, 2008, p. 67), seemingly of huge interest in 17th-century France. He based his system on who was ahead and by how much when the evening's proceedings were brought to an untimely finish and, most important, that any unfinished games would be randomly decided, i.e., binomially or 50/50.

For example, in a best-of-seven game of cards (or any other gambling venture), Pascal's method shows that one will go on to win 66% of the time after winning the first game, 81% after the first two, and 94% after winning the first three, practically a dead certainty.[13] From these odds,

Pascal calculated who owed what. Such logic applies to any series of random events, such as roulette spins, stock performance, and sports playoff series.

Of course, winning isn't necessarily a random process—say, in sports, where winning is affected by home advantage or ability. But Pascal's method still applies if we know or can estimate the odds. Let's say Roger Federer is playing a low-ranked player in a tiebreak at Wimbledon (first to seven) and his odds of winning any point are 3 to 1. According to Pascal, Federer will win 97.6% of the time and will even win 60% of the time after losing the first 4 points![14] To be sure, he can still lose and will almost 1 in 40 times if the odds of winning each point are 3 to 1, which explains why poorer players occasionally win tiebreakers against better players, how lower-ranked teams win playoff series, how some gamblers can win at roulette through blind luck, or, as we will see next, how some stock pickers seem to be more successful than others.

Pascal was credited with a great many things, especially the principle that pressure in a fluid is transmitted equally throughout a liquid (which defines the basics of hydrostatic or hydraulic pressure), as well as adding machines, number theory, and of course calculating the probability of binomially distributed events. Thankfully, he didn't get around to the mail-order business or the 17th-century equivalent of spam. If he had, we would all be losers.

7.3. Is He or Isn't He? A Not-so-Random Guru World

Taleb (2004) called a similar version of the previous con "the mysterious letter" (pp. 157–158), where bullish/bearish predictions are made by a trickster and sent to a group of unsuspecting possible investors each month (10,000 in this case)—for example, on January 1, February 1, March 1, etc. Although it still seems like a lot of work (stuffing envelopes, licking stamps, following up each month), the principle of the ruse is the same and again shows up the not-so-mysterious properties of the geometric series. Taleb referred to such mechanisms as "survivorship bias" (pp. 93, 146), since the dwindling numbers after each outcome—the resultant "successful," though talentless gurus—end up calling the shots, solely because of probabilistic attrition. History written by the winners?

However, he incorrectly used the same logic with another 10,000 traders, whose success he based on a 50/50 coin flip (Taleb, 2004, pp. 152–153). In his fictional accounting, the traders also work for George Soros, who fires any "losers" at the end of each year, leaving 5,000, 2,500, 1,250, 625, and finally 313 traders still standing, although such exact numbers cannot come out so neatly. In fact, the likelihood of an exact halving can be shown to be impossible using the simplicity of Pascal's triangle (for only four throws, the probability of getting exactly two heads and two tails is 37.5%, or 3/8, not a perfect 50/50 split). And the difference gets worse as the numbers increase—much worse (not as unlikely as that famous monkey typing *Hamlet,* but still quite remote).

Furthermore, using Pascal's triangle[15], we can see that the likelihood of an even split dwindles as *n* increases (for *n* = 2, 4, 6, 8, 10: 1/2, 6/16, 20/64, 70/256, 252/1,024, shown in Figure 7.3 for *n* = 2, 4, and 6). Continuing to higher *n*, the odds of 144 traders producing exactly 72 successful and 72 unsuccessful traders (as far as my computer's Pascalian combinatorics would take me) is about 1 in 15 or 6.6% (1.48×10^{42} / 2.23×10^{43}), and the odds of 10,000 producing exactly 5,000 good and 5,000 bad traders is 0.0000798, an exceedingly unlikely occurrence. The odds of 10,000 producing exactly 5,000 followed by exactly 2,500, 1,250, 625, and 313 is essentially 0 ($\sim 1 \times 10^{-15}$).[16]

| Figure 7.3 | Pascal's triangle: The mean is not as likely as the rest |

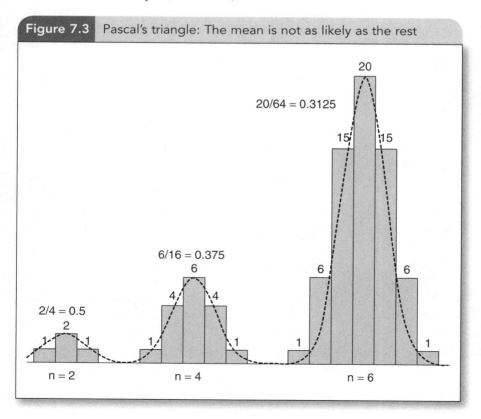

Indeed, we get a 50/50 split only for two throws, otherwise an even split is much more unlikely. Note also how the continuous Gaussian distribution (aka bell curve or normal curve) wonderfully approximates the discrete binomial distribution.

Taleb (2004) did, however, correctly describe how cancers are more likely to form clusters, given that a randomly distributed attribute is much more likely to be unevenly spread out in a population, in the same way darts thrown randomly will show a skewed or nonaverage behavior (pp. 169–170) or, as Mlodinow (2008) noted, the way raisins clump in a box of cereal (p. 13)—or, as seen above, how binomial distributions cluster about the middle.

Happily, such an analysis also shows that the average is *not* the most likely occurrence, which I think may be heartwarming to many, including the willfully provocative Taleb. Although the mean is the most likely *single* occurrence (most probable), it is highly unlikely that any one person will be exactly average over a large population—in fact, most of us are *not* average (*ne pas hommes moyen*).

Nonetheless, Taleb (2004) did correctly note that a successful trader will not necessarily be successful the following year if such success was randomly based—although, according to his simple, trader-game logic, 166 out of 10,000 will be successful after 6 years (p. 153). Furthermore, such traders shouldn't be trusted to make future decisions, because randomly directed strategies will be right only half the time, and one might unfortunately fall in the bad half:

> We tend to think that traders were successful *because* they are good. Perhaps we have turned the causality on its head; we consider them good just because they make money. One can make money in the financial markets totally out of randomness. (p. 93)

As Taleb also noted, Russian roulette players rarely live to their 50s (p. 24), typing monkeys producing *The Iliad* shouldn't be expected to type *The Odyssey* next (p. 135), and battle winners who won by a crazy decision won't likely continue to succeed (p. 256).

Such logic is similar to a random or so-called drunkard's walk, where in any two-pronged, binomial outcome (random stock picks, coins, roulette colors, or drunken steps from a pub door), there is an equal chance of a win/loss, head/tail, red/black, or left/right step.[17] Random walks elegantly explain why bad stock traders can make money, in the same way red occasionally comes up on a roulette wheel 10 times in a row—a nonetheless highly unlikely occurrence ($1/2^8 = 1/256$).

The same can apply to supposed hot streaks in Hollywood, where Paramount Pictures followed three great financial successes (*Forrest Gump, Braveheart, Titanic*) with some not-so-great financial successes (*Timeline, Lara Croft Tomb Raider: The Cradle of Life*; Mlodinow, 2008, pp. 13–14). Subject to the uncertainties and vagaries of the movie business, this may be purely random and have nothing to do with talent (a theory not likely to be of any consolation to Paramount's fired CEO).

Similar thinking can extend to why some people are more successful than others—for example, Bruce Willis, an unknown actor who took a trip to Los Angeles to watch the Olympics and while there auditioned for a role in *Moonlighting*, which then led to a career in movies; or Bill Gates, whose small software company was hired to supply DOS to IBM after a chance meeting with an employee, which then led to him becoming the richest man in the world (Mlodinow, 2008, pp. 206–209). Similarly, Barack Obama was a relatively unknown first-term senator from Illinois when he was picked to give the keynote speech at the 2004 Democratic Convention in Boston, without which he likely wouldn't have ended up as president.

Precedence or cumulative advantage, or as some call it, path dependence, is at the root of many storied successes, from child actors Elizabeth Taylor and Mickey Rooney to Lana Turner, who skipped high school typing class to buy a drink and was famously "discovered" at a Los Angeles pharmacy. Likewise for a book becoming a best seller, as when 22 million copies of then-unknown author Jacqueline Susann's *Valley of the Dolls* were sold in 1966 after Bantam rolled out a relentless publicity bandwagon, prompting the publisher to remark that "the only thing you could turn on without getting Jacqueline Susann was the water faucet" (Frank & Cook, 2010, p. 140).[18] Is the art of Damien Hirst, the music of Lady Gaga, or the success of many manufactured pop stars today any less susceptible to the viral nature of media exposure in the buildup of careers? Although other factors clearly determine the binomial multiplying in such "domino determinism," success can and does follow from multiplying the effect of an initial result without which nothing would otherwise have followed (something we'll see more of in Chapter 10 with regard to feedback loops and winner-take-all markets in economics).

Furthermore, an increase followed by an increase or a decrease followed by a decrease can be entirely random, despite the law of large numbers (or regression toward the mean), first noted by the English mathematician Francis Galton in successive sweet pea generations and then in humans, where large peas and taller humans tend to have smaller and shorter offspring, and small peas and shorter humans tend to have larger and taller offspring.[19]

Quiz games such as Trivial Pursuit, *The Weakest Link*, and *Who Wants to Be a Millionaire?* can similarly depend on luck, i.e., who gets asked what, and not necessarily who is the smartest (as we saw for overlapping content and knowledge of tests in Chapter 4). Furthermore, future success can have nothing to do with past results, depending again on the luck of the draw and false impressions rather than on a player's knowledge.

But we have to be careful with those who like to oversimplify a belief in randomness, refuting things we know to be causally related. In his attempt to never be too sure, Taleb (2004) called himself a "naïve falsificationist" (p. 128) and glossed over much. Sure, some people bet their lives and win while many more lose, and we have to live with the lucky ones who then assert their luck as talent. But much of life is not random; it is biased, whether in favor of the house winning 5.3% for every roulette spin and 7% at the blackjack table, or in the more than 99.99999% chance in the lottery,[20] not to mention a trader, developer, or politician taking advantage through insider knowledge or connections.

Taleb (2004) did make the important point, however, that past data doesn't necessarily contain information about future data—owing, as he argued, to a prevailing randomness (pp. 160–162). His point is best summed up in one line: Past success does not equal future success, particularly if the past is random. It should be noted, however, that he didn't mind stating, "We have been getting things wrong in the past and we laugh at our past institutions; it is time to figure out that we should avoid enshrining the present ones" (p. 243), suggesting that the present does depend on the past, as most of us would think.

Taleb's (2004) subthesis, however, is more palatable: that our habitat has dramatically changed in 3,000 years, yet our mathematical abilities (or tools) have not (pp. 198–200), especially probabilistic reasoning, which presumably was in short supply in caveman times—although I would not relate such a lack of probabilistic reasoning to evolution. Nonetheless, a good simplification, if simplifications are needed, is that science and discovery are a mixture of the old and the new, the reliable and the untested, the signal and the noise, which must be verified at every turn. As remarked by stock-broker-turned-painter Paul Gauguin, we must ever balance the revolutionary with the plagiarist. And we must ever protect ourselves against numerical magicians.

7.4. Not All Numbers Are to Plan: Noise and the Art of Confusion

Numbers don't always go according to plan—for example, with birthdays, which at first thought should be randomly distributed through the year, each month having the same number of birthdays as any other, i.e., 1/12 or 8.3%.[21] It is possible, however, that more births occur 9 months after a great emergency (say, a storm or war) or after a large sporting event (for the winning city or nation anyway). Or perhaps more conceptions occur during colder months, when people stay inside more, suggesting that births are more likely in October/November in the northern hemisphere or in April/May in the southern hemisphere. In Europe, particularly France, where everyone goes on holiday in August, the birth register may even see a rise 9 months later in May and June.

Sports teams' rosters again provide us with readily available data, from which we can put such a theory to the test and try to assess if professional athletes are born, not made, i.e., a greater number born in a particular month. As before, when we looked at the number of people in a room to ensure a 50/50 chance of a double birthday, data from the NBA, NHL, MLB, NFL, and soccer World Cup were analyzed (a sample set of almost 5,000 players[22]). From the data, we can see if a randomness of birth applies and whether there is a marked difference in birthdays by month.[23]

As illustrated in the histograms in Figure 7.4, the data shows a reasonably flat spread, although not exactly 8.3% per month. Here, from our data set of almost 5,000 professionals—a seemingly large enough sample—the "winning" birthday month isn't so clear-cut, showing a preference for June in basketball, March in hockey, August in baseball, January in football, and May in soccer (not shown). But can this be attributed to anything other than chance, or is it due entirely to the randomness of numbers?

Gladwell (2008) questioned whether sporting success is based on individual merit and talent or whether it has to do with "relative age," that is, being born in the early school cutoff months of January, February, or March. He cited a greater number of players in Canadian hockey born in the first 3 months of the year and suggested this was because of the eligibility date of January 1, meaning that a child born earlier in a year will develop faster and be selectively advanced throughout life compared with one born later in the year (pp. 15–34). However, he based his analysis on only two teams, both with relatively small sample sizes

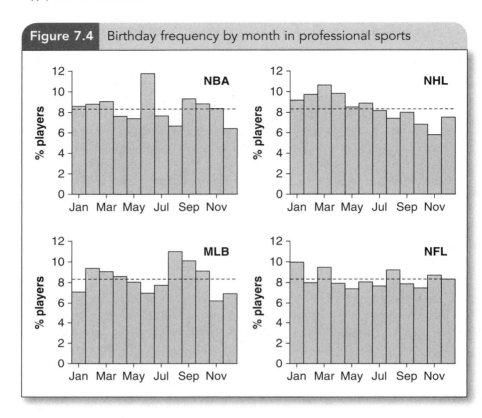

Figure 7.4 Birthday frequency by month in professional sports

(25 and 21)—the 2007 Memorial Cup-winning Medicine Hat Tigers hockey team and the 2007 World Cup-finalist Czech National junior soccer team—as if cherry-picking the data to fit his conclusion.

So let's try a more recent professional winning hockey team (2011 Boston Bruins) for a similarly small sample of 26 players. Here, however, the data doesn't quite show Gladwell's supposed Matthew effect (a biblically inspired leg up, labeled by sociologist Robert K. Merton), highlighting the danger of picking small data sets, and suggests that the conclusions are specious. In fact, only two players were born in January, the same as in November. In a slightly larger sample (the four 2011 NHL semifinalists, Boston, Vancouver, San Jose, and Tampa Bay), eight players were born in December, the same as in January, and more were born in April (14) than in any other month (from a total sample of 113 players). Even with a larger sample, e.g., the 2006 NHL season (984 players), more players were born in April, May, and June than in January, February, and March.

The most apt refutation of the so-called "birthday lottery" thesis, however, may come from hockey's best-ever player, Wayne Gretzky—born January 26, 1961—who broke every scoring record, winning the MVP nine times, the scoring title 10 times, and scoring more than 200 points four times in his 20-year career, a feat not replicated once by any other player. Gretzky could arguably be considered the best athlete ever in any sport from his statistics relative to those in other sports (e.g., comparing a normalized points total by percentage to the rest of the league).

At first look, Gretzky's birth date seems to validate the supposition, *except* that he played his entire junior hockey career with players up to 4 years older than him, starting as a 6-year old in an under-10 novice league in Brantford, Ontario. When he first started turning heads on the ice, he was only 8 years old, 2 years younger than most of his teammates, yet he racked up more than a point per game and finished the season as the top scorer. Furthermore, throughout the rest of his junior years, he was typically the youngest and smallest on any team and was only 17 when he turned professional, joining the NHL at the age of 18.

Even as a professional, Gretzky was on the small side, although he eventually grew out and up enough to handle the higher degree of physicality in the NHL. Thus, it seems that his success is because of talent and practice, *not* maturity as Gladwell and other birthday lottery proponents have assumed. But of course, my data is based on a data set of only one, and it certainly doesn't follow that all younger, smaller, January-born Brantford natives named Wayne will succeed at hockey. To be sure, something is up—the numbers have been picked to support a thesis. With small data sets, one can make the numbers do anything one wants.[24]

Nonetheless, why does the data show a greater number of players born in some months than in others? Well, if there is a systematic bias, there could be any number of reasons and much more data is required for a proper analysis. Here, the cutoff age is not the same for each sport because leagues start at different times during the year and players come from different countries with various customs. Furthermore, relative age is much less likely on teams with older players (Levitt & Dubner, 2010, p. 63)—for example, a December-born 20-year-old could be much bigger than a January-born 20-year-old, a rather unlikely occurrence, comparatively, among 4-year-olds. To be sure, all sorts of reasons are possible with limited data sets that suggest order when there is only randomness.

But if one wanted to infer from the data that the best chance of succeeding as a professional athlete is to be born in January, one would be wrong. Just because more professional basketball players are born in June, hockey players in March, baseball players in August, football players in January, and soccer players in May (for the given years surveyed), it doesn't follow that being born in a particular month increases one's chances of playing in a professional sport. More likely, a tall person has a better chance of making the NBA (with an average player height of 6'7"), someone from a northern country is more likely to be a hockey player (53% Canadian and 23% American versus about 5% each of Swedish, Czech, Finnish, and Russian), a big person is more likely to play in the NFL (with average player height and weight of 6'2", 247 pounds), or a person with a professional player for a father is 800 times more likely to play in baseball's major leagues than is a random player (Levitt & Dubner, 2010, p. 65).

So what's up with randomness, and can it be explained? In the example above, we expect that if birthdays are completely random, then 8.3% of the population should be born in each month ($1/12 \times 100\%$), yet in our sample, 9.6% were born in March and 7.7% in December. In this case, the difference between the expected and the real relates to the natural error or noise of numbers, where the square root of the sample size provides the plus/minus error, such that for our sample size, any month falls in a range from 6.9% to 9.8% (i.e., between ± 70 of the expected value). What's more, as the number increases, the relative error decreases (as in the law of large numbers). Figure 7.4 especially shows how there is less variation in the monthly differences the larger the sample (NBA: 409 players; NFL: 2,302 players). As the numbers increase, the variance correspondingly decreases.

Furthermore, as we saw in Pascal's triangle, which showed an extremely low probability for an exact 50/50 split of coin tosses, it is highly unlikely to get an *exact* split of birthdays throughout the year (i.e., exactly 8.3% for each month). In fact, one would be extremely surprised to find an exact split in any data set, which would immediately call the sample into doubt.

Betting Strategies: How to Go Broke Over the Long Run

Randomness is hard to explain to the numerically fearful (something we'll look at more in Chapter 8) but is often tied up in superstition and small-number occurrences that suggest a pattern where there is none. For example, if six people choose a different number on a roll of a die, the one who wins feels lucky, as if the number was meant to be. But the number 1 will come up just as often as the numbers 2, 3, 4, 5, and 6 *over the long run* within a quantifiable statistical plus/minus error, no matter if 6 came up 10 times in a row in any prior sequence of throws (less than once in more than 60 million rolls). Red will also appear as often as black on a roulette wheel, even if it occasionally appears 10 times in a row (about once in every thousand spins).

For dice rolling, the odds of ten 1s in a row are $(1/6)^{10} = 1/60,466,176$.
For roulette, the odds of 10 reds in a row are $(1/2)^{10} = 1/1,024$.[25]

Even an octopus can sometimes pick multiple winners in a row, as did the famous Paul, who tried his luck on the outcome of each of Germany's seven 2010 World Cup soccer matches (a 1-in-128 likelihood of getting all right). An unbroken sequence of 1 million zeroes will even appear at least 10 times in a series of $10^{1,000,007}$ randomly distributed ones and zeros, a truly extraordinary example (Mlodinow, 2008, p. 174). Indeed, the *appearance* of bias is often purely random, or, as Paulos (1996) noted, "Randomness on one level of analysis constitutes a kind of order on a higher level" (p. 180).

Don't believe me? Try a strategy of betting the outcome of a roulette wheel by choosing the color that didn't come up on the previous turn. Over the long run, you will lose. Taleb (2004) referred to such traders in the financial world as "dip buyers," who dominated the heady bull market times between 1992 and 1998 but were shown up when the markets dramatically slumped (p. 93). Eventually, the randomness of events swamped the apparent order in a small number of occurrences.

One can win, however, by applying the niceties of binomial probabilities, using a simple betting strategy—provided one has the nerve and a large enough bank account to cover the occasional losing streak. Start with a bet of one unit ($1, $10, whatever you can afford), betting another unit after each win but doubling the previous bet after any loss (arithmetic wins, geometric losses, minus the house edge). Despite the patterns that emerge, the losses are covered.

Alas, such "good money after bad" strategies will eventually make huge losses, thanks to the power of geometric doubling—as in the investment banks we hear about from time to time, where a rogue trader keeps betting that the market will "turn" but runs out of time and money (aka gambler's ruin). Just ask Nick Leeson, the derivatives broker who broke Britain's Barings Bank when he continued to bet that the Japanese markets would recover after the Kobe earthquake, or Jérôme Kerviel, who almost did the same to France's Société Générale when he lost €5 billion by hedging on European index futures. Both were little more than gamblers who instantly became famous for "doubling down" their losses and losing more.

Gambler's ruin will always happen when you don't have enough money to cover the doubling, just like a collapsing pyramid scam that runs out of friends.[26] What's more, the more you play, the higher the chance of apparent rarer events, as Benoît Mandelbrot showed years ago when he made chaos theory famous. Clearly, butterflies flapping their wings in Brazil don't lead to hurricanes every day, but they have to do so only once to cause a lifetime of damage.

Or, as we'll see with regard to the causes of the credit crisis (in Chapter 12), a whole series of unexpected events in the stock market can happen all at once to blow up the best-laid trading plans. As Patterson (2010) noted, "Risk management is about avoiding the mistake of betting so much you can lose it all—the mistake made by almost every bank and hedge fund that ran into trouble in 2007 and 2008" (p. 297). Indeed, if we've learned anything about probability, we know that unlikely events can and do happen.

So what can we do to better prepare for the unlikely? In probability—· whether the stock market, games of chance, or birthday distributions— what is important is not to look at a result and work backward, such that one event caused (or will cause) another, but to understand that any outcome is one possibility, however likely or not, on which you might want to bet given a fair system and favorable odds. However, randomness is just that, and occasional patterns appear (or don't, although we may see them anyway) aided by our desire or inclination to see patterns—whether in the movement of pollen seeds suspended in water (the original Brownian motion); in the spread of heights, weights, or birthdays in a population; or in the ups and downs of stock market ticks.

The taxman knows all about such randomness and can check whether your statement shows a natural disorder or betrays a too-tidy tinkering. Try rounding your expenses to the nearest dollar on your next tax return and see how fast you get a call from the auditor, cited as No. 2 in "Ten Common Tax Audit Flags" (Bailey, 2008).[27] Remember, randomness is just that—random.

Unfortunately, chance can never be one-upped, other than in the world of fiction—as in *Rosencrantz and Guildenstern*, Tom Stoppard's play about determinism, where in the opening scene a series of coin tosses comes up heads 20 times in a row, an event that would take more than 1 million tosses to happen *once* by chance ($(1/2)^{20} = 1/1{,}048{,}576$), which may well have happened given three performances per day since the play first opened in 1967. Nonetheless, such an occurrence is due entirely to chance.

As for one's lucky stars increasing the chances of playing for the New York Yankees in a World Series or for the Pittsburgh Steelers in a Super Bowl—which underlies the idea of birthday lotteries and the like—unfortunately, birth month is not a predictor of future success. The mother's biological stress (smoking, drinking, and pressure), however, *is* a significant predictor of future success, as noted by Nobel prize-winning economist James Heckman (2006b). Heckman also suggested that birth dates should start 9 months before the actual birth to account for the important development time during pregnancy, an interesting thought that emphasizes prenatal development, not to mention standing the whole of astrology on its head (more on that in the next chapter).

The scariest thought of all, though, is whether genes are a predictor of future success. Now, that's something I'd be far more worried about than in which month I was born or, for that matter, my astrological sign.

7.5. Monopoly: What Would Aesop, Smith, Rousseau, and Marx Think?

To end this chapter, we look further into the nature of chance and probability, starting with the board game Monopoly, which simulates the economics of the marketplace with dice and from which we can begin to see how a computer simulates real-world phenomena, from physics to stock markets. Some even believe that mathematicians and physicists are to blame for increased stock market instability, first by using computers to predict trends and then by automating a lemming-like sell process when a particular economic marker is reached. It doesn't take a mathematics degree to know that if everyone has the same software,[28] the same decisions will be made across the board, often with disastrous results.

We will then try our luck at a computer-simulated version of Monopoly and other games to compare analytical solutions (i.e., using equations) to numerical solutions (i.e., using computers). Today, with ever-faster computers, it has become easier to calculate hard-to-work-out probabilities, using numerical or Monte Carlo techniques that crunch out millions of "runs." In some cases, when the mathematics is too complicated, there is no other way to solve a problem.

But, first, let's look at Monopoly, a simple economic simulation created in 1930 by Charles Darrow after he lost his sales job following the

1929 American stock market crash, and sold to the established game makers Parker Brothers, who released the version we know today.[29] As most of us know from our game-playing youth, in Monopoly one buys properties and collects money from rent and passing "Go" on a repeating four-street board, a model not too dissimilar from life if properties are substituted for possessions and "Go" represents a weekly (or monthly) paycheck. Over the course of the game, some succeed and live it up, whereas others don't and find it hard to make ends meet when the bills come around, whether Electric Company or landing on Boardwalk with a hotel. Suffice it to say, Monopoly life gets worse for most players as Monopoly life goes on, and the moral of Monopoly is that there is only one winner, which may be great for a vacation game but is not so good in life.

One can also see that at the start, players coexist in an easy way, not worrying from one pass of "Go" (paycheck) to the next, until the high-rent monopolies (sets with hotels) kick in, making it impossible to sustain the earlier balance. Spikes soon appear as rents and penalties increase, and savings become more uncertain. Interestingly, as some families play the game, trading properties is discouraged and the balance is prolonged for hours, one of the usual complaints of Monopoly (for the older players anyway). But when played more aggressively, as in a real marketplace, Monopoly typically lasts no longer than an hour.

As an analogy for our own communities, we have already seen that the benefit to the player of Community Chest is better than that of Chance, although the benefits of both diminish when the freewheeling begins and Chance (and chance) becomes more of a contributing factor to the outcome of the game (more redirects and higher rents)—where one roll can win or lose all, not unlike the increasing monetary load of high-risk mortgages, casinos, and lotteries today. Of course, the early honeymoon phase in Monopoly is valid only because labor is creating wealth as each player participates in the building up of the bank. But by hard work or luck, some end up with more, and as soon as the capital becomes organized to build high-cost homes and wealth is no longer shared (as the money in the bank originally was), excessive capital starts to swamp labor and the weekly paycheck no longer covers expenses. When rents and mortgages are many more times income, $200 just doesn't cut it. As private control over the public wealth increases, more and more of the weekly (now monthly or yearly) paycheck covers less and less of the individual's and the public's holdings.

Is this the price of progress, accompanied by a rise of greed and excessive moneymaking that leads to devaluing the community and an unhealthy preoccupation with wealth, where making money becomes the primary or even sole goal in life? Galbraith (1991) wondered, with regard to safeguarding the economic system, whether "something, perhaps much, must also be attributed to the pleasure of winning in a game where many lose" (p. 220), suggesting that economics is seen by many as a game—in this case, a winner-take-all game. He also noted that there must be something wrong in a society with so much private wealth in the midst of so much public squalor (Galbraith, 1958/1999, pp. 70, 101, 191).

Well, it is only a game, but perhaps Parker Brothers was on to something with its dice-rolling economic model. Odd, though, that luxury tax doesn't increase as the game goes on. At a flat rate of $75, luxury tax is irrelevant to the high rollers with property sets and hoards of money.[30]

What would some of the great economists and moralists say about Monopoly and its comparisons to reality? Well, for fun, let's imagine a Christmas game played by Thomas More, Jean Jacques Rousseau, Adam Smith, Karl Marx, George Santayana, Laurence J. Peter, and William Shatner ("Do you want to make more money? Of course, we all do."), if only to compare competing ideas about money and morals.

In our make-believe game, Thomas More is the dad, regularly topping up everyone's eggnogs, although sadly low on rum. He is forced to play only because it's a holiday, and always rolls out of turn. In *Utopia*, he wrote, "'We'll never get human behaviour in line with Christian ethics,' these gentlemen must have argued, 'so let's adopt Christian ethics to human behaviour. Then at least there'll be some connexion between them.'" As the top hat, he goes out first, making a poor trade, which produces the first set of the game for Shatner.

Rousseau, the older college-student brother, looks down his nose at such wanton abuse of private property and debased games. Nevertheless, he is the thimble, refusing to trade his properties to anyone, counting only on luck, and seeming to land on "Free Parking" more than his fair share, though only Santayana notices. In his *Social Contract*, Rousseau wrote, "The first man who, having fenced in a piece of land, said, 'This is mine,' and found people naive enough to believe him, that man was the founder of civil society." Rousseau's luck is quite unbelievable, as he lands on all three orange properties in a single turn after getting out of Jail (double 3, double 1, followed by a 4 to land on "Chance" and a "Go backwards three spaces" redirect).

Adam Smith is the wheeler-dealer, the car, who will swap with anyone, saying again and again that it's win-win all around. He buys everything he lands on, but when Santayana accuses him of stealing from the bank with his invisible hand, he goes off in a huff, yelling about his corollary from

The Wealth of Nations: "Consumption is the sole end and purpose of production; and the interest of the producer ought to be attended to only so far as it may be necessary for promoting that of the consumer." No one seems to understand.

Marx is the intellectual uncle, a bit tipsy on wine, who is stuck playing banker. He casts a knowing eye when an argument breaks out about who owes what to whom, but nonetheless ensures that everyone pays their fines. He is the wheelbarrow and plays with care. He wrote,

> The modern bourgeois society that has sprouted from the ruins of the feudal society has not done away with class antagonisms. It has but established new classes, new conditions of oppressions, new forms of struggle in place of the old ones. (Marx & Engels, 1848/1992, p. 3)

Laurence J. Peter is the iron and is knocked out second, denying that he lost because he was saddled with the utilities and two greens and passed on Park Place because he didn't want to mortgage Pacific Avenue (duh?). He shouts to anyone who will listen, "In a hierarchy every employee tends to rise to his level of incompetence."

As the cutthroat game continues, the rich get richer and the poor get poorer, as in any winner-take-all market (which we'll look at more in Chapter 9). Marx eventually leaves, however, to join the wives' card game, saying that he doesn't have the stomach for all this ruthless capitalism (taking care to return his property and money to the bank). As the game continues, Shatner plugs his various products (Priceline.com, Kellogg's All-Bran Wheat Flakes, Chrysler, etc.), and Santayana finally wins when Shatner and then Rousseau land on his Boardwalk with a hotel. As Rousseau counts out his last dollar, Santayana smugly announces, "Those who cannot remember the past are condemned to repeat it," at which point Smith pours his eggnog over him while an argument breaks out about the working conditions of the Roman slave, Shatner questioning under what circumstances a slave could own property or whether they were, in fact, only property. Smith gets angry again and shouts at Shatner for saying that the slave was better off than the working man in Britain in the midst of the Industrial Revolution, which he points out permitted 19-hour workdays and corporeal punishment for children until the Factories Act of 1802 (aka the Health and Morals of Apprentices Act). "Do you want to make more money? Of course, we all do," Shatner giddily yells, before Smith runs off again, muttering to himself that no one understands.

As for Aesop, who has been quietly reading a book by the fire the whole time, he is asked who would fare better at Monopoly, the tortoise or the hare. Courtesy of his new half-off, no-money-down-for-a-year, Christmas-sale computer, he excitedly shows them a graph he has made, telling them that since the biggest increase in rent is from two to three houses, the best strategy is to be like the hare and race to put three houses on a property but then to chill out like the tortoise on Easy Street (see Figure 7.5).

They all smile, even Shatner, who asks how he can get a franchise. After more discussion about Keynesian fiscal policy and the creation of the welfare state by Bismarck, the game breaks up and they all go down to the local Pig and Whistle for pints—all except Marx, who seems to have become somewhat of a hit with the wives.

As for predicting where one might land in Monopoly, the probabilities cannot be worked out analytically, and so we turn to modern computing to simulate a large number of throws, called a Monte Carlo analysis, the ultimate in modern number crunching.

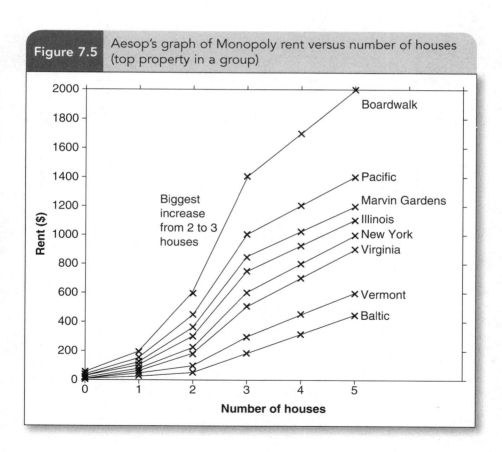

Figure 7.5 Aesop's graph of Monopoly rent versus number of houses (top property in a group)

The relatively simple Monte Carlo calculation shown in Figure 7.6 took about 30 seconds on an off-the-shelf, 3GB RAM PC and gives the probability of landing on any square as a percentage over 1 million dice throws, a suitably large number to see any possible trends (excluding the relatively minor effect of Chance and Community Chest). We can simulate the results for any number of throws with a sufficient random-number generator, subject to available computer time.

Here, we can see that the Jail square (10) is the most likely square to land on (roughly 5%, and about twice as often as any other square), as one might expect given the "Go to Jail" (30) redirection and as can also be worked out quasi-analytically using Markov chains.[31] Without the "Go to Jail" redirection, the percentages are 2.5% or 1/40 for each square. Note that the squares after Jail from 14 to 33 are more frequently landed on than the other squares from 34 to 13, again because of the "Go to Jail" redirection. From this simple analysis, it seems that properties on

Figure 7.6	Monte Carlo-calculated frequencies for Monopoly properties

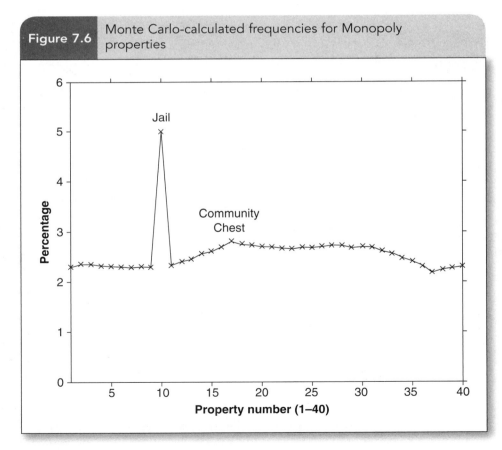

the streets after Jail, especially from orange to red (squares 16, 18, 19 and 21, 23, 24),[32] are the best properties to own, with the oranges perhaps more so if the lower costs to buy and develop them are factored in (although, it seems, not for Rousseau in our one-off Christmas game).

It can also be seen that square 17 (Community Chest) is the most likely square to land on after Jail (7 being the most likely throw from Jail). Note that the same results are found for repeated simulations, thus reducing the chance that chance (as in the distribution of throws straying from theoretical expectations) played a part in the results. At 10 million throws, the chance of chance playing a part is reduced to almost zero.

More Monte Carlo Solutions

Monte Carlo simulations can also calculate odds for blackjack, for example, determining whether one should hit with 16 when a dealer shows a 9 up (yes), Texas Hold 'em strategy for all possible pairs of down cards, or other scientific calculations for which the excessively large number of atoms in even the smallest-sized matter makes analytic solutions impossible.

The Monte Carlo calculation in Figure 7.7 shows whether to stick or hit in blackjack against any up card from the dealer. Here, 25,000 hands were dealt and took 2 minutes to run on the same off-the-shelf computer as before. Although 25,000 games is a small simulation, the results in some cases are nonetheless decisive (e.g., 5 to 1 against for a player 16 versus a dealer 9). Note that the percentages close to 50% are anybody's guess

| Figure 7.7 | Blackjack stick strategy versus the dealer's up card |

		Dealer's up card									
		2	3	4	5	6	7	8	9	10	A
	12	.42	.40	.44	.52	.47	.30	.29	.31	.23	.26
	13	.41	.38	.41	.49	.48	.29	.21	.23	.24	.23
	14	.40	.46	.46	.48	.46	.27	.20	.24	.21	.22
	15	.38	.45	.48	.48	.46	.30	.24	.31	.23	.25
	16	.42	.44	.44	.48	.51	.30	.24	.20	.22	.28
Player's card total	17	.51	.47	.46	.53	.57	.50	.28	.22	.26	.29
	18	.53	.70	.64	.66	.59	.75	.66	.38	.38	.54
	19	.76	.79	.72	.81	.66	.88	.92	.65	.56	.54
	20	.87	.87	.85	.86	.88	.92	.90	.92	.81	.62
	21	1	1	1	1	1	1	1	1	1	1

(stick strategy > 50% are highlighted in grey), but the results clearly show that one should hit with 16 against a dealer who shows anything from a 7 to ace, however counterintuitive that may seem.

Monte Carlo analyses have also become a bona fide mathematical technique for working out many calculations that cannot be solved analytically, including nuclear reactor models, molecular simulations, and, as is more common today, the wired world of social networking. A Monte Carlo calculation is the ultimate trial-and-error perseverance method, where all one needs is a standard random-number generator[33] to simulate a dice throw, calculate the motion of a nucleus in a DNA sequence, or model temperatures in a climate-change projection calculation.

Weather patterns can also be modeled using Monte Carlo statistics and the simplest mathematics of Pascal's binomial method, where the flipping of a coin is used to model a future event, the so-called random or drunkard's walk, each step having an equal probability of left or right (recall the binomial distribution from Pascal's triangle). Once again, randomness rules, as can be seen in the two-dimensional random walk shown in Figure 7.8, where an equal chance of a step north, east, west,

Figure 7.8 The drunkard's walk and climate change

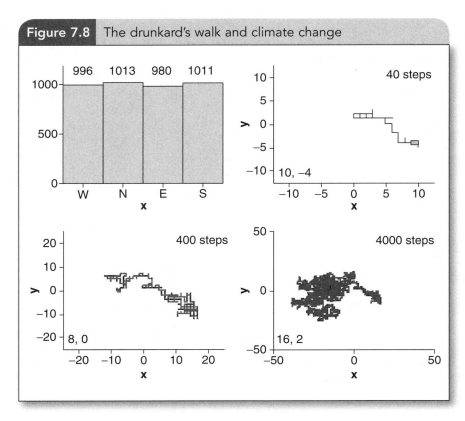

or south shows a relatively flat distribution (top left) and the position of a random walker after 40, 400, and 4,000 steps.

Such logic is often used to simulate the dynamics of a complex system. Fleming (2005) even likened long-term weather prediction to predicting the destination of a drunken man from his first few steps after exiting a pub, stating that "weather is a very variable phenomenon, hourly, daily, monthly, yearly. Extracting long-term trends is difficult." In such complex, many-bodied interaction systems, computers are invaluable and often the only means to study how these systems work.

To be sure, advances in computing over the past few decades have been extraordinary, providing almost real-time simulations for car design, oil exploration, medical imaging, financial analysis, genomic research, social networking, and the granddaddy of them all, weather prediction. At the same time, the change to the consumer has been staggering, where in less than 10 years the supercomputer of yesterday has become the desktop of today, selling at knockdown prices. That's a whole lot of power available to the average number cruncher.

With computer simulations, general car safety has improved without the need for crash-test dummies, which aren't particularly good analogs of the human body; racing cars can be redesigned to fit the next track (and expected weather conditions) in less than 3 days without the need for expensive wind-tunnel tests; and air flow simulated in planes has reduced the need for wind-tunnel tests to improve flight safety. Simulating oil explorations reduces the chance of drilling a bad well, with savings in the millions, whereas 3-D CAT scans can be presented faster, from 2 to 3 hours today to 6 minutes (at 100 TF[34]) to almost real time (at 1 PF). Satellite navigation rendering, Internet fraud checking, DNA sequencing, portfolio risk analysis, hurricane touchdown locations, the diffusion of information, climate-change predictions, and even potato chip design are all treated to the mathematics of today's supercomputer.

What's more, high-performance parallel supercomputing, whether in a multicore laptop, cluster, or grid (including cloud computing), can reduce turnaround times by simultaneously running calculations as separate tasks, particularly good for Monte Carlo simulations, parameter sweeps, and back fitting. High-performance computing is now so commonplace—from the very small (10^{-21} or atto) for probing molecular structures to the very large (10^{21} or zetta) for simulating global weather predictions—anyone can try their hand at previously restrictive numerical analysis. And, of course, financial markets are now run by ever-more-complex software, indicating when to buy and sell based on complicated statistical analysis. No wonder a few mathematicians and physicists have been recruited into the financial world,[35] as well as being blamed for computer-driven spikes in the market.

Nonetheless, here's a poser for the philosophically minded: How random is a random number, such as the result of a dice throw, if there are only a thousand or so throws in a game or in a limited computational analysis (which can also explain why Rousseau lost)? And even more interesting from a stochastic perspective, how random is a random-number generator that selects numbers not previously selected over the long run to maintain a spread in the data?

Despite such dilemmas, George Boole could never have imagined the computational power of today, one that uses his basic logic both to run a computer (the "and" and "or" logic of 0 and 1 bits) and to search data (the intersection and union of sets to restrict databases). With the advent of ever-faster computers, mathematical understanding is improving by the day, where spreadsheets can be used as easily as word processors and the simplest of numerical methods can be manipulated to dupe us. In today's wired world, no one can afford not to know.

Notes

1 **Ancient astronomical accounting**: In Ireland, a Neolithic passage tomb built more than 3,500 years ago at Newgrange is aligned to mark the winter solstice. In England, Stonehenge marks changing seasons and eclipses. The pyramids at Giza may also be oriented to stars in the Orion constellation.

2 **Pythagoras's theorem using similar triangles**: A young Albert Einstein elegantly proved Pythagoras's theorem using similar triangles.

3 **Multiplication and addition**: Multiplication and addition operations are related, as can be seen if one thinks of multiplication as repeated addition: e.g., $3 + 3 + 3 + 3 + 3 = 5 \times 3 = 15$.

4 **Accuracy of logarithms**: The actual answer, 838,102,050, depends on the number of digits used in the exponent.

5 **Newton and Kepler**: Newton verified Kepler's quasi-empirical third law (the radius squared of a planet in orbit varies with its period cubed) using the inverse-square law and calculus (called "fluxions" by Newton), which he had created to represent a planet as a point mass at its center.

6 **Babbage's Difference Engine No. 2**: The London Science Museum constructed a working replica of Babbage's Difference Engine No. 2 in 1991, and it is on display there along with portions of his unfinished Analytical Engine and Difference Engine No. 1. Napier's bones and Pascal's Pascalina are also on display there.

7 **The first integrated circuit**: The first integrated circuit was constructed in 1957 by Jack Kilby and was wired on a single piece of silicon instead of discretely cutting pieces and stitching them together. It was composed of one transistor, one resistor, and one capacitor on germanium, although silicon was used soon after. Texas Instruments and Intel have both been credited with its invention.

8 **University of Cork**: The University of Cork was known as Queen's College while Boole was professor of mathematics there from 1854 to 1864.

9 **Birthday frequencies**: Technically, the odds are a little more or less than 1/12 (8.33%) depending on the month. For 30-day months, the odds are slightly worse (8.22%), and for 31-day months, the odds are slightly better (8.49%).

10 **The so-called "birthday paradox"**: Number of people in a room such that the odds are 50:50 that two share the same birthday: $0.5 = 1 - (364/365 \times 363/365 \times 362/365 \times \ldots \times (365 - n) / 365)$ or $365! / (365 - n)! \times 1/365^n = 0.5$, which gives $n = 23$. A quick computational approximation is $(364/365)^{n(n-1)/2} = 0.5$.

11 **Pascal's triangle**: The triangle was previously known to the 10th-century Persian mathematicians Al-Karaji and Omar Khayyam. Another simple way to determine the coefficients for the first four rows is from 11^n, e.g., $11^2 = 121$, $11^3 = 1,331$, $11^4 = 14,641$ (after which the simple product fails).

12 **Advanced combinatorial odds**: The odds are $6! / (6 - n)! / n!$ for $n = 0$ to 6, where the exclamation mark means "factorial." Thus, "6 choose 6" $= 6!/6!/0! = 1$, "6 choose 5" $= 6!/5!/1! = (6 \times 5 \times 4 \times 3 \times 2 \times 1) / (5 \times 4 \times 3 \times 2 \times 1) / 1 = 720/120/1 = 6$ (just as in the triangle), 6 choose 4 $= 6!/4!/2! = 720/24/2 = 15$, 6 choose 3 $= 6!/3!/3! = 720/6/6 = 20$, reproducing the coefficients in the triangle. Note that Pascal's triangle is symmetric, which makes sense since the number of ways to get one wrong out of six is the same as getting five wrong out of six. For a more advanced 6/49 lottery game, the odds are $49!/43!/6! = 13,983,816$ to 1.

13 **Best of seven odds**: Ahead 1 to 0: $(20 + 15 + 6 + 1) / 64 = 42/64 = 0.66$; ahead 2 to 0: $(10 + 10 + 5 + 1) / 32 = 26/32 = 0.81$; ahead 3 to 0: $(4 + 6 + 4 + 1) / 16 = 15/16 = 0.94$, as can be seen from Pascal's triangle for rows 6, 5, and 4, where four coefficients are used to determine at least three wins in six games, two wins in five games, and one win in four games.

14 **The Federer solution**: 1) Federer wins 7 to 6, 7 to 5, 7 to 4, 7 to 3, 7 to 2, 7 to 1, or 7 to 0 over a maximum 13 points.

2) Federer wins 7 to 2, 7 to 1, or 7 to 0 over a maximum 9 points.

$$1)\ \sum_{n=0}^{7} \binom{13}{n} p^{13-n} q^n \quad 2)\ \sum_{n=0}^{2} \binom{9}{n} p^{9-n} q^n$$

15 **Calculating Pascal's triangle**: Pascal's triangle can be calculated combinatorially for any coefficient using $n! / r! / (n - r)! / 2^n$, where n is the row and r is the row coefficient, from 1 to n. For an even 50/50 split, the probability is thus $n! / (n/2)! / (n/2)! / 2^n$, and so for $n = 4$, the probability of an even split is 37.5% ($4!/2!/2!/2^4$), for $n = 6$, 31.25% ($6!/3!/3!/2^6$), for $n = 8$, 27.34% ($8!/4!/4!/2^8$), and for $n = 10$, 24.61% ($10!/5!/5!/2^{10}$). Note that even splits apply only for even numbers.

16 **Beyond Pascal's triangle**: The results of a 50/50 split for 10,000 traders and the further splittings were calculated by approximating the binomial distribution with a Gaussian curve and integrating. The Gaussian curve was discovered in just this fashion by the French mathematician Abraham de Moivre, who wanted to work out higher-order tiers in Pascal's triangle.

17 **The Brownian motion of matter**: In 1873, James Clerk Maxwell used a random walk to predict the size of a hydrogen molecule (~ 6 Å). In his first great paper of 1905, Albert Einstein confirmed the atomic theory by showing that atoms move in Brownian motion (i.e., a random walk), comparable to particles suspended in a liquid, as first seen by the Scottish botanist Robert Brown.

18 **Best sellers and feedback loops**: Frank and Cook (2010) noted that Donald Trump bought thousands of copies of his own book to keep it on the best-seller list (p. 192).

19 **Correlation and the law of large numbers**: Galton also used the same analysis to devise the idea of correlation after plotting parents' versus children's heights, which for a perfect correlation would produce a straight line at 45 degrees (Mlodinow, 2008, pp. 162–163).

20 **Roulette, blackjack, and lottery odds**: On a standard American roulette wheel, there are 38 numbers, of which 1 to 36 win a player 35 times the bet and 0 and 00 win nothing. The house edge (or expected value) is thus $35 \times 1/38 - 1 \times 37/38 = 0.053$. In Europe, there is no 00, and so the house edge is only $35 \times 1/37 - 1 \times 36/37 = 0.027$. In blackjack, the house advantage is much harder to determine and must be calculated numerically, whereas the probability of winning a typical lottery is more than 1 in 10 million (more on this later in Chapter 8).

21 **Random calendar distributions**: Some things that aren't randomly distributed through the year are holidays, daylight hours, and prom nights. As Levitt and Dubner (2010) noted, murders and sexual assaults peak in the summer (p. 46), while Muslim babies *in utero* for the first month during Ramadan (when the mother is fasting during daylight) can suffer fetal distress, an effect that can be especially detrimental during summer months in high latitudes where nights are much shorter (p. 62).

22 **Sports league sample**: The 4,964-player sample was made up of 409 NBA, 744 NHL, 750 MLB, 2,302 NFL, and 759 World Cup soccer players.

23 **Small sample sizes**: We have seen before that an outcome can show preference due entirely to chance when there aren't enough data points in the sample (i.e., people or players).

24 **My own Matthew effect**: Of the 22 professional athletes mentioned in this book, only one was born in January—Wayne Gretzky (which is more like January minus 48 since he was 2 years younger than most of his teammates). The data also showed a marked preference for the second half of the year, particularly August, October, and November. And so, according to my limited sample set, it seems that being born later in the year is best to ensure sporting success and mathematical fame.

25 **Roulette wheel red and black odds**: With the inclusion of the green numbers (0 and 00) on an American wheel, the odds are somewhat less than $(1/2)^{10}$ at $1/(18/38)^{10}$ or $1/1,758$, although the odds of red versus black stay the same.

26 **Optimal betting strategies**: Since doubling up will eventually end in gambler's ruin, the best (or optimal) strategy is to bet everything on the first go and then stop. If you lose, you are no worse off than if you had played for hours (or days) and lost all. If you win, you have doubled your money. Of course, the hard part is to bet and stop.

27 **Randomness and financial data**: Numbers in some distributions are not randomly distributed and appear more often than others, for example, the wear pattern on a keyboard or ATM pad or the leading digit in a collection of financial data. According to Benford's law, the frequency of each leading digit decreases from 1 to 9: number 1 appears about 30%, 2 about 18%, 3 about 12%, 4 about 10%, . . . , and number 9 about 5%. If the data were randomly distributed, each number would appear 11.1% or 1/9 (in this case, numbers don't begin with 0). Mlodinow (2008) noted a fascinating story about an enterprising forensic accountant who compiled a list of 70,000 numbers from a businessman's transactions and was able to help return a fraud conviction using Benford's law, although a similar analysis of Bill Clinton's tax returns showed the expected distribution (p. 84).

28 **Machine-run stock market equipment**: The Bloomberg terminal, created by New York Mayor Michael Bloomberg, is one such system that provides real-time financial data on a desktop or handheld device. Updated data for bonds, shares, indices, and currencies are available.

29 **Monopoly the game**: The Landlord's Game, patented by Elizabeth Magie in 1904, may be the basis for Monopoly. Influenced by her Quaker background, Magie wanted to show how ownership creates monopolies. An updated version of Monopoly for our time could be called Oligopoly.

30 **Reducing wasteful luxury spending**: Frank and Cook (2010) noted that a progressive tax on consumption would reduce luxury spending as well as create more efficient consumption practices (pp. 213–214).

31 **Markov chain**: A Markov chain is a continuously iterative system that nonetheless has a finite countable sample space. Markov chains have numerous applications, including random walks, stock market modeling, thermodynamics, and information sciences.

32 **Monopoly squares**: In the American Monopoly version, the oranges are St. James Place, Tennessee Avenue, and New York Avenue and the reds Kentucky, Indiana, and Illinois. In the standard London Monopoly version, the oranges are Bow Street, Marlborough Street, and Vine Street and the reds Strand, Fleet Street, and Trafalgar Square.

33 **Random-number generators**: A computer random-number generator returns a series of digits that lacks any perceivable pattern. Many take a digit from a computer's internal clock to seed a randomize function. Others use π's decimals, thermal variations, radio noise, and even atomic states (quantum cryptography).

34 **Flop**: A flop is a bit per second. A teraflop (TF) is 10^{12} flops, a petaflop (PF) 10^{15} flops, an exaflop (EF) 10^{18} flops, and a zettaflop (ZF) 10^{21} flops.

35 **Scientists in the financial industry**: There is little data on this, but one Irish survey noted that 5.6% of physicists work in the finance sector (Martin & Associates, 2011). Moyo (2011) noted that "engineering graduates and biotechnicians are all heading to Wall Street" (p. 99), adding that only 31% of engineering and science graduates were employed in their related fields (p. 99).

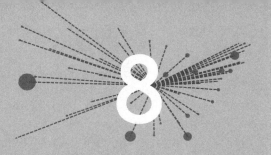

8

WHO TO BELIEVE?

Numbers and the Art of Deception

A better understanding of numbers helps us see how various systems have been based on error, superstition, or ignorance. Indeed, many "legacy" systems have been created through the ages, based on incomplete information, and are with us today only because of custom or tradition rather than validity. In this chapter, we look at a few such systems to see how astrology was erroneously devised using pre-Newtonian concepts, the days of the week were incorrectly named, and so-called lucky numbers came to be.

We also analyze the basics of codes and number scams before looking at the lottery, a highly questionable means of raising funds that has undermined centuries of communal values and replaced them with a seemingly innocuous, get-rich-quick ethos and pro-gambling culture, based on little more than our superstitious fascination with numbers. Better methods that do not unfairly burden the less well-off or undermine a well-intended work ethic can achieve the same goal.

8.1. Debunking the Mystery Behind the Meaning

Our fascination with the unknown is regularly encouraged by a misuse of numbers and fuzzy logic. For example, do heavenly objects influence our lives? Can life, love, and happiness be explained by the movement of the sun, moon, and planets relative to the zodiac constellations at one's birth, as many believe?[1] These are typical questions asked today, but are there quantifiable answers?

In 1543, Nicolas Copernicus proposed a heliocentric view of the solar system in his landmark book *The Revolution of Heavenly Bodies.*

Published the year he died, *De Revolutionibus* changed our world like none before and spawned the use of *revolution* to mean radical change. The story goes that Copernicus received the first copy on his deathbed, not wanting to publish the unsettling news while he was alive.[2] But the deed was done, man was no longer at the center, and the lowly earth was just one of six planets whizzing around a massive sun. Uranus (1781), Neptune (1846), and Pluto (1930) hadn't yet been discovered (and in the case of Pluto, "undiscovered" in 2006 after being reclassified as a dwarf planet in the Kuiper belt).

Of course, the obvious question is, if astrologers believed that the sun revolved around the earth, as was commonly thought prior to Copernicus, how could their so-called charts be right? It is true that much of astrology became popular to the masses after Copernicus, as did popular science and science fiction as a result of Galileo's telescopic discoveries of extra-terrestrial worlds (Kuhn, 1957/1997, p. 225); nonetheless, astrology's most basic assumption about the heavens—that the sun revolves around the earth—was wrong. It's like basing one's belief in the Greek gods of Mount Olympus or a leprechaun at the end of a rainbow. Three undiscovered planets also weren't included in the analysis, nor was a 10th "planet," dubbed Xena and found outside Pluto in 2005—putting a rather large wrench in any predictions.

To be sure, astrology is plain old astro-tomfoolery and contains many celestial misconceptions, some of which are nonetheless worth illuminating, if only to help us understand the canvas of the astronomer and cosmological thinker. Interestingly, Copernicus was against astrology for religious reasons, yet Brahe and Kepler made a living doing people's charts, a favored parlor game in 17th-century aristocratic Europe (Kuhn, 1957/1997, pp. 93–94).

The zodiac or "zoo circle"—all are animals except for Libra—refers to the 12 constellations on the ecliptic.[3] All are about 30 degrees wide (30° × 12 = 360°) and within about 8 degrees above and below the ecliptic, roughly corresponding to 1 month in the night sky, as intended for timekeeping and navigation purposes. Alas, the "constellation" stars are no more than made-up patterns that helped order the night sky for early travelers, chief among them the Big Dipper (in the northern hemisphere), a seven-star asterism that is part of the constellation Ursa Major (aka The Big Bear). Constellation stars are, in fact, many light years away from each other—for example, in the Big Dipper, ranging from 78 light years (Mizar or Ursa Major ζ) to 124 light years (Dubhe or Ursa Major α). Connecting distant stars is like saying that Los Angeles and London, seen at night from a distance, are next-door neighbors.

As for planets influencing our lives, it is reasonable to ask how this could be. Presumably, the influence is based on proximity and mass, such that a near object affects one more than a far object and a large object more than a small one. We can even quantify such effects with some help from Isaac Newton, who first calculated the force that all objects exert on one another.

According to the famous story, Newton discovered gravity in a eureka moment when an apple fell near where he was sitting under a tree, from which he immediately understood that the moon and the earth attract each other in the same way as any two objects do. His equation, which he derived from calculus and Kepler's third law, is $F \propto Mm / r^2$, where F is the force, M is the first mass (e.g., the earth), m is the second mass (e.g., the moon or another planet), and r is the distance between them.

His gravitational equation is an inverse-square law, where the force between two objects is inversely proportional to their separation squared and proportional to their masses. In an inverse-square law, if the separation doubles, the force diminishes 4 times, if the separation triples, the force diminishes 9 times, etc., showing how little effect a tiny, distant planet such as Pluto has compared with the many times closer and more massive moon.

Coulomb's equation is also an inverse-square law that measures electric force between two charges separated by a distance ($F \propto q_1 q_2 / r^2$), as is the body mass index formula: $BMI = W / H^2$ (W = body weight (in kg) and H = height (in m), where a value < 18.4 is considered underweight and > 25 overweight). The importance in an inverse-square law is that the denominator has a greater effect on the result because of the squared dependence.

So which planets exert the greatest attraction on earth and, by extension, on us? Table 8.1 gives the mass and distance from earth for each of the nine planets, the moon, and the sun, from which we can easily calculate their relative force using Newton's law of gravitation (mass divided by distance squared).

Here, we can immediately see that the sun exerts the greatest force, which is probably what most of us would expect, followed by the moon, Jupiter, Venus, and then Mars. Note that we're only interested in relative strengths, so we don't need to worry about constants, i.e., the actual force.[4]

Any number of comparisons can be made. Although 10 times closer to earth, Mercury has an attraction almost 10 times less than Saturn's (it is more than 1,000 times less massive). Jupiter, which is about 10 times farther from the earth than is Mars, Venus, or Mercury, nevertheless has a

Table 8.1	Planetary mass, distance from earth, and relative force			
Planet	Mass (kg)	Distance (km)	Mass/Distance² (kg/km²)	Rank
Sun	2.0×10^{30}	1.5×10^{8}	8.9×10^{13}	1
Mercury	3.2×10^{23}	9.2×10^{7}	3.8×10^{7}	7
Venus	4.9×10^{24}	4.1×10^{7}	2.8×10^{9}	4
Earth	6.0×10^{24}	—	—	—
Moon	7.4×10^{22}	3.8×10^{5}	5.0×10^{11}	2
Mars	6.4×10^{24}	7.8×10^{7}	1.0×10^{9}	5
Jupiter	1.9×10^{27}	6.3×10^{8}	4.8×10^{9}	3
Saturn	5.7×10^{26}	1.3×10^{9}	3.5×10^{8}	6
Uranus	8.7×10^{25}	2.7×10^{9}	1.2×10^{7}	8
Neptune	1.0×10^{26}	4.4×10^{9}	5.4×10^{6}	9
Pluto	1.1×10^{24}	5.8×10^{9}	3.3×10^{4}	10

much greater attraction because of its large mass. The sun exerts 180 times more force on earth than does the moon, Jupiter exerts 5 times more force than Mars, and Mercury is as insignificant as the more massive but much more distant Uranus. Pluto (the size of Australia and 6 million km away) might as well not be there.

As can be seen, however, one of the problems with astronomical (or highly varied) data is the size of the numbers. Large absolute values are messy, especially if the range is also large, but if we normalize the data (i.e., divide all by the same amount), the results can be presented more manageably. Table 8.2 shows the same data with mass and distance normalized, where mass is normalized relative to the mass of the earth (mass = 1) and distance is normalized relative to the distance between the earth and the sun (defined as 1 astronomical unit or 1 a.u.). Two columns have also been added to show the effect of the massive sun and nearby moon, both of which swamp the other planets. In these two columns, the relative effects from the planets excluding the sun and the moon can be more easily seen.

Again, only the sun and the moon show any significant effect (we haven't changed the data, just normalized it), almost all of which is from the sun. The sun (at a whopping 99.43%) exerts the most force on the earth, followed by the moon (0.56%) and Jupiter (0.01%). The other planets clearly don't amount to much, although their relative influences can now be seen in the last two columns, where the effect of the sun and then

Table 8.2	Normalized planetary mass, distance from earth, and relative force				
Planet	Mass (normalized)	Distance (normalized)	M/d^2 (%)	M/d^2 (%)	M/d^2 (%)
Sun	332,998.8	1.0	99.43	—	—
Mercury	0.1	0.6	0.00	0.01	0.42
Venus	0.8	0.3	0.00	0.56	31.21
Earth	1.0	—	—	—	—
Moon	0.0	0.0	0.56	98.20	—
Mars	1.1	0.5	0.00	0.21	11.51
Jupiter	317.9	4.2	0.01	0.95	52.85
Saturn	95.1	8.5	0.00	0.07	3.83
Uranus	14.5	18.2	0.00	0.00	0.13
Neptune	17.2	29.1	0.00	0.00	0.06
Pluto	0.2	38.5	0.00	0.00	0.00

of the sun and moon have been removed. Thus, the moon exerts 100 times more gravitational pull on the earth than Jupiter does. After that, Jupiter, Venus, Mars, and Saturn attract earth the most (53%, 31%, 12%, and 4%, respectively).

But can such data debunk astrology? Well, astrology is based on celestial objects affecting us at the time of our birth. And although astrologers don't say why they affect us—there's no astrology manual—it seems reasonable to assume that size and closeness are important. As such, by quantifying the attraction in a straightforward way, one can easily see that other than the sun and the moon, celestial objects don't affect us. More thorough double-blind experimental debunking has been done—for example, by Carlson (1985) at Berkeley—but here we have easily debunked such claims based only on a simple mathematical relation. Basically, only the sun and the moon have any quantifiable effect, which agrees well with the very real gravitational effects of the tides, the result of the interplay of the sun and the moon on our oceans (Cole & Woolfson, 2002, pp. 383–386) and, indeed, on us.

It would seem, then, that "Saturn in Libra" or "Jupiter in Aquarius" is quite meaningless, not to mention the highly selective and self-centered approach of using the earth as a reference point in the cosmos and, thus, as an arbiter of personality. However, the moon aligning with the sun can create a very real effect, as in large spring and neap tides.[5]

It is worth noting that the science writer John Gribbin believed that the combined attractive forces of Jupiter, Mars, and the moon—aligned in the same direction relative to earth, where their forces add most—would be so important on January 23, 1972, that he felt compelled to warn the world of impending doom. He dubbed his doomsday scenario the "Jupiter effect," presumably to sell lots of books, and, of course, he sold lots of books. Yet January 23, 1972, came and went without disaster.

From our simple analysis, using readily attainable data and the basic mathematics of inverse-square laws and normalized data, we have easily debunked such stupidity. Furthermore, as noted by Nave (2011) on his excellent Georgia State University hyperphysics website, "Don't lose much sleep worrying about the Jupiter effect. You change the gravity force on yourself by taking one step up a stairway more than the combined gravitational effects of both Jupiter and Mars if they were perfectly aligned!"

Jupiter does, however, have a profound effect on the earth's safety, by sucking up asteroids and other passing objects that otherwise might strike it with potentially disastrous results. Jupiter's large gravity (24.9 m/s^2 or 2.5 × that of earth[6]) can vacuum up, deflect, or even redirect toward earth passing interplanetary stuff. As noted by Cox, 400 such near-earth objects are out there on possible collision courses (Olding, 2010), including the next possible strike of any significance, asteroid 1950 DA, which according to calculations has a 1 in 300 chance of striking earth sometime around March 2880 (Giorgini et al., 2002).

Impact craters in South Africa, Canada, Mexico, and Russia that measure more than 100 km in diameter and numerous others of more than 20 km around the world are grim reminders of earth's precarious location in space. Some of these craters have been identified only since the advent of satellite technology in the 1960s (e.g., one near Portnoo in remote Donegal, Ireland) and even using Google maps (e.g., a 0.27-km crater in northwest Australia found in 2008 and named the Hickman Crater after its diligent discoverer; Cain, 2008). The massive 170-km Chicxulub crater in the Yucatán is also thought to have been responsible for the extinction of the dinosaurs 65 million years ago (Bell Burnell, 2010).

Not only does a habitable planet such as earth need to be positioned at a precise distance from its parent star—the so-called Goldilocks zone, not too far away to lose essential heat and not too close to be overheated, and where water is found predominantly in a liquid state—it must also have an outer, planetary protector such as Jupiter, i.e., a planet with high gravity (Perryman, 2006), although such planets can also redirect potentially destructive objects *toward* earth (Olding, 2010).

To be sure, the effect of Jupiter's gravity can be examined and measured and has real consequences in our lives. I just wouldn't bet on it changing your personality.

Understanding the solar system and our place in the celestial clockwork is ever changing as we learn about the seemingly immeasurable scales (see Figure 8.1). Perhaps that's the scariest bit of all as we contemplate our being—that we are just a small part of a big universe and that the earth is as small as small can be. Sadly, astrologers and other charlatans would have us think otherwise.

| Figure 8.1 | Planetary scales (Mercury, Venus, Earth, Mars, Jupiter, Saturn, Uranus, Neptune, Pluto) |

SOURCE: NASA/Lunar and Planetary Institute

8.2. How Old Are We? A Simple Calculation

In the late 16th century, James Ussher, the archbishop of Armagh and primate of all Ireland, calculated the date of creation by counting backward from life spans given in the Bible. As stated in his *Annalium pars Posterior*, creation began "at nightfall preceding 23 October, 4004 BC," a very precise calculation, to be sure.

Of course, it is easy to debunk such thinking using advanced techniques available to us today, such as radioactive decay of uranium samples found in the earth (8–11 billion years); cooling rates from the oldest white dwarves, including formation time after the Big Bang (11.5 billion years); or a cosmological model that uses the Hubble constant and the density of matter and dark energy (13.7 billion years; May, Moore, & Lintott, 2007, p. 22). But other, more accessible methods also show that such a biblical reckoning is not quite right.

One indication that the earth is older than first thought is the fossil record. The great science populizer Thomas Huxley (1868/1995) deduced from the thousand-feet deep calcified animal skeletons (i.e., chalk) found in the coasts off the south of England that "the time during which the countries which we call south-east England, France, Germany, Poland, Russia, Egypt, Arabia, Syria, were more or less completely covered by a deep sea, was of considerable duration" (p. 143). Although Huxley doesn't give an exact time for the creation of the earth, he does state that because of the thickness, "the chalk can justly claim a very much greater antiquity than even the oldest physical traces of mankind" (pp. 144–145).

One can also measure time from the movement of the earth's tectonic plates, which are slowly jolting past one another, as every earthquake attests. Measured in millimeters, the earth's continents have drifted apart from one single aggregate mass (called Pangaea), readily seen in the shapes of South America and Africa, which fit together like two pieces of a giant geological jigsaw puzzle.[7] Judging from the separation of the continents today, the shift must have taken a very long time indeed. In the same way, the Himalayas were created by the uplift of land caused by the collision of two continents, elegantly explaining the presence of 6-million-year-old fish fossils at the top of Mount Everest.

It is, of course, hard to verify these numbers without access to advanced scientific methods, but we can make a rudimentary guess at a modern societal age, i.e., when human populations began doubling. Here, we work backward from a current known population of 7 billion and assume a doubling rate of 30 years, not unreasonable given the age at which people have children and the oft-quoted statistic of 2.3 children per family.[8]

Calculating backward by halving is the inverse of doubling by twos, similar to calculating the age of radioactive material from its half-life, from which nuclear waste lifetimes are measured and the oldest fossils and galaxies have been dated.[9] The same method applies, whether calculating the age of the earth, determining the formation of Niagara Falls, or dating the Shroud of Turin,[10] where carbon-14, the radioactive isotope present in all living matter, is used to "carbon-date" the past.

And so $2^N = 7,000,000,000$, where N is the number of generations. Thus, $N = \log_2(7,000,000,000)$ or 32.7 generations.[11] From our doubling birthrate of 30 years, we get 32.7 generations × 30 years/generation or 981 years (1031 AD). Of course, we haven't accounted for wars, famine, natural disasters, etc. that will significantly affect the doubling rate, but we have come up with a reasonable time frame to account for the origins of growth-based societies.

Determining the past is never easy and will continue to fascinate anyone with an interest in our origins. The latest cosmological theory on the block, however, suggests that time and space may have no beginning at all and that an ever-expanding universe is not stable, because the size of the universe is too big for its age using the conventional Big Bang accounting. According to string theory, a cold and empty universe existed in a pre-Big Bang period, not the local singularity from which all expansion is believed to occur. To be sure, the unknown persists, as does the uncertainty of our creation, which is the basis for much misinformation.

8.3. Modulo Arithmetic: How the Days Got Their Names

The heavens were notoriously difficult to decipher, and it took the work of Copernicus and Kepler to determine that the planets travel in elliptical orbits in a heliocentric solar system, contrary to the confusing geocentric and fitted paths of yore. Making decisions based on partial knowledge, early man made fact of superstition and gods of mystery to explain the heavens.

Even the days of the week were muddled. As most of us know, the days were named in English from a mixture of planets' names (Saturn's day, Sun's day, Moon's day) and Norse gods' names (Tiw's day, Woden's day, Thor's day, and Frigg's day). But why were they so ordered? That is, why does Thor's day follow Woden's day and not the other way around? It all has to do with Claudius Ptolemy and the mathematics of cycles or modulo arithmetic.

First, let's take a look at the Romance languages, where the original Roman planet names are more easily seen, since the Roman gods who ruled each day weren't supplanted by their corresponding Norse-based Teutonic gods for Tuesday to Friday.[12] In French, *mardi* is Mars's day (Tuesday), *mercredi* is Mercury's day (Wednesday), *jeudi* is Jupiter's day (Thursday), and *vendredi* is Venus's day (Friday). Italian and Spanish are similar except that Sun's day is God's day and Saturn's day is the Sabbath (see Table 8.3). It would seem that the days are somehow ordered according to a celestial mapping. But how?

Following from there, in Mesopotamia it was believed that each of the seven ancient "planets" ruled the earth in turn for 1 hour. A different planet thus ruled the first hour of each day for 24 hours through 7 days, thus "naming" the corresponding day, a practice continued by the Romans. Alas, the astronomer of the day, Ptolemy, believed that the earth was at the center of the known universe. Furthermore, the order of the

Table 8.3	Days of the week as planet/god names			
Planet/God	*Latin*	*French*	*Italian*	*Spanish*
Sun	Soli	dimanche	domenica	domingo
Moon	Lunae	lundi	lunedi	lunes
Mars	Martis	mardi	martedi	martes
Mercury	Mercurii	mercredi	mercoledi	miércoles
Jupiter	Jovis	jeudi	giovedi	jueves
Venus	Veneris	vendredi	venerdi	viernes
Saturn	Saturni	samedi	sabato	sábádo
Planet/God	*Teutonic*	*German*	*Dutch*	*English*
Sun	Sun's day	Sonntag	zondag	Sunday
Moon	Moon's day	Montag	maandag	Monday
Mars	Tiw's day	Dienstag	dinsdag	Tuesday
Mercury	Woden's day	Mittwoch	woensdag	Wednesday
Jupiter	Thor's day	Donnerstag	donderdag	Thursday
Venus	Frigg's day	Freitag	vrijdag	Friday
Saturn	Saturn's day	Samstag	zaterdag	Saturday

planets was out of whack—i.e., according to Ptolemy's reckoning, the heavens were arranged from the earth out as follows: the moon, Mercury, Venus, the sun, Mars, Jupiter, and Saturn, which relates to their orbital periods as measured from earth (see Figure 8.2).[13]

But how does that *order* the days? It all comes down to the gods who ruled the day and modulo arithmetic, in this case, 24 mod 7. For example, expressing each calendar day by an ordinal number (January 1 = 1, February 1 = 32, March 1 = 60, etc.), one can work out that October 23 (the 296th day of the year) and Christmas (the 359th) are on the same day, since 296/7 = 42 with remainder 2 and 359/7 = 51 with remainder 2, where the remainder gives the day of the week, which in this case falls on the same weekday as the second day of the year. In modulo arithmetic, this is written as 359 *mod* 7 = 296 *mod* 7 = 2 *mod* 7, which states that Christmas and October 23 always fall on the same day of the week every year and also on the same day as the second day of the year, excluding leap years.

Figure 8.2	Ptolemy's universe (circa 100 AD)

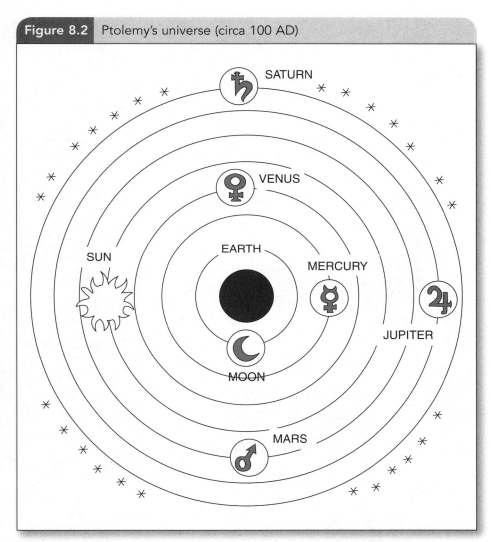

SOURCE: http://en.wikipedia.org/wiki/File:Ptolemaic_system_2_%28PSF%29.png

Modulo arithmetic: October 23 (day 296) and December 25 (day 359): (359 − 296) / 7 = 63/7 = 9 with 0 remainder and, thus, day 296 and day 359 fall on same day of the week. A Monopoly board is similar, "wrapping around" after Boardwalk (square 40) to Go (square 41 or 1), thus 41 *mod* 40 = 1 *mod* 40.

A 24-hour, 7-day table is used in the same way to determine the ruling "planets," where the modulo format (24 *mod* 7) can easily be seen cycled through each day according to Ptolemy's believed geocentric order (see Table 8.4). The table starts backward from the sun and follows each planet from Day 1 for 7 hours, repeating until Hour 24 on Day 7, which thus gives us the sun (Day 1, Hour 1) at the top of Day 1 and the moon (Day 2, Hour 1) at the top of Day 2, followed in the same way by Mars, Mercury, Jupiter, Venus, and Saturn. Voilá, we have the days of the week in the order we know them.

It is highly informative to see how the days got their names from a mixture of mistaken theories about the heavens and a simple numerical

Table 8.4 Days of the week ordered according to ruling planet

Hour	Day 1	Day 2	Day 3	Day 4	Day 5	Day 6	Day 7
1	**Sun**	**Moon**	**Mars**	**Mercury**	**Jupiter**	**Venus**	**Saturn**
2	Venus	Saturn	Sun	Moon	Mars	Mercury	Jupiter
3	Mercury	Jupiter	Venus	Saturn	Sun	Moon	Mars
4	Moon	Mars	Mercury	Jupiter	Venus	Saturn	Sun
5	Saturn	Sun	Moon	Mars	Mercury	Jupiter	Venus
6	Jupiter	Venus	Saturn	Sun	Moon	Mars	Mercury
7	Mars	Mercury	Jupiter	Venus	Saturn	Sun	Moon
8	Sun	Moon	Mars	Mercury	Jupiter	Venus	Saturn
9	Venus	Saturn	Sun	Moon	Mars	Mercury	Jupiter
10	Mercury	Jupiter	Venus	Saturn	Sun	Moon	Mars
11	Moon	Mars	Mercury	Jupiter	Venus	Saturn	Sun
12	Saturn	Sun	Moon	Mars	Mercury	Jupiter	Venus
13	Jupiter	Venus	Saturn	Sun	Moon	Mars	Mercury
14	Mars	Mercury	Jupiter	Venus	Saturn	Sun	Moon
15	Sun	Moon	Mars	Mercury	Jupiter	Venus	Saturn
16	Venus	Saturn	Sun	Moon	Mars	Mercury	Jupiter
17	Mercury	Jupiter	Venus	Saturn	Sun	Moon	Mars
18	Moon	Mars	Mercury	Jupiter	Venus	Saturn	Sun
19	Saturn	Sun	Moon	Mars	Mercury	Jupiter	Venus
20	Jupiter	Venus	Saturn	Sun	Moon	Mars	Mercury
21	Mars	Mercury	Jupiter	Venus	Saturn	Sun	Moon
22	Sun	Moon	Mars	Mercury	Jupiter	Venus	Saturn
23	Venus	Saturn	Sun	Moon	Mars	Mercury	Jupiter
24	Mercury	Jupiter	Venus	Saturn	Sun	Moon	Mars

calculation, one that cycles or loops through a list, especially since modulo arithmetic is one of the basic filtering techniques found in computer programming today, used in anything from Internet casino programs that number a 52-card deck into four suits of 13 (52 *mod* 4) to modern cryptography.

What would the order of the days be if the correct planetary order had been used?—Monday, Saturday, Sunday, Friday, Thursday, Wednesday, Tuesday, which can be worked out by substituting the correct order of planets into the 24/7 table above. Not that there would be any difference. Well, we might say TGIW and sing "I don't like Saturdays" and "goodbye, Ruby Sunday."

8.4. More Debunking: What's in a Number or Code?

Numerology is as old as numbers themselves, where a connection between two numbers or the symmetry in the greater whole is seen to divine a cause. As Voltaire's Pangloss might say, numbers and numerology fit like a nose to spectacles. But why should arbitrary numbers have such relevance?

Oddly, we didn't start counting in 10s, as might be suggested by our fingers, but in 12s and 60s, since both are factors of 360, the closest whole number of days in the year, which divides other cyclical measures of time such as 12 months and 30 days. And so, we get 360 degrees in a circle and 60 minutes in an hour, but there is no rule that a circle must have 360 degrees or an hour 60 minutes—it is an arbitrary choice, like 7 days in a week or 24 hours in a day, or even the length of a second, our most arbitrary of measures tied to the length of a human heartbeat. Highlighting its arbitrariness, the International System of Units defines the second as exactly 9,192,631,770 cycles, corresponding to the transition rate between two energy levels of the ground state of cesium-133.

Many centuries removed, the reasons for our arbitrary measures of time have been lost and it is harder to see that time standards were made by man. But we could just as easily use three 10-day weeks to make a 30-day month, although then the weeks wouldn't follow the phases of the moon so nearly. We could also have 10 months of 30 days, with an inter-calendar period thrown in at the end of the year now and again when the seasons become misaligned, as in the Roman calendar, which gave us the excess of drunken Bacchanalian festivals to mark the year end, and may be why they were discontinued.

As it is, we have our own cyclical system in the West (the Gregorian calendar) that tries to keep the seasons in line so we don't eventually celebrate Easter at Christmas, although that annual clock is imperfect and must be adjusted by adding in leap-year days every 4 years, but not century years that aren't also millennium years.[14] In Islam, where the calendar is tied to the moon, the months aren't aligned to the seasons, and the months (for example, Ramadan) rotate through the seasons, starting earlier each year by about 10 days. Other calendars (e.g., the Chinese and Hebrew calendars) are lunisolar and fall in and out of time with the Western calendar, such that holidays are more or less but never exactly at the same time each year.

How we got our calendar and how different calendars came to be is a fascinating story, one we can't explain in detail here. But the basics are that the earth orbits the sun incommensurately and the moon orbits the earth incommensurately (not a whole number of days); furthermore, since the moon orbits the earth independently of the earth orbiting the sun, we can't fit the "monthly" moon into the "annual" sun—no matter how hard we try. We can't even fit the number of days exactly into a year.

What does all this have to do with numerology? Well, if one sees that the number of seconds in a minute, the number of minutes in an hour, the number of hours in a day, the number of days in a month, and the number of months in a year are all numbers that have been made by man, it is easier to see that there is no underlying significance to the number 7 or the number 12 or the number 60. They are only approximations of incommensurate things that hold unwarranted importance in our day-to-day lives, even into the 21st century.

To be sure, we will always try to make associations with numbers. It is in our nature to ask why. Given a puzzle, we turn it upside down and spin it around until it fits just right. But the number of days in the week was arbitrarily chosen to match a presumed perfection in the sky—the seven known planets, two of which were the moon and sun, and which had been incorrectly ordered by the leading astronomer of the day. When multiplied by the four phases of the moon, this number gave the nearest whole-number approximation to the number of days in the orbit of the moon and when rounded up to 30 and multiplied by 12 gave the nearest number of easily divisible whole-number days in the year. That's all. And yet, from such arbitrary methods, we have constructed all sorts of ordained meaning.

The Meaning of Numbers and Words

Not only numbers have been associated with hidden meaning, but ancient texts have also been gleaned for translations of the divine or glimpses into the past—for example, the Koran, the Bible, and Egyptian hieroglyphics. Perhaps the best known of such artifacts is the Rosetta Stone. In 1799, a soldier in Napoleon's army found in the Egyptian city of Rosetta a granite stone fragment on which both hieroglyphics and Greek were inscribed (incidentally, announcing a tax amnesty for temple priests). From this great find, ancient Egyptian could at last be translated, which led to decoding hieroglyphics throughout the ancient Nile valley.

In mathematics, this is called a mapping, and if the mapping is one-to-one, the translation can go either way—in this case, from Greek to Egyptian or from Egyptian to Greek.[15] The Rosetta Stone is a great link between today and ancient times and gives us much to think about regarding how to decode messages from the past and even our own daily communications or miscommunications. Deciphering meaning is inherent in the human condition, as we constantly attempt to answer the question, why?

An example of a mapping that isn't one-to-one is two homophones (words that sound the same but have different meanings), such as *hare* and *hair*, where the meaning can be misinterpreted if the context isn't given. Or words and expressions with double meanings, such as "sorry" to mean "excuse me" or "Are you okay?" to mean "May I help you?" as is the custom in Ireland. Two people living at the same address receiving a letter addressed to "occupant" is also a mismatched mapping. In file compression, the term is *lossy*—used especially with photographs, where pixels are removed to reduce large file sizes. In a many-to-one encoding, some meaning or information will always be lost.

Mappings are like a kid's decoder ring, where although the same language is used, each letter is shifted to garble the message. The Caesar wheel was the original decoder ring, devised by Julius Caesar to encipher messages during battle, where each letter was shifted by three—for example, the letter A enciphered as the letter D, B as E, C as F, and so on (as shown here in English). The first transmitted telephone message, uttered by Alexander Graham Bell when he spilled acid on his hand, shifted to the right by one letter, is "XBUTPO DPNF IFSF J OFFE ZPV." When shifted back at the receiving end, the original message is deciphered to reconstruct "Watson come here I need you."[16] In the same way, HAL of *2001: A Space Odyssey* fame is an encoding of "IBM," shifted by one, Arthur C. Clarke's cryptic wink to his readers.

In World War II, much computational work at the Government Code and Cypher School in Bletchley Park north of London went into breaking the German encoder machine Enigma, using Colossus, one of the first modern computers. In today's banking system, similar codes are essential for electronic transactions to keep our personal data hidden. Financial stock pickers also use code-breaking methods to determine minute non-random patterns in stock prices, where ultra-secretive algorithms troll through mounds of data, as do speech recognition programs that use past data to determine what comes next.

Note that an encoding doesn't have to be encrypted. Morse code encodes (or maps) each letter by a combination of dots and dashes, where the most common letter E is a simple dot. The American Standard Code for Information Interchange (ASCII) table encodes alphabetic, numeric, and other symbols for use in computing (as shown in Figure 8.3), which are then converted to binary encodings of 0s and 1s for electronic transmissions. Resistor color codes, the Dewey Decimal System, Internet addresses, and bar-code price stickers are also encodings.

Mnemonics are also a type of encoding, where an easy-to-remember saying is substituted for a hard-to-remember pattern, such as "Every good boy deserves fudge" for the EGBDF notes on a piano or "Richard of York gave battle in vain" for the seven perceived colors of the rainbow: red, orange, yellow, green, blue, indigo, violet (ROYGBIV).

In computing, ASCII and binary encoding are both used. For example, "Watson come here I need you" is encoded as "**087**097116115111110 099111109101 104101114101 073 110101101100 121111117" in ASCII

Figure 8.3		The ASCII code												
65	A	80	P	95	_	110	n	125	}	140	î	155	¢	170	¬
66	B	81	Q	96	`	111	o	126	~	141	ì	156	£	171	½
67	C	82	R	97	a	112	p	127	□	142	Ä	157	¥	172	¼
68	D	83	S	98	b	113	q	128	Ç	143	Å	158	P	173	¡
69	E	84	T	99	c	114	r	129	ü	144	É	159	ƒ	174	«
70	F	85	U	100	d	115	s	130	é	145	æ	160	á	175	»
71	G	86	V	101	e	116	t	131	â	146	Æ	161	í	176	_
72	H	**87**	**W**	102	f	117	u	132	ä	147	ô	162	ó	177	_
73	I	88	X	103	g	118	v	133	à	148	ö	163	ú	178	_
74	J	89	Y	104	h	119	w	134	å	149	ò	164	ñ	179	¦
75	K	90	Z	105	i	120	x	135	ç	150	û	165	Ñ	180	¦
76	L	91	[	106	j	121	y	136	ê	151	ù	166	ª	181	¦
77	M	92	\	107	k	122	z	137	ë	152	ÿ	167	º	182	¦
78	N	93	]	108	l	123	{	138	è	153	Ö	168	¿	183	+
79	O	94	^	109	m	124	|	139	ï	154	Ü	169	_	184	+
80	P	95	_	110	n	125	}	140	î	155	¢	170	¬	185	¦

(e.g., *W* converted to 087) or, as seen in Figure 8.4, in binary packets of 0s and 1s (e.g., 087 further converted to 000010000111). In binary, the same message can then be sent at high frequency by laser or LED transmitters along millions of meters of fiber optics that make up the Internet.[17]

Figure 8.4	"Watson come here I need you" in binary

W	a	t	s	o	n
087	097	116	115	111	110

```
000010000111000010010111000100010110000100010101000100010001000100010000
000010011001000100010001000100001001000100000001000100000100000100000001
000100010100000100000001000001110011000100010000000100010001000100010001
000100000000000100100001000100010001000100010111
```

Sent as a secret message, however, such an encoding would be easy to decipher because of the inclusion of the dead-giveaway, one-letter word *I*—shown boxed in Figure 8.4 (octal 073, binary 000001110011)—and the frequency of the letter *E*—shown underlined for each of the five occurrences in Bell's call for help (octal 101, binary 000100000001)—indeed, as we saw before with letter patterns in Shakespeare's plays.

Of course, messages now are mostly digital, allowing for easy replication. Where we once had an analog encoding of a physical measurement—such as voice encoded by carbon filings on an electrical current in a phone, transmitted by copper wire, and reconstructed at the other end in reverse—we now have digital encoding—for example, 1s represented by optical pits in a CD or DVD and scanned by a laser or the billions of on/off field-effect transistors in an EPROM flash memory or drive. The more information there is, the more 1s and 0s can be encoded from the original music, photograph, or data file.[18] As well, digital information doesn't degrade as long as the physical medium is updated as needed.

Nonetheless, with easier copying—whether small-time sharing of music files or wholesale pirating of CDs and DVDs—the need for safe encoding increases. The reputation of banks and credit card companies, who make billions of electronic transactions per day, depends on it, and they spend a fortune to encode each transfer and ensure the safety of each message, using ultra-complex public-key cryptography to keep away prying eyes.

But does that mean we can't crack the code? Not necessarily. If you can find the prime factors for the following 232-digit number in a useful amount of time (i.e., a second or so), you can crack anything:

12301866845301177551304949583849627207728535695953347921973224521517264005072636575187452021997864693899564749427740638459251925573263034537315482685079170261221429134616704292143116022212404792747377940806653514195974598569021434 13[19]

Some Codes: Old and New

To many, the most intriguing code of all is that of Revelation, the last book of the New Testament and one of the oft-cited riddles of our past. Many have tried to pry loose its meaning and that of the infamous mark of the beast. Others have attempted to glean meaning from patterns in the Bible, such as the ridiculous book *The Bible Code*, which claims that the assassination of John F. Kennedy, the rise of Hitler, and the plays of Shakespeare can be found in skip patterns of the Pentateuch (Wheen, 2004, p. 152). Here, we try a simple debunking of such attempts.

Anyone who has glanced at Revelation will find the material hard going, full of riddles and numbers and not at all easily interpreted or decoded. The number 7 is prominent (seven lamps, seven stars, seven spirits), as is the number 4 (four lions, oxen, humans, eagles, each with six wings and eyes), the number 24 (24 thrones for 24 elders), and, of course, the famous mark of the beast, 666. As stated in the text: "This calls for wisdom: let anyone with understanding calculate the number of the beast, for it is the number of a person. Its number is six hundred and sixty-six."

To early decoders, the year 666 was considered a harbinger of the Apocalypse, as was the year 1666, when the Great Fire of London must have seemed like the end of the world to some. Yet that doesn't jibe, because Western dates are out of whack due to the incorrect calibration of Christ's birthday by Roman Abbot Dionysius Exiguus, who was working out Easter tables in the fifth century. Because of his miscalculation, the birth of Christ actually occurred in 4 BC. Besides, even a cursory reading of Revelation, as above, reveals that the number of the beast is that of a man, not a year.

There have been many stabs through the ages at possible beasts and, indeed, the meaning of 666. Some believe that 7 is perfection, and, thus, 6 is one less, but why three 6s? Another explanation is that 666 is DCLXVI in Roman numerals (the numbers of the Bible), which contains each of the six Roman numbers, but the equivalent in our Hindu-Arabic number system, 9,876,543,210, is hardly scary. Perhaps the number refers to the 9,876,543,210th person born, which may have already occurred. Someone suggested that a child born on the 6th of the 6th of the 6th (June 6, 2006) is the one, the beast, the Damien. Twentieth Century Fox even released the remake of *The Omen* on the same devilish date, their public relations machine salivating at the delicious timing—although no one told Fox that Christ was born in 4 BC or that we are now in the 21st century.

In Reformation days, Martin Luther was thought to be a possible beast. And it's no surprise that apocalyptic numerologists found meaning in Ronald Ray-gun's moniker, a name right out of science fiction. Add his middle name and 666 seems a perfect match for Ronald Wilson Reagan. Who could mistake the message? Interestingly, both Adolph Hitler and Joseph Stalin had six letters in their first and last names, although Stalin's real name was Joseph Vissarionovich Dzhugashvili—tough, but not beastly.

Of course, Martin Luther, Adolph Hitler, Joseph Stalin, and Ronald Reagan have all come and gone. A more recent beast, Saddam Hussein, misses out with six and seven letters, as did Albert Einstein, who was considered by some for his role as father of the atomic bomb. Perhaps the most ridiculous suggestion of all was Mikhail Gorbachev, suspected because of the wine-stain birthmark on his face (Wheen, 2004, p. 160), highlighting the absurdity of such numerical gibberish. Indeed, something may have been lost in the translation.

To be sure, any mapping that encodes letters to numbers can be cooked up with a little numerological trickery, such that Ronald Wilson Reagan, Mohandas Karamchand Gandhi, and even Betty Crocker can become one's very own beast. By summing the letters in "Betty Crocker is alive, Betty Crocker is alive, Betty Crocker is alive," where each letter is encoded as a number from 1 to 26, one gets, you guessed it, 666 (see Figure 8.5).

This is also how books such as *The Bible Code* that use skip patterns of words or letters from various texts can claim to concoct futurist messages, such as "Kennedy will be killed" or "Hitler is a monster." But as British journalist Francis Wheen (2004) succinctly noted, "Select every third letter from 'generalization' and you get 'Nazi'" (p. 153), although if one were to use the British spelling "'generalisation," one would get "nasi," a Malay and Indonesian word for rice. Gobbledygook to be sure. In the same way, apparent messages can be found in songs played backward if one looks hard enough or listens long enough, such as Led

Figure 8.5	How Betty Crocker became the devil		
B	2	i	9
e	5	s	19
t	20		
t	20	a	1
y	25	l	12
		i	9
c	3	v	22
r	18	e	5
o	15		
c	3	Σ =	222
k	11		x 3
e	5		= 666
r	18		

Zeppelin's "Stairway to Heaven," which Singh (2008) noted contains the backward lyric, "*There was a little toolshed where he made us suffer, sad Satan*"—if one is listening for it.

Oddly, 666 is the sum of the first seven primes squared and also the back-and-forth sum of the first six cubes,[20] revealing a tantalizing numerological palindrome. The first 144 digits of π also sum to 666, but, to be sure, any trivial coincidence can pop out if one tries hard enough. Even within the digits of π (recently worked out to 2.7 trillion decimal places), one will find one's phone number or any encoded message one wants if one looks hard enough and long enough, which, as we saw earlier, can be explained by purely random probabilities.

Using the same strained logic to see or make up patterns that aren't there, my own suggestion for the mark of the beast is the Tim Berners-Lee–created interface to the Internet, the World Wide Web. The first three letters all start with *W*, which is the 23rd letter of the alphabet, and $2 \times 3 = 6$. If one thinks of the Internet as a living organism, it's a slam dunk. In the future, who will be able to shop without the Internet? WWW, beware the mark of the modern beast.

My favorite encoding, however, was created by Arnold Schwarzenegger, who as governor of California sent the following cryptic note to a Democratic assemblyman who had heckled him at a prior event:

To the Members of the California State Assembly:

I am returning Assembly Bill 1176 without my signature.

For some time now I have lamented the fact that major issues are overlooked while many unnecessary bills come to me for consideration. Water reform, prison reform, and health care are major issues my Administration has brought to the table, but the Legislature just kicks the can down the alley.

Yet another legislative year has come and gone without the major reforms Californians overwhelmingly deserve. In light of this, and after careful consideration, I believe it is unnecessary to sign this measure at this time.

Sincerely,

Arnold Schwarzenegger

Although the message is hidden (check out the first letter of each line), one can imagine that Governor Schwarzenegger wanted the meaning to be decoded, yet for obvious reasons wanted to be able to deny any surreptitious slight. When asked about the encoded message, a spokesman wryly noted, "My goodness. What a coincidence. I suppose when you do so many vetoes, something like this is bound to happen" (Li, 2009). Intentional or unintentional—who is to know?

For the final word on the meandering meaning found in the Book of Revelation, however, Thomas Jefferson had this to say: "The ravings of a maniac, no more worthy, nor capable of explanation than the incoherences of our own nightly dreams" (Wheen, 2004, p. 125). As for numerology and numbers meaning more than they seem, we will always see patterns that aren't there. The secret is to see man's additions to the celestial tapestry of the heavens and our own handiwork therein.

8.5. The Doubling Game and the Superstition of Choice—as Advertised

As belief in religion continues to diminish, belief in fate and fortune increases, manifest by the ever-increasing use of horoscopes, tarot, psychic lines, and the like, all peddling happy futures. Indeed, many think that a prescribed personal meaning rules our way, as if the future is preordained, never mind the Catch 22 of trying to see and change a future that is already determined—i.e., if a psychic sees a client's death and warns him or her, the client will presumably not die, and how then could the psychic have seen the death?

It's hard to explain to the believer how much hokum superstition is. But one doesn't avoid walking under ladders because one and all one's descendants will forevermore be unlucky but, rather, because one doesn't want paint spilled on one's clothes or a hammer dropped on one's head. Admittedly, lesser superstitions are nothing more than playground games, such as throwing salt over one's shoulder after spilling something (?), knocking on wood lest what one is talking about should happen (??), or 7 years' bad luck for breaking a mirror (???). They make no sense and assume a causal relation between unlinked events.

Paulos (2001) cited some excellent examples of supposedly related occurrences that are nothing more than chance—meeting someone who knows someone you know on a trip (p. 38), a dream premonition (p. 73), and the famous monkey typing Shakespeare (p. 75)—all of which can be explained by multiplying small probabilities by large numbers (e.g., people, dreams, and infinity). In fact, randomness is built into the large number of permutations of events in our daily lives. As Paulos reminds us, it would be a miracle if there were no "miracles," deliciously stating with regard to those ever-precocious monkeys someday typing *Hamlet*, "Though some have taken this tiny probability as an argument for 'creation science,' the only thing it clearly indicates is that monkeys seldom write great plays" (p. 75).

Organized superstitions are more dangerous, however, and advise us to do something according to arbitrary prescription, which removes responsibility from ourselves. No less than blind faith, horoscopes, psychic lines, and the lottery are harmful practices that take away from our being in the world. Taken as innocent fun, astrology is as harmless as a kid's Ouija board, but as an alternative to thought, it is dangerous. But, more important, because such practices are incessantly advertised to the unwitting, they plague us. "Love, money, happiness, the answer is in the stars," sings the television ad. "Call the number on your screen now for an appraisal. Psychics are standing by to take your call. Find out the future now."

Such shams are snake-oil hucksterism at its most shameful and should be banned in the same way cigarette ads and alcohol ads portraying irresponsible behavior are banned. At the very least, small-print charges and any possible extras should be clearly displayed in lettering the same size as the hook. Better yet, how about a psychic accreditation? If a psychic wannabe can tell the shape of an object on a card in a blind test (say, four shapes, 10 cards) and score better than random (25%), then he or she gets a psychic license. Otherwise, no license, no advertising, and no "life, money, happiness."

In Ireland, one such practice is called *Psychics Live* and receives more than 100,000 calls a year at €2.40/min. The service has even sought to introduce a monthly fee, even though its former owner freely admitted, "You can get people dependent on anything" (Lyons, 2006), a telling statement about the real motivation behind such practices. I wonder if *Psychics Live* was owned by a millionaire. You don't need to be a psychic to figure that one out.

Today, television call-in shows and phone contests are also keen to offer monetary prizes for answers to the silliest of questions, where the caller is direct debited immediately upon calling the premium-rate number, but no information is offered on how unlikely winning the prize is. One such show offered prizes up to €10,000 for answering 10 multiple-choice questions, where each question had four possible answers. Of course, the quiz starts off with the easy questions, such as "Who was the first man on the moon?," before working up to the harder questions, such as "What is the population of Timbuktu?" And you have to answer all the questions to get the top prize, although getting one wrong sends you back to the start.

The odds of randomly answering all 10 questions is easy enough: $(1/4)^{10}$ or 1,048,576 to 1, which with a bit of knowledge is a bit reduced, or so it would seem given the impossibility of the last few questions. The real odds are immeasurable. Alas, they also play music between questions, the kind of music you get when you're put on hold, only you're still paying for the premium-rate telephone line. And each question is laboriously drawn out, such that there is a hefty bill to pay even before the impossible questions roll round. Interestingly, in these loathsome games, the better player is punished more than the poorer player, who loses earlier and thus pays less—no wonder more than a few people complained (13,000 according to an independent regulator). Furthermore, these shows are not considered as games of chance and thus are not required, as lotteries are, to give a percentage to charity. In reality, they are nothing more than for-profit swindles.

Interestingly, the ads for these shows tell you that you're going to lose straight off. Not content to take your money in their electronic three-card monte games, they rub it in with their seemingly cautionary commercials.

One such ad featured an actor dressed up as Benjamin Franklin and flying a kite and a quirky voiceover asking, "Who discovered electricity?" with four choices listed on the screen (e.g., Bon Jovi, Ben Franklin, Winston Churchill, Moses) and a reminder, "Call now to win." Another had the same actor dressed as Isaac Newton, sitting under a tree, and the question, "Who discovered gravity? Bill Clinton, Mahatma Gandhi, Bobo the Clown, Isaac Newton. Call now to win." After the spiel about how well-off you'll be when you win, the commercials end with old Benjamin being electrocuted and a bird doing its business on poor Isaac's head. No need to play, it would seem, when you know you'll be electrocuted or shat on in the end.

Such pervasive rip-off practices have even made their way to children's television. As Turner (2007) noted, her children would cry out to her to call in their answers to such moronic brain teasers as "Is the British prime minister called a) Tony Hair b) Tony Blair or c) Tony Carebear?" all charged at the usual, small-print, premium rate, although with an asterisk stating callers must have the fee-payer's permission. She questions the legitimacy of such programs but hopes at least that her children are being taught "the valuable life-lessons of skepticism" and a mistrust of television executives who call such games a "moron tax." But, really, why do we permit such cynical abuse of children?

What's more, are these games and call-in lines only harmless fun or should we be worried about such questionable practices? Other versions now running regularly on nighttime television are clearly bogus, low-end lotteries that rip off callers by asking inane questions (e.g., name a movie with a color in the title; name an animal; name a David), all the while billing callers at a premium rate.

Furthermore, do such "psychic" lines and lottery games aim to empty the bank accounts of the more troubled? Is any of this new-age and new-technology media properly regulated, or are they, instead, permitted by uninformed and lazy governments that spend little on consumer advocacy and fail to protect the unsuspecting from being taken in? Hasn't anyone figured out that the snake-oil salesman is constantly reinventing himself, shedding old cons for new cons just as the snake itself sheds its skin? Come on—name a David?

We should be asking why the government encourages us to spend money so irresponsibly. We don't allow snake-oil salesmen through our doors. Why should the government? Alas, if such games of chance are no more than simple fun—the pre-information-age, fairground equivalent of "guess your age" or "toss the hoop"—shouldn't they at least be fair or properly taxed? But television quiz games and call-in lines are not designed to be fair; they are designed to make money, and without necessary safeguards against minors playing. Regulation seems nonexistent, particularly for games shown on state-run channels or aimed at children.

All this, of course, raises questions about why we are so keen to encourage gaming, which has fast permeated our everyday lives, not least because of the advent of sophisticated electronic technologies and in no small part because of the alarming increase in the number of casinos, online betting sites, and lotteries the world over. What's more, the numbers are growing: Total gaming revenue in the United States is more than $90 billion per year, a 50% increase in only 7 years. What's more, 30% of adults visited a casino last year, 1% of whom are considered "pathological gamblers" (American Gaming Association, 2011). A British gambling boom has produced "a 20 percent increase in the combined wealth of the country's super-rich," where half of new entries for more than £500 million in a list of richest Britons made their money from casinos or online gambling ("Gambling Boom Boosts British," 2009). As for the premier cash-cow game of chance, the lottery—run by governments everywhere with little thought to social impact—it accounts for more than $200 billion spent annually worldwide (which we look at next).

So what can we do to protect ourselves, our children, and those around us who don't seem to know any better? Given the lack of government action, the only appropriate course is not to play. The informed consumer should know that there is no such thing as a free lunch and not to trust the man in the suit selling success. He is a shill. As the ads themselves divine, not only will you lose your money, you will be electrocuted or shat on.

8.6. While We Were Sleeping: The Growth of the Modern Lottery

"Imagining the freedom," as the lottery ad goes, has never been so popular. What started as a curious diversion is now big business, with more than 200 million people wagering up to $200 billion per year on any number of games, known the world over as Mega Millions, Fantasy Five, Lotto 6/49, Money Madness, Lucky Stars, Bono Lotto, Beach Blanket Bucks, Win for Life, and on and on.

Winning, however, is highly unlikely, no different from *La Lotto de Firenze* in 1530, the first lottery to use money for prizes. In fact, the chance of winning a six-ball, 49-number lottery is about 14 million to 1, or 6 times less likely than being struck by lightning, according to J. Laurie

Snell, Dartmouth professor of mathematics and numbers oracle at Chance News, a web-based statistics forum.[21] A program of lesser prizes does attempt to offset the unlikely odds of winning, where "winning" includes a 4-to-1 chance to win another ticket and so on.

Many trumpet the Irish Hospitals Sweepstakes as the first modern lottery, offering payouts based on horse finishes, but the electronic clamoring for today's millions really began in New Hampshire with microchips and secure telephone lines. A revamped American Dream, posting multimillion dollar prizes Monday to Sunday of every week, has done the rest.

The lottery wasn't always so bold. Banished by the United States Supreme Court for 70 years because of a "demoralizing influence upon the people" (Heberling, 2002, p. 604), New Hampshire (1964), New York (1967), and New Jersey (1969) cautiously started the ball dropping again. Little debate has since ensued over any negative influence. Today, one can even buy a "virtual" lottery ticket on credit on the Internet, with winnings deposited directly into a Swiss bank account.

Though half of all sales revenue is given back to the "players," as lotteries dub their ticket-buying population, governments take enormous pride in how they spend *their* windfall, as though justifying the lottery through philanthropy, which of course could just as easily be achieved with traditional taxes. According to the Ontario Lottery Corporation Act, almost one third of each lotto dollar must be used for "the promotion and development of physical fitness, sports, recreational and cultural activities and facilities, the activities of the Ontario Trillium Foundation, and the protection of the environment." A science center, a music hall, a museum renovation, and numerous community projects have been financed in part by lotto money. The act further states that any profits "not so appropriated in the fiscal year shall be applied to . . . the operation of hospitals." Other jurisdictions similarly apportion their philanthropic largesse, although no one questions whether the same monies could be raised in other ways.

Using lotteries to fund public works is nothing new. Lotteries helped finance such landmark buildings as the British Museum and the Sydney Opera House. The Continental Congress in 1776 voted to establish a lottery to raise funds for the American Revolution, although the scheme was soon abandoned. Other public works programs are more imaginative—for example, in Kansas, which puts 10% of lottery money aside for prisons, and in Nebraska, where a percentage of lottery revenue is spent on a solid-waste landfill and a compulsive gamblers' assistance fund. "Interlotto," the world's first Internet lottery, lets players pick their charity from a drop-down list before playing. This is all a far cry from Louis XIV, who had different ideas about lotto funds when he and a number of his ministers won top prize in his own state lottery.

In Ontario, after returning 7% for retailer commissions, the remaining 13% of each lotto dollar goes to operating costs, including 1.6% for advertising. Each year, the government spends tens of millions of dollars telling people how to "imagine the freedom," "live dangerously," "take the plunge," and "escape the jungle for an instant," making the game "as popular as possible for the player." In New Hampshire, the state has even simplified its $1 fork-over to two best-selling scratch tickets called "One of These" and "One of Those." No need to think, let alone imagine.

To date, the biggest lottery payout is $380 million, won in a U.S. Mega Millions draw. Elsewhere, prize winnings regularly top the multimillion-dollar mark (see Table 8.5). Few seem to question the value of giving away millions for nothing.

Given its complete integration into daily living, one wonders why the odds aren't better known. In a 6/49 lottery, six balls are chosen from 49, and, thus, the odds of choosing any six balls in the order they drop are 1 in $49 \times 48 \times 47 \times 46 \times 45 \times 44$ (a little more than 10 billion to 1). But since it doesn't matter in which order the balls fall, there are $6 \times 5 \times 4 \times 3 \times 2 \times 1$ (720) ways to draw the same six balls. Thus, the odds of choosing any six balls from 49 is 1 in $(49 \times 48 \times 47 \times 46 \times 45 \times 44) / (6 \times 5 \times 4 \times 3 \times 2 \times 1)$, or 1 in 13,983,816.

Table 8.5	Biggest lotto winners around the world			
Country	Name	Odds/Numbers	Prize	Date
United States	Mega Millions	56 to 5 and 46 to 1: 1 in 198,711,536 4, 8, 15, 25, 47 (42)	$380 million	January 4, 2011
Europe	EuroMillions	50 to 5 and 9 to 2: 1 in 76,275,360 9, 21, 30, 39, 50 (1, 3)	€183.5 million	February 3, 2006
United Kingdom	National Lottery	49 to 6: 1 in 13,983,816 2, 3, 4, 13, 42, 44 (24)	£42 million	January 6, 1996
Canada	Lotto 6/49	49 to 6: 1 in 13,983,816 5, 11, 20, 30, 37, 43 (31)	$54.3 million	October 26, 2005

The only way to increase the odds, however, is to buy more tickets or play in groups, as many do. "Systems" abound and are especially designed to separate one's dollar from one's pocket. Most are based on random numbers, past performance, or the usual appetite for numerology and lucky numbers, where number 7 seems to be everyone's favorite—although *lucky* number 7 is no luckier than *unlucky* number 13 when it comes to lottery balls. Random number systems have as much chance as any homespun system, such as picking birthdays or wedding dates. Typically based on "past performance," such systems compile results from a wealth of frequency data to show what's "hot" and what's not, and are all the rage. HOTLOTT, HitLotto, and Lottomatic are but a few.

In truth, past-performance systems prey on the ignorance of those who believe in "hot" and "cold" numbers, as if the most basic laws of chance are invalid. In Powerball USA, a 5/59-plus-bonus ball game, number 20 is currently "hottest," popping up on 13% of draws in the past 5 years. The "coldest" is number 29, falling on only 6% of the same tickets. Should one then stay with "hot" number 20 or dump it in favor of "overdue" number 29? Sadly, many players, subscribing to their own ideas about the law of averages in the grand scheme of the numbers cosmos, think that number 29 is ripe for the picking. Of course, it makes as much sense to stick with the "hot" number 20, if it's that hot. In theory, hot 20 and cold 29 should pop up about 11% of the time or once in every nine draws (to see why this doesn't always happen, see Chapter 7).

In poker, most players believe that luck depends on where you're sitting; in the international lottery game, luck comes down to where you're living. In the UK National Lottery, 39 is currently the "winningest" number (86 of 522 draws) and 7 the "losingest" (43 in 522 draws). Only miles away on the Emerald Isle, where luck is cultivated as an art not a science, number 2 is "luckiest" and number 43 "unluckiest," at least according to 5 years of the Irish National Lotto. In Canada, 40 beats 16 hands down, popping up about 70% more in the past 5 years of Lotto 6/49. None of this can be attributed to any bias.

Numerous schemes attempt to beat the odds. Not only clubs but organized syndicates exist, perhaps winning up to a third of all draws. For a fee, an administrator plays an array of numbers and distributes the winnings evenly. Payouts come in bunches, as favorite numbers are "boxed" in multiple combinations. "Wheeling" is the jargon of the serious, syndicate lotto gambler. But wheeled or not, one ticket is just as likely—or unlikely—to win as any other. Mind you, the only way to increase the payout is if no one else plays the same numbers—impossible in a syndicate. If at all reputable, syndicates are for the more seasoned suckers of the lottery sham.

When asked about lottery clubs or syndicates, the Ontario Lottery Corporation states that "players should purchase their own tickets through an authorized lottery retailer only." For the multiplayer office lottery pool, they advise an agreement on paper drawn up by the players, just to be sure you know who your friends are when the millions pour in. For the increasingly popular multiplay and out-of-country syndicates, however, nothing can be done to ensure their legitimacy or legality. Charging a fee for such syndicates may in fact be illegal since a lottery ticket cannot be purchased above its retail value.

Out-of-country syndicates also contravene mail and tax laws in some countries. Swiss Lotto, Lotteria Nazionale Italia, and the Irish Lotto are just a few lotteries for which syndicates are easily found on the Internet. A Swiss Lotto syndicate even issues a "Swiss quality guarantee" that winnings will be paid immediately into a Swiss bank account. Payment is made by faxing or e-mailing credit card information, international wire transfer, international money order, or by cash—U.S. cash only, mind you, which they advise you to send by registered mail. An Irish Lottery syndicate boldly stated that "the pool of players is not as large as other international lotteries, so you stand a better chance of winning the jackpot," a statement that contradicts the most basic understanding of chance. They further add, "No one knows that you have bought a ticket and no one knows that you have won . . . except you!"

The first Internet lottery, Interlotto, was a 6/40 lottery, claiming the world's largest percentage payout at 65% of ticket sales. It is still operated by the "International Lottery in Liechtenstein Foundation," a charitable organization supervised by the Liechtenstein government and audited by an independent accounting firm. For legal purposes, players are deemed to be *in* Liechtenstein when they play. Draws are made in public and results passed along the Internet within hours. Everything is more or less the same as at the corner store, except you can't buy any milk.

One can list the wrongs of a state-sponsored lottery quite rationally—most obviously, that it is a regressive tax. "A tax on the poor," Manitoba MLA Jim Maloway stated in response to the increasing number of video lottery terminals in his province ("Debates and Proceedings," 1995). He further questioned the propriety of government involvement in such lottery schemes:

> What we have done is basically legalize and organize another vice by doing this with gambling. The fact of the matter is that it is here, but the whole

fiscal image, the whole fiscal position of the government rests largely on this gambling revenue that they have managed to put aside for the last couple of years. ("Debates and Proceedings," 1995)

Heberling (2002) further noted that when the money-raising euphoria wears off as ticket sales decline, coercive action is needed to keep up the revenues (slyly referred to as a "voluntary tax"); as such, governments end up raising taxes, increasing the number of ads, and "devising new, more exciting (and addictive) versions of the game" (p. 598). What's more, lottery spending is greater the poorer the income group (those earning less than $10,000 per year spend at least $500 per year on the lottery), and, thus, lottery ads are targeted specifically at the poor (pp. 598–599).

One can also argue that selling unlikely dreams is contrary to fostering a work ethic many of us subscribe to each day or that government-run lotteries should not be sold as tantalizing entertainment. Scratch and *lose* is what they should be saying, informing us of the unlikely event of ever winning, not encouraging our use with seductive advertising. Of course, if a lottery is essential to scratch an irresistible itch or to keep the numbers rackets and organized crime out, why not a simple lottery with smaller and more reasonable payouts? Or is it all just Money Madness?

At the very least, there should be a disclaimer on every ticket, not unlike that on cigarette packages: "Playing lotteries decreases your chances of saving" or "Player not likely to win in 10,000 lifetimes." No society should count its lottery as entertainment, and no government has any business selling dreams as sour as 14 million to 1.

Notes

1 **Belief in the unknown**: Wheen (2004) quoted polls indicating that one in four Americans believe in astrology and almost one in two believe in UFOs, and a study that almost 50% of Wall Street brokers consult horoscopes when deciding what to buy and sell (pp. 103, 116, 125).

2 **Copernicus and the Catholic Church**: The Church didn't allow printing of books claiming the earth's motion as real until 1822 (Kuhn, 1957/1997, p. 199).

3 **The ecliptic:** The ecliptic is the apparent path of the sun around the earth (apparent, because the earth travels around the sun).

4 **Gravitational force**: To calculate the actual force in newtons, multiply by the universal constant of gravitation, G, $= 6.67 \times 10^{-11}$ m³/kg/s². The 10^{-11} factor shows how extremely small the force is between two 1-kg objects 1 meter apart.

5 **Tidal force**: The shape of the coastline and water depth also greatly affect the tides. At mid-ocean, where the effect of the coastline is nonexistent, high tide is about 1 meter above

average sea level. The tide-raising force is also proportional to m/r^3 (not m/r^2), resulting in the moon having a greater effect than the sun, which more realistically relates to experience: $F_m/F_s = m_m/m_s \times (r_s/r_m)^3 = (7.4 \times 10^{22}$ kg $/ 2.0 \times 10^{30}$ kg$) \times (1.5 \times 10^8$ m $/ 3.8 \times 10^5$ m$)^3 = 2.28$.

6 **Jupiter's gravity = 2.5 times the earth's**: $g_J = g_E \times M_J/M_E \times (r_E/r_J)^2 = 9.8$ m/s$^2 \times$ $(1.9 \times 10^{27}$ kg $/ 6.0 \times 10^{24}$ kg$) \times (6.4 \times 10^6$ m $/ 71.5 \times 10^6$ m$)^2 = 9.8 \times 2.54 = 24.9$ m/s^2.

7 **Africa and South America**: Darwin first observed the related continental species of the two continents, which has now been verified by DNA evidence.

8 **Population doubling**: 30 years is not a bad first guess. In less-developed regions, birthrate doubling will be faster but is offset by infant and child mortality. Note that below a so-called "subreplacement fertility" (about 2.1), a population will shrink.

9 **Half-lives**: U-238 $t_{1/2}$ = 4.468 billion years; Pu-242 $t_{1/2}$ = 380 thousand years; C-14 $t_{1/2}$ = 5,570 years.

10 **Dating Niagara Falls and the Shroud of Turin**: Niagara Falls was formed when Lake Erie joined Lake Ontario after the recession of the Laurentide ice sheet, as measured by carbon-dating shells found at the base of the continuously eroding falls. The Shroud of Turin was carbon-dated and thought to be a medieval hoax; however, the tests may have been performed on a patch and not the original relic.

11 **The logarithm function is the inverse of exponentiation**: To get rid of a 2^N in an equation, take the log of both sides; $\log(2^N) = N$.

12 **Teutonic gods**: In German and Dutch, the Teutonic gods replaced the Roman gods only for Thursday and Friday, and English added the two gods for Tuesday and Wednesday.

13 **Solar system periods**: Moon 29.5 days, Mercury 88 days, Venus 224 days, sun 365 days, Mars 687 days, Jupiter 12 years, and Saturn 29 years.

14 **Calendar imperfections**: By adding leap years, the Gregorian calendar corrects the earlier, more imperfect Julian calendar. Because of the precession of equinoxes, the first point of Aries that marked the Roman year-end is also out of whack and is currently in the month of March.

15 **Rosetta Stone**: Both Demotic and hieroglyphic Egyptian are inscribed on the Rosetta Stone, providing three one-to-one language mappings (between Classical Greek, hieroglyphic, and Demotic).

16 **Enciphered text**: Enciphered text is capitalized to differentiate it from the original plain text.

17 **Bell's photophone**: One of Bell's early inventions in sound reproduction, the photophone, transmitted speech on light, a precursor to today's laser/LED Internet.

18 **Storage size**: One interesting change as a result of modern digital recording is that information is now stored in indeterminate lengths, unlike records, tapes, and CDs, which are constrained by the size of the physical medium. Most digital storage devices today are much bigger than any single file.

19 **The two prime factors are the following 116-digit numbers**:
33478071698956898786044169848212690817704794983713768568912431388982883 79387800228761471165253174308773781446 7999489 and 36746043666799590428244 63337996279526322791581643430876426760322838157396665112792333734171433 96810270092798736308917 (Matson, 2010).

20 **Odd numerical beasts**: The biblical mark of the beast, 666, is the sum of the first seven primes squared $(2^2 + 3^2 + 5^2 + 7^2 + 11^2 + 13^2 + 17^2)$. The number 666 is the back-and-forth sum of the first six cubes $(1^3 + 2^3 + 3^3 + 4^3 + 5^3 + 6^3 + 5^3 + 4^3 + 3^3 + 2^3 + 1^3)$.

21 **Chance News**: http://www.dartmouth.edu/~chance/chance_news/news.html.

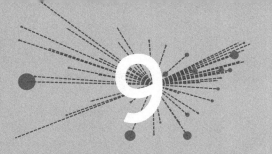

TWO KINDS OF PEOPLE

Those Who Divide and Those Who Don't

Z ero-sum thinking is the basis for most games and the starting point of economic theory as used by the likes of mathematicians John von Neumann and John Nash. Of course, strategy plays a part in the outcome of any game, but is a better strategy (or strategies) obvious or counterintuitive, as in the Hollywood movie *A Beautiful Mind*, where the seemingly maladroit Nash attempts to woo a woman in a bar not by starting from the "top," as it were, and possibly ruining his chances down the line if unsuccessful, but by starting from the "bottom," thinking such a strategy more rewarding in the long run.

Basically, not everyone can win when competition is the prevailing attitude, from dating to sports to economics. Competition is deemed essential, whether or not the playing field is free from bias or error—which, as we have seen, is not always the case—and politicians, statisticians, and economists particularly like to divide the world into two kinds of people: winners and losers, those above the mean and below the mean, the haves and have-nots. The list is endless, but all can ultimately be thought of as competitors in a game.

The zero-sum game recognizes the inherent inequality in the competition between players with opposing interests, where everyone cannot win. In this chapter, we look at a few examples, from ecotourism to the music business, from subsidies to bailouts, and from taxes to the national debt—perhaps the best example of how our world is defined by the balance between what is private and public.

9.I. The Forward Stampede and the Home-Comers

Schumacher (1993) called the two categories of people the "forward stampede" and the "home-comers." He quoted a prominent past chief of the European Economic Community with the watchwords of the forward stampede: "more, further, quicker, richer" (p. 128), which has an almost Olympian quality of "faster, higher, braver," whereas the home-comer has come back to older ways:

> For it takes a good deal of courage to say "no" to the fashions and fascina-
> tions of the age and to question the presuppositions of a civilisation which
> appears destined to conquer the whole world; the requisite strength can be
> derived only from deep convictions. (p. 128)

One wonders why we divide. Can we not see the other in ourselves: male, female; for, against; empire, rebel; tortoise, hare? If, as in "The Tortoise and the Hare," it makes better sense to plot a steady rather than aggressive burnout course, is there still nothing of value in the hare's ways? He is enthusiastic, has a carefree attitude, is charming, and is full of confidence. These are good qualities. It is telling to assume that his dominant trait is haughtiness, but he may well have been a modicum of enthusiasm, insouciance, and confidence. He just blew a gasket, overes-timated his charm, tried his hand at one too many hedges. He may have done everything right other than judge his resources.

As the prime example of excessive consumption versus production, the United States is not so much arrogant and haughty—as some critics suggest—as brimming with the enthusiasm and confidence of a great industrial power, ever expanding its frontiers and seemingly endless limits. If growth is not connected to resources or real needs, however, its purpose is questionable. Why go to the trouble of cooking a meal only to throw out half? To what end—to indulge one's sloth? And as we reach our limits, we may soon be competing for increasingly diminishing resources and arable land, especially as the developing world adopts first-world ways.

GNP is generally given as the measure of national success, but why do we assume that GNP is the only measure—what Jackson (2009) called the "busy-ness" of an economy (p. 179)—condemning "poorer" nations because they don't have our first-world incomes and highly industrialized GNPs? Why do we want to continue increasing income and GNP at oth-er's expense, when GNP, as Veseth (2005) noted, "is in any case a very limited measure of the human condition" (p. 176)? Why are we in such a hurry to grow without considering the connection to the environment or our own needs, especially given the inequalities that often occur? All such questions are related to economic competition.

Slower improvement, such as the sure-and-steady ways of the tortoise rather than the up-and-down spikes of the hare, can make sense economically and spiritually. For example, double-digit unemployment is as debilitating to the idle worker as it is to the affected families and communities. Why not more jobs at a lower wage, such as a 4-day workweek, if it results in more employment and everyone participating in society? Frank and Cook (2010) suggested more public holidays (p. 228) and noted that "if *everyone* worked less we would *need* less" (p. 143).

Krugman (2009) noted that workers in France work 14% less than American workers and yet are only 10% less productive (p. 254), an actual savings. Jackson (2009) noted that reduced workloads are "the simplest and most often cited solution to the challenge of maintaining full employment with non-increasing output" (p. 136). And, as Schumacher (1993) reminded us, Gandhi saw that *real* revolution demands "production by the masses rather than mass production" (p. 57).

Furthermore, it is not heresy to slow down if overconsumption is destroying our world. Building housing estates that satisfy continued growth without considering demand, the environment, or the social fabric of an area is mad. Building massive homes in gated communities next to tenement slums overvalues money and undervalues community. Commuting 2 hours to work in traffic-jammed, single-occupant cars is a failure of economic planning, not an achievement. Yet we are told that the economy must keep growing, that our tried-and-true capitalist ways keep us in good stead, and that a great invisible hand guides us in our choices.

Alas, few of us want to live like this, so why should we fashion our economy this way? Adam Smith, the father of the modern economic world, perhaps said it best: "Consumption is the sole end and purpose of production; and the interest of the producer ought to be attended to only so far as it may be necessary for promoting that of the consumer." Smith clearly included balance in the midst of his economics of production and consumption. Using the language of error, he stated that we must err on the side of the consumer, for the greater error is to let the producer produce without concern for the consumer, and that the best of capitalism must be protected from the worst of its excesses. We have done much to improve our living standards, including reducing disease and hunger and the folly of underproducing and underconsuming, but to overproduce and overconsume can be even more disastrous and is what we are choosing for ourselves now.

228 DO THE MATH!

Perhaps more important, should success be measured by winning or losing a game and the size of one's economic achievements, such as Lee Raymond, who earned $686 million during his 13-year stint as chief of the ExxonMobil oil supply (more than $6,000 an hour every hour; Maass, 2009, p. 117), while communities around the world were laid waste by relentless oil extraction and corruption? Is money the sole measure of the man?

In a chapter titled "Technology With a Human Face," Schumacher (1993) wrote,

> What is quite clear is that a way of life that bases itself on materialism, i.e., on permanent, limitless expansionism in a finite environment, cannot last long, and that its life expectation is the shorter the more successfully it pursues its expansionist objectives. (p. 121)

Does that sound like the doubling game? As we have seen, *the faster we become, the faster we lose.*

We must ask ourselves, are we becoming victims of our own success and speeding to the end? Some estimates put current oil supplies at 40 years and deem some fish stocks depleted beyond return. Have we overextended ourselves like the hare, stuck in a manufactured, expansion-only economic model, focused solely on production and market levels? In his lust to win, the hare overextended himself. In our lust to rush into the future, we are destroying the world. And worse, if we lose, there may be no return.

Many of us grew up with a "waste not, want not" ethos, as did many raised by parents who lived in war-ravaged times. Few would disagree. So why have we become a world of wasters, advancing no farther since the 1970s, when President Carter (1977) noted during the Gulf oil crisis that "we waste more energy than we import." Is it right that we use electricity to heat water or machine-dry clothes in southern climates, that housing developments are built without considering the direction of the sun, or that few consider the impact on the world of a disposable lifestyle, as though the one among us does not count. We are meant to be building societies, not statistics. Not everything is a commodity—individual lives do count.

Galbraith (1958/1999) noted that "the ancient preoccupations of economic life—with equality, security and productivity—have narrowed down to a preoccupation with productivity and production" (p. 97). The quest for more and the belief that more is better are at the core of modern economic thinking. He further stated,

> The importance of production is central to [the economist's] calculation. All existing pedagogy and nearly all research depend on it. Any action which increases production from given resources is good and implicitly important; anything which inhibits or reduces output is, *pro tanto*, wrong. (p. 115)

The fascination with speed is everywhere in our culture: faster, higher, braver. We are like junkies, pushing ourselves to find new fixes. But selling cannot be the sole end of production and must be tied to production *and* consumption. There is no point to production if it is not connected to consumption, other than artificially to create need that ultimately leads to manufactured cycles of boom and bust (more on that in Chapter 10).

Few would disagree that we are overvaluing the producer and undervaluing the consumer (machine versus man?) and praising short-term gain, e.g., first-quarter earnings over long-term goals. Jackson (2009) noted that the freedom "to find meaningful work at the expense of a collapse in biodiversity or to participate in the life of the community at the expense of future generations" (p. 44) may be a freedom too far. Naisbitt (1984) wrote that advocating short-term gain after a period of poor performance is unwise, stating that "long-range plans must replace short-term profit or our decline will be steeper" (p. 82). Echoes of Schumacher. Echoes of the hare. Lessons that are not being learned.

Frank and Cook (2010) noted that "the inability to set one's sights on larger, more distant rewards is associated with, among other difficulties, criminal behavior, alcohol and other substance abuse, marriage dissolution, and pathologically low savings rates" (p. 203). They further noted that valuing long-term goals over what appears better in the short term is similar to developing the ability to defer instant gratification:

> A job flipping hamburgers after school, for example, holds the immediate attraction of providing money to buy a car, but it also entails having less time to qualify for admission to a good university, and hence a lifetime of diminished opportunity. (p. 203)

Schumacher (1993) referred to the "economic calculus" being more important than "meta-economics," where man produces even at the expense of "health, beauty, and permanence." Jackson (2009) called it a "fetish with macro-economic labour productivity . . . a recipe for undermining work, community and environment" (p. 132).

A simple example of how our lives have become ruled by competition and disconnected from shared values suffices. As a pedestrian, it is in my interest that cars go slow, but as a driver, it is better for me if cars go fast.

Given that there are different and conflicting goals, depending on my perspective, it is a telling abstraction to think of a collective understanding rather than that of the individual. The mix is not solely utilitarian, where statistics rule, because that smacks of an inhumane numbers game, and, indeed, the individual will rarely sacrifice his own good for the whole. No, it is something else. When one considers the collective good as part of his own good, he transcends himself, transcends the individual, and aspires to a higher good, the one we all know from poetry and song and fable and heroic myth that tells us no one is an entity unto himself but part of a couple, a family, a community, a society, a species. Even as an individual, one is a part of the world, with which we silently communicate in every waking moment.

So, as a driver, I must stop for the pedestrian, even when he or she is crossing against the red. As a driver, I sit atop a pinnacle of engineering achievement—the internal combustion engine—having benefited from the sweat and blood and inventiveness of the countless people who came before me, without whom I could never have created it in a million years. Can I not recognize the benefits that have accrued to me? Can I not wait 2 seconds for a pedestrian to cross in front of me? Am I the sole criterion in my judgment?

> A hare one day ridiculed the short feet and slow pace of the tortoise, who replied, laughing: "Though you be swift as the wind, I will beat you in a race."

I did not make the car, and I cannot claim it as my own or treat a pedestrian with such contempt or, worse, honk at him or her to hurry. But, sadly, I do. Nor did I make the house I live in, the glasses I need to see, or the food I eat. And yet, more and more, I am constructing the world as an extension of me and not of others as well as me—although, ironically, the benefit from behaving as part of a society is that both the individual and the group can prosper.

Other examples show the same misguided strategies to success, such as bankers' obscene bonuses tipping the scales of fairness, supported by taxpayer relief; an athlete receiving bonuses for scoring goals, which may increase his goal production but decrease that of the team if he plays more selfishly; or congressional representatives making millions or a minister for justice owning 14 properties, while society crumbles.

Similarly, where production or love is withheld, an overworked worker may quit or an abused lover leave, because the relationship between the contracted parties has become unfair. Even in an unequal relationship between two parties, where the inequality seems to serve both, by not giving into one's worst self and by curtailing one's worst excesses, a passable amount of servitude can continue, provided one doesn't go too far.

But when the balance is tipped too far, the system breaks down. A lover can take the occasional haranguing, a worker will put up with the occasional slight, a populace will not question its lot—but not as a matter of course. We all recognize that life deals out the occasional unfairness, yet from the state of affairs in our modern, have-and-have-not world and rich and poor first-world cities, one must wonder how the basic interplay between people has been forgotten, the very reason we banded together—for freedom and security.

Man's great leaps forward were to walk upright and talk. The great leap forward for mankind is to recognize the collective as well as the individual need and that the collective need *is* the individual need. Many hands make light work, as the old adage goes, but only when the load is shared can we all prosper. Worse, as we shall see in this and the next chapter in examples that itemize the reality of the zero-sum game, seeking out that which benefits only the self can be gravely detrimental. What's more, if we continue in our ways, we may not only lose a race, we may lose a world.

9.2. A Simple Zero-Sum Game: Keep Your Eye on the Birdie

In the early 20th century, sea eagles, ospreys, and kites were hunted to extinction in England, Scotland, Wales, and Ireland. Most of the damage was done in the name of science, as the birds were stuffed for display in museums along with their unborn eggs, without much argument from the local farmers, who believed that such birds of prey were responsible for numerous farm kills.

In the 1960s, however, the birds were reintroduced into Scotland and Wales as a pilot project by bird enthusiasts who transplanted young birds from Norway and Sweden into the dormant countryside. Success was slow, but there is now a burgeoning population in such remote areas as Mull in northwest Scotland, where the magnificent sea eagles are once again part of the diverse wildlife scenery.

At first, there was some hesitation on the part of local farmers, who saw the eagles as potential predators to their animals—and, indeed, some eagles did attack and kill young lambs—but the number of tourists interested in sea eagles soon increased, providing a boon to the local economy. People also started coming in greater numbers to see the ospreys and kites in their renewed habitats, and the new attractions were accepted and welcomed as the economic benefit from tourism increased—"more bums in beds," as one local put it.

However, what is good for Mull must be *less good* elsewhere. Tourist money spent in Mull cannot be spent in other ecotourist spots, a straight-forward example of a *zero sum* in a limited-supply condition, where money spent in one place equals money not spent elsewhere. Life is a trade-off, as seen in this simple example, where only so many tourist dollars are available.

But that's life. In fact, it is the basis of classical economics. As Adam Smith might have put it, "everyone does their thing, spends where they see fit, and the great, good invisible hand guides the economy." And as long as consumer interests don't change overnight, everything is hunky-dory. In the case of ecotourist dollars, the change in the overall economy was minor (although significant in Mull) and no one minded or noticed decreased takings elsewhere.

Moving from Scottish birds to baseball birds, namely, the Toronto Blue Jays, we see the effect of economic planning on a larger scale. In 1977, professional baseball returned to Toronto, reintroduced like the sea eagles to the fourth-largest city in North America. From that snowy April day when the Toronto Blue Jays defeated the Chicago White Sox 9 to 6, the turnstiles didn't stop turning, and the Blue Jays were soon posting record crowds, winning the World Series 15 years later. They were the talk of the town—fitting, it seems, since baseball's most storied player, Babe Ruth, hit his first professional home run there in 1923. But, as with tourism in Mull, only so many dollars can be spent, and if they are spent on base-ball tickets, hot dogs, sports paraphernalia, and the like in Toronto, they can't be spent elsewhere.

All economics is a trade-off, no matter the size—whether ecotourism in the Hebrides or the entertainment sports dollar in North America. In the same way, Wal-Mart swamps local market sellers, Starbucks displaces local cafés, and the government lotteries put the local lotto charities to pasture. That's the way economics works, on back to the central tradition that placed "a premium on efficiency," as espoused by English political economist David Ricardo, where a more efficient means of production results in better business.

Economist Joseph Schumpeter called it "creative destruction," where innovation continues to revamp the old in a "relentless pursuit of nov-elty." In her analysis of city economies, Jane Jacobs noted that increased efficiency comes from "import replacements" that provide easier access to local manufacturers to manufacture products cheaper and are the life-blood of a functioning city. But whatever the name—import replacement, creative destruction, more-efficient production—they are all restatements of the old adage, "Build a better mousetrap and the world will beat a path to your door"—natural economic selection at its best.

Except it isn't a selection if the rejigging is unfair, where government economic planning supports or advantages one group while neglecting or disadvantaging another. The whole of the Toronto economy improved because of the Blue Jays, but others lost out, such as local baseball, the city of Dallas (or some other city for which a franchise wasn't awarded), and scores of other less-developed sports. The same applies to any protected or subsidized industry, from wine growing in Ontario to the multi-billion-dollar American automobile industry bailed out to the tune of tens of billions of dollars. Furthermore, if deep-pocketed companies are given incentives to relocate, should they then be able to do as they please, cutting prices and operating at a prolonged loss to flush out the competition before reinventing themselves elsewhere to do the same?

There is nothing wrong with subsidizing indigenous industries—preferential treatment has spurred on many renowned products, from vodka and whiskey to airlines and autos—but the problem is that we end up with a game of musical chairs as support is offered and then removed or companies move to new locales, changing the landscape in one area at the expense of another. Lower corporate tax rates, relocating manufacturing to low-wage countries, outsourcing work to India at a loss of 2 million American jobs (Zielenziger, 2003) are all part of the same government-supported economic tinkering. In Ireland, such dramatic rejigging occurred when Dell left Limerick for Poland, taking 4% of the Irish GDP with it while directly axing 1,900 workers and putting various others out of work in one fell swoop (Purves, 2010), yet the heads of the company were feted as local heroes in their new locale (having previously been feted in Limerick).

The same local-hero syndrome was at play in President Obama's historic $938 billion health care bill, as the government attempted to extend its coverage to greater numbers. Here, the debate was not that one *should* receive health care but *how many* should receive health care, as some of the already insured (83%) hoped to keep their slice of the pie by opposing increased insurance to those not covered (12% more, as proposed), who were primarily poor at that. In the end, after much debate, health care was extended to another 32 million people by government law (Pear & Herszenhorn, 2010).

Providing subsidies also fuels demand at arbitrary prices, when increased demand would lower prices under normal circumstances. But putting the supply cart before the demand horse leads to bubbles, as seen with disastrous effects in the housing sector because of poorly engineered interest rates and bad loans. Soros (2008) noted that countries with the fastest-growing demand, such as China and other major oil producers, "keep domestic energy prices artificially low by providing subsidies" (p. 234).

A particularly apt example of the zero-sum game at work in today's globalized world is seen in the consumer oasis of Dubai. If one shops or swims in Dubai, where tourism accounts for 30% of GDP, one can't shop or swim elsewhere, and now vacationers make their way to the playground attractions in the remodeled emirate by the planeload instead of visiting other holiday resorts. Called Dubai, Inc., for its no-tax, no-foreign-exchange, no-capital restrictions, Dubai is a perfect example of an artificial economy, built on exploited migrant workers with minimal labor laws, sidestepping basic human rights to provide its modern tourist Mecca. As Ali (2010) noted, "Construction workers in Dubai regularly work six-day weeks of eleven-hour days, and often another half day on Fridays. They work in dangerous conditions with more than seven hundred deaths on the job and ninety suicides per year" (p. 83). Such practices are hardly fair to the competition, not to mention the horrendous abuse of so many put-upon, exploited workers.

Illegal immigrants in the United States, some of whom have lived and worked in the country for decades, also provide cheap labor—numbering 10.8 million in 2009 according to the Department of Homeland Security (Hoefer, Rytina, & Baker, 2010)—while full-fledged citizens earn more and receive protected benefits. In fact, such immigrants are disenfranchised, low-wage workers who live on the fringes of society, fundamentally undermining the ideals of American democracy by living without any right to vote. However, the economy benefits too much from their labor, providing no political will to legalize or at least offer amnesty, although in the wake of increasing unemployment, there has been renewed interest in curbing immigration in some of the most affected states.

It is hard to fault a company if it can employ under-the-table workers at a fraction of the usual cost, trim expenses by moving to cheaper labor markets, or obtain preferential government subsidies at the expense of other competitors. But the real costs are astronomical, as a vast citizenry is kept from fairly participating or competing in the everyday exchanges of life. Sharing is the issue, as in any kindergarten sandpile—two of us simply can't have the same toy in a limited-supply game—and is little different from the changed landscape in Mull or Toronto. Whether the players are competing countries or competing companies, the zero-sum game is at the core of everyday economics.

The Economic Landscape and Government Engineering

Engineered economies are nothing new. Governments have been changing the economic landscape for centuries. During the 17th century, the British navy was expanded by Royal directive to support war in Europe, creating new jobs, new direction, and national pride, and bankrolled by the newly established Bank of England, which offered government bonds at 8%.[1] As a result of World War II, the U.S. government directly controlled parts of the U.S. economy, particularly wages, even maintaining such controls after the war ended (Krugman, 2009, pp. 52–53). Income tax has also done its part to aid different sectors of the economy in its collection and disbursement, begun in Great Britain in 1799 to finance the Napoleonic wars and in the United States in 1862 to support the Civil War (Atwood, 2008, p. 136).

At the end of World War II, sweeping changes were implemented, with vast amounts of public money spent to foster the move from war to peace. In the United States, the G. I. Bill (or Serviceman's Readjustment Act of 1944) was introduced, which still exists today at a cost of almost $2.5 billion per year (Shakir et al., 2008). The Marshall Plan cost $12.5 billion from 1947 to 1951 (Williams, 2008b, p. 206), or about $100 billion in today's money, and greatly stimulated industry and commerce in war-ravaged Europe. Thereafter, public spending increased throughout the world (especially nationalized public health services), with corresponding greater tax takes. President Obama's 2009 trillion-dollar stimulus package included a $50 billion top-up to build new roads, railways, and airport runways—an obvious attempt to direct public spending to create jobs—whereas the total liability to the U.S. Treasury as a result of the credit crisis amounted to $3.6 trillion in outlays and $16 trillion in guarantees (Mason, 2009, p. 195), including quantitative easing[2] by the Federal Reserve (so-called QE and QE2), which has pumped another $1.2 trillion into the economy in the hopes of stimulating growth.

The granddaddy of all government-engineered work programs (or in this case, back-to-work programs) was Franklin Delano Roosevelt's 1933 New Deal, his response to the Great Depression, which cemented modern government as part of the everyday landscape (like a fabled Superman incarnate: faster than a speeding bureaucracy, more powerful than a privately built locomotive, able to leap seemingly insurmountable planning hurdles in a single bound). The New Deal put unabashed controls on the economy, including worker relief, economic recovery, and financial reform, setting up national social security, insurance, and power generation for decades to come.

The New Deal economic development program included building the Hoover Dam, a national interstate highway system, and the Tennessee Valley Authority (TVA), the nation's largest public utility, which brought industry to one of the poorest regions in the United States and today provides 174 billion kWh from 75 power-generating sites, helping "attract or retain more than 40,500 jobs" (TVA, 2011). On the other hand, the TVA prevents competition from other private utility companies, bringing condemnation from conservative opponents, including Barry Goldwater and Ronald Reagan. Depending on which side of the fence one stands on, the TVA was (and is) either all that is wrong with government—top-down planning and restricted competition—or all that is right with government—organized effort and economic development.

Times Square is another perfect example of a controlled market regulated by government. Remodeled from a seedy, crime-ridden peripheral attraction of the 1980s to a major shopping and entertainment area, it now attracts millions of tourists each year. This wasn't achieved by the free market but, rather, by city planning that permitted buildings to be knocked down and offered preferential lease arrangements and low-interest loans to the likes of Morgan Stanley and Disney. As noted by the president of the Times Square Authority (TSA), "Its transformation is due more to government intervention than just about any other development in the country" (Bagli, 2010). Similar to the TVA, the TSA remade the playing field according to its own agenda, aiding commerce to increase tourism while providing more than an estimated $1 billion in property tax abatements and zoning changes (Bagli, 2010).

Government direction and intervention occurs everywhere. In Germany, the government bars intercity bus travel to protect state-run train travel; in Canada, government laws on foreign investment keep a potash company in a price-fixing cartel; in Britain, a global wine industry was revamped overnight when licensing regulations were changed to allow supermarkets to compete with specialty shops. As Veseth (2005) noted, "Broader distribution networks, scale economies, and longer opening hours" changed the wine-selling landscape forever (the economic *terroir*), which "triggered the global wine avalanche" (pp. 153–154).

In China, the government subsidizes green technology for export, undermining competitiveness in other countries, contrary to the tenets of the World Trade Organization. Bradsher (2010) noted that "the central government in Beijing and China's provincial governments have used land grants, low-income interest loans and dozens of other measures that violate W.T.O rules." Such violations are giving Chinese companies the edge as they unfairly "expand their share of the world market for wind turbines, solar panels, nuclear power plants and other energy equipment at the expense of jobs in the United States and elsewhere" (Bradsher, 2010).

What's more, China's new-styled statism is short on basic freedoms, preferring economic efficiency to public liberty and winning more economic battles with command structures that care little for consensus to compete unfairly in the now global economic sandpile. China's 2008 stimulus package (at 14% of GDP) was "implemented in a matter of days" (Moyo, 2011, p. 172), whereas airports, highways, and industrial parks there are built in months (Zakaria, 2009b, p. 134). As Moyo (2011) noted, "In the emerging world the state's reach seems to know no bounds; the collective state is paramount and takes precedence over any individual, contravening the Western dogma that the individual is king" (p. 147).

In fact, politics is at the core of all world economy, which is anything but free. As Beattie (2010) noted,

> The development of the world economy may look like an onward march of impersonal market forces which lays all inefficiencies waste before it. In truth, . . . some industries, and most particularly agriculture, are shaped as much by politics as economics. Their sustenance owes much to the fact that small groups of producers who will throw everything into protecting their livelihoods can often win out over much larger interests who care much less. (p. 151)

Given the history of such market control, it seems that government is not in the business of ensuring fair competition, as many think. The idea of a free market is a myth, and government routinely designs and redesigns the playing field to reflect the prevailing agenda in accordance with its own policies, which in most Western democracies is typically that of the right that seeks cheap labor or of the left that seeks to curtail excessive exploitation. In reality, it is the difference between the individual and the collective, as in any zero-sum game.

In a libertarian[3] world with minimal government intervention or control, the highest bidder wins—not at all what is intended by government and citizenship. A neutral enforcement agency is thus essential to keep the market from being eaten by the rich, an agency that naturally falls to government to provide. One can only hope that more than just economic interests and zero-sum thinking are considered.

9.3. Laissez-Faire Illusions: *Cui Bono?*

Prices are also manipulated by today's businesses, where laissez-faire economics—the supposed hallmark of the free market that suggests a natural balance between predator and prey—doesn't apply and rarely includes considerations other than profitability. Everything from price-fixing to bogus offers to monopolies scars the landscape of fair practice.

In the United States, 16 airlines were caught colluding on prices in the air cargo business and were fined $1.6 billion after a Department of Justice probe. In Europe, a price-fixing cartel between rival detergent giants Unilever and Procter & Gamble was found after operating illegally for almost 3 years, resulting in more than $300 million in fines. Computer chip makers in California paid $173 million for their price-fixing scheme. One wonders how many others exist undetected.

A simple example shows the raison d'être of many businesses operating today, and the lack of government bite to protect everyday consumers. Worried about reduced profits during the 2009 downturn, supermarkets turned to downsizing packaging on items such as beer, cigarettes, and processed dinners—sometimes by up to 20%. The prices, however, remained the same. Worse, supermarkets offer "reduced" prices directly after increasing them. One could argue that such practices are smart marketing, but why are consumers not informed? Or is it just an economic sleight of hand, aided by governments uninterested in monitoring such bogus trading practices, let alone enforcing fairness or applying penalties—not to mention price-fixing as part of a de facto oligopolistic marketplace.

Furthermore, various companies routinely establish de facto monopolies, some created by government license, others for which competition is restricted—from phone and cable companies to transportation and utility companies—which is recognized as a flaw in the capitalist system (Galbraith, 1958/1999, p. 33). That the small cannot compete fairly with the large is one of the greatest weaknesses of the supposed free-market system. When competition is restricted, free-market capitalism vanishes, resulting in modern economies that have nothing to do with Adam Smith and the so-called classical orthodoxy of supply and demand, despite a casual belief that they do.

What's more, inequality increases as monopolies begin to take shape and dominate the landscape, marked by feedback loops in today's speed-of-light electronic information exchange, where the Internet of today is already creating the monopolies of tomorrow. Fair competition is always the loser in such a viral, mousetrap-building world (more on feedback loops in Chapter 10).

Alas, the market is not the great regulator of economic life in perfect competition as Adam Smith believed and as we are taught. It does not reward producers of well-made, affordable products and punish producers of poorly made, expensive products (Galbraith, 1977b). Smith's invisible hand is a strapped-on prosthetic, used when needed to explain an idea of capitalism as an effortless rule of nature, as though preordained, where talent and determination alone win out.

In fact, markets are made according to an agenda, and, thus, the question one should be asking is, who benefits from government decision making (or lack thereof), or, indeed, who benefits more than they should from government decision making that should be in the general public interest? More public transit or more highways? More pedestrian-free areas or more traffic? More human-scale businesses with integrated wages and community enhancement or more low-cost superstores with poor wages and questionable marketing practices? More Wall Street profits or more jobs? Economics is not free and is a constant trade-off between who is financed and who is not.

If change or investment is based on cooperation involving all interests, everyone wins, but rarely are special interests not considered first as supported by the government of the day. The lobby system is particularly egregious, where private interest groups openly solicit political favors with money and privilege. Ordinary interests are lost as politicians support their own projects, typically based in their own constituencies. Zakaria (2009b) noted how a "highly dysfunctional" (p. 211) political system has taken root, "captured by money [and] special interests" (p. 212). "By every measure—the growth of special interests, lobbies, pork-barrel spending—the political process has become far more partisan and ineffective over the last three decades" (p. 212).

As for the birds in Scotland, on the heels of the transplant success in Mull, more ecotourist areas have been planned throughout Britain. Although laudable, such uncoordinated planning highlights the problems of laissez-faire economics. In zero-sum mathematics, everywhere (and everyone) cannot be successful. Indeed, the birds of Scotland will soon have to compete with a nearby turbine wind farm, highlighting the uncoordinated and yet interconnected competing needs in any economy.

The truth is that we live in a forced economy, where everyday decisions are made—good for some, not so good for others—seemingly regulated by fair-minded governments. However, government is not a neutral arbiter—fairly deciding who gets what, as it should be—but, rather, is engaged on behalf of some and not others. We have only to think of ecotourism in Mull or sports in Toronto to see the effects of such decisions. It seems extraordinary to say, but government must represent all: "Of, by and for," as in Lincoln's great reminder to the weary at Gettysburg, are not just words; they are a call to fair-mindedness and inclusiveness.

9.4. Economies of Speed: The Ever-Expanding Financial World

The Tennessee Valley was remarkably poor in the 1930s, although, as Jacobs (1985) noted, "neither the poverty or economy nor its backwardness was actually surprising, considering the fact that the entire region lacked an import-replacing city and never had one" (p. 112). Jacobs defines import-replacing cities as those that freely trade with one another, exchanging basic needs to create functioning, vibrant economies that grease the wheel of prosperity. In effect, we grow from the bottom up and out.

As such, Jacobs (1985) believed projects such as the TVA are ineffective since they do not address the root cause of poverty or economic inactivity. To her, subsidies, welfare, and make-work contracts are recipes for disaster (pp. 191–196)—particularly for underdeveloped supply regions that lack working city economies—and are instead "transactions of decline" (p. 206): "They mitigate regional inequalities but can't eliminate the causes" (p. 206).

To Jacobs (1985), subsidies are Band-Aids that result in booms that go bust when funding ends, as in Seattle, where the loss of Boeing contracts in the 1970s resulted in a localized depression, as well as adding to the national debt (pp. 185–186). What's more, nations can be dragged down by inappropriate responses to poorly performing cities, where "nothing else in their grab bags of economies suffices. . . . In the end, both cities and the nation itself come to ruin" (p. 212). In Limerick, when 4% of Irish GDP was removed overnight with the departure of Dell, the Irish economy entered a tailspin from which it has not recovered, the knock-on effect of which continues to hamper the entire eurozone. The heydays of Detroit's Motown, Seattle's Airtown, and Limerick's Delltown fueled their demise because of unrealistic ideals of permanent growth. What's more, the problems were exacerbated when real competition from elsewhere appeared. As Jacobs noted, "Economic competition is a hazard" (p. 66).

But doing nothing is no solution. In today's engineered economies, which attempt to create more efficient specialized markets (via tariffs, subsidies, tax breaks, etc.), governments have a responsibility to protect citizens when conditions change and reduced demand makes everyday markets obsolete. Change is not just economic but cultural, where livelihoods and ways of life are at stake.

So what can be done? Jacobs (1985) noted that breaking into smaller sovereign regions with meaningful exchange rates where currencies are used as intended—as measures of economic performance—would work (p. 215) but can be theoretical in the extreme (although smaller, city-based exchanges can work). Currencies provide feedback information about the strength (or weakness) of economic activity. As she noted about the collapse of Detroit's economy after its exports began to dry up, "It got no feedback, so Detroit merely declined, uncorrected" (p. 163), which is true of many former colonies and politically engineered economic unions, as seen especially in the recurring problems with the single-currency, multifederal European Union (EU). Instead, bottom-up economics sustained by volatile and creative free trade is required, which promotes real need in a real-time economic system—the opposite of mismatched, one-size-fits-all, macroengineered solutions.

Common sense also prevails, as in the adage, "Give a man a fish and you feed him for a day; teach a man to fish and you feed him for a lifetime." In Detroit, as elsewhere, it seems that most people want to work and the problem lies in economic engineering that supports extreme movements of labor and do-nothing welfare payments instead of better education and organized, bottom-up thinking—not to mention military expenditure, which could solve a host of real-world problems if appropriately applied closer to home.

Arrighi (2010) takes the argument further, however, examining in detail the root cause of the failure of the four main capital-accumulating powers in history—the Genoese-backed Spanish (1450–1648), the Dutch (1628–1784), the British (1776–1914), and the Americans (1917–present)[4]—each of which advanced the capitalist world system through a "systematic cycle of accumulation" (p. 129). In his meticulous analysis, he noted (after Braudel) that "the maturity of every major development of the capitalist world-economy is heralded by a particular switch from trade in commodities to trade in money" (p. 111). As such, the transfer from a manufacturing economy to a financial-services economy fuels its own demise, where old-fashioned usury ultimately ends up playing havoc with the economy.

Arrighi (2010) noted that the "socially polarizing effects of 'financialization'" were in evidence in Renaissance Italy (p. 326) and that as far back as the 16th century, capital was becoming more important than production, where "the largest profits were made not in the buying and selling of commodities but in exchanging currencies for one and another" (p. 131). And, although stock markets existed prior to the Amsterdam Stock Exchange, what was new under Dutch dominance was the volume and speculative freedom of transactions coupled with government-chartered joint-stock companies (pp. 142–143).

Great Britain ruled the waves next, after "they had robbed the Spanish, copied the Dutch, beaten the French, and plundered the Indians" (Ferguson, 2007, p. 51), primarily through their navy and the East India

Company, which was bringing increasing volumes of cheap food and raw materials from India to the markets of Europe through the Suez Canal. Jacobs (1985) noted that "throughout Britain's period of imperial expansion, England was unremittingly investing, publically and privately, in the conquest and shaping of far-flung supply regions" (p. 198), which ultimately undermined British production. At the same time, business was receiving preferential government treatment through an expanded circle of merchant bankers—such as the Rothschilds—who, as Polyani noted, "were subject to no one government" (Arrighi, 2010, p. 171). Furthermore, "the massive relocation of surplus capital from industry to finance resulted in unprecedented prosperity for the bourgeoisie, partly at the expense of the working class" (p. 178). Ferguson (2007) noted that of the 347% increase in per capita British GDP during almost 200 years of Indian rule, "a substantial share of the profits which accrued as the Indian economy industrialized went to British managing agencies, banks or shareholders" (p. 217).[5] Over the same time, the Indian GDP per capita grew by only 14% (p. 217).

When world power changed again, from Britain to the United States at the end of World War I, business was already firmly entrenched as the de facto power of the state, its newfound prosperity solemnized by President Calvin Coolidge's remark that "the chief business of the United States is business." As Arrighi (2010) noted, the British-conquered world "was consolidated into a system of national markets and transnational corporations centered in the United States" (p. 225). He further noted (after Chandler) that the American system changed the previous British system of "economies of size" to "economies of speed" by internalizing transaction costs (p. 247). In essence, the corporation had become king.

In many ways, business was more important than the state. As Arrighi (2010) noted, the rise of transnational corporations in the United States (from 10,000 in 1980 to 35,000 in 1993) was "the single most important factor" in undermining the exclusiveness of the state (pp. 74–75), reorganizing humans as never before and establishing corporate extraterritoriality that could roam the globe as seen fit according to "world power policy" (p. 129).

According to Arrighi (2010), the change from trade and production to financial speculation and intermediation marks the end of the productive phase of material expansion, where competition becomes more cutthroat, "the primary objective of which is to drive other organizations out of business even if it means sacrificing one's own profits for as long as it takes to attain the objective" (p. 233). The dilemma of *real* capitalism thus became excessive competition, which drives prices and, hence, profits to intolerably low levels, as originally predicted by Adam Smith (p. 295).

In the final phase, financial expansion becomes subject to violent up-and-down swings, marked not by buying and selling but by long-term investment and risk, out of which corporations work not to enhance the market but to authoritatively determine prices to stay alive (Arrighi, 2010, pp. 240–243, 263, 297). A stacked-market economy is then realized, controlled by a highly efficient vertically integrated corporation or group of corporations, a de facto monopoly that restricts and excludes competition. As Arrighi noted, "The market is suspended when the planning unit enters into contracts specifying prices and amounts to be provided and bought over long periods of time" (p. 297). In the process, risk and uncertainty are reduced so that profits can be orderly maintained, contrary to the intention of a free market and real capitalism. As Hilferding noted, "Capital . . . detests the anarchy of competition" (p. 299).

At the precise point when money trumps production and trade, an economy sows the seeds of its own demise. According to Arrighi (2010), the turning point of the American cycle of accumulation came in the 1970s, when "the volume of purely monetary transactions carried out in offshore money markets already exceeded the value of world trade many times over. From then on the financial expansion became unstoppable" (p. 308). As Phillips noted, "Excessive preoccupation with finance and tolerance of debt are apparently typical of great economic powers in their late stages. They foreshadow economic decline" (p. 325).

Since 1960, financial companies in the United States have grown substantially: The number of employees at Morgan Stanley has increased more than sixtyfold (Davidoff, 2011), the profit of the financial industry as a percentage of corporations has almost tripled from 14% to 39% (Mason, 2009, p. 128), and by 2008, "the financial industry accounted for a third of total domestic profits—about twice its share two decades earlier" (Krugman, 2010a). In 2007, before the financial meltdown, derivatives amounted to 8 times the size of the real economy (much of which resided in a complex and shady off-balance-sheet banking system) and currency trading was 17 times world GDP (Mason, 2009, p. 66). One could say that in the race for more profits the American Dream has become leveraged to the hilt, where paper fortunes are more important than work and by which the wonders of multiplicative financial investment seek to make money only for those who have money (or access to money) and not for those who can provide only labor.

In the process, inequality has become enshrined as the ethos of modern economics. The old ideals of hard work and equality were traded for overblown financial usury, aided and abetted by a reformed Wall Street casino that learned how to play the percentages better, first in hedge funds from 1949 onward and then in more advanced real-time quantitative trading. The pioneering days of yore became unrecognizable as a new world conquest emerged, that of taming the gaming tables and beating back the uncertainties of chance. In the process, modern economics bears no resemblance to any classical orthodoxy.

Bayley (2010) argued that we have turned our back on making things, the real source of our wealth that "recognises the crucial link between effort and reward," and said he regrets the casino world we now inhabit: "A system that gives priority to an engagement with products over a lust for quick returns is a more stable and wholesome one than a system where derivatives are a more reliable source of wealth than making a teapot." Could gambling with our futures be a problem, where we believe that such fortuitous decisions make us all the richer and smarter for our avowed success?

The problems of today's economics may also be the result of reduced demand, because we already have what we need—who needs a new toaster, kettle, or car when the ones we have work perfectly well? Thus, economies shrink, putting excess pressure on debt-ridden governments reliant on growth. Business models that consider only one way in their financial futures—relentless growth—can no longer compete.

We are also reaching geological limits, which in frontier days went unnoticed because resources were plentiful and space seemed infinite as the economy continually expanded, ever postponing a future reckoning. But the frontier days are long gone. As such, the ultimate Ponzi scheme may well be the economy itself, which must continuously grow to hide its faults—whether account imbalances or social inequality—all the while sold by governments in the name of productivity and national worthiness.

Simply put, we can't all be rich, and all the more so when money becomes the means for making money. Furthermore, as manufacturing turns more to finance and finance in turn becomes more of a game, the gap widens, creating ever more winners and losers, as we might expect in any zero-sum–run economy.

9.5. The Private and the Public: Spend Now, Ask Questions Later

Nowhere is the idea of the zero-sum game more hotly debated than with taxes and the public debt accrued from the regular shortfall of the annual tax take. The argument is essentially one of scale and whether we should be individually or collectively totting up the balance each year, which is what taxes are for, after all—a group collection to manage our individual wills.

During his presidency from 1980 to 1988, Ronald Reagan liked to popularize the phrase "trickle-down economy," where the stalwarts of business created employment for others through their massive wealth, with a little help from their tax-slashing friends in high places. In fact, such preferential treatment is a "zero-sum, trickle-*up* economy," since not only do the spoils *not* trickle down, but those at the bottom end up with less. In the heat of the Cold War, Reagan's trickle-down economy was more like the Politburo with a smile, where those in political favor (with their hands on the tap) got more—not to mention the unprecedented postwar increase in national debt under his tenure (as we will see below).

Even his election contender George Bush Sr. didn't get the trickle-down thing, calling Reagan's supply-side chicanery "voodoo economics"— that is, before Bush got on the supply side of the trickle-down spigot as Reagan's presidential running mate. Reagan even had a formula to back up the questionable accounting—called the Laffer curve after Arthur Laffer, a colleague of Reagan's main economics man Milton Friedman at the University of Chicago—which suggested that continually raising taxes *reduces* the tax take because people stop paying, thus *increasing* the deficit. Conversely, lowering taxes *decreases* the deficit.

Limited tax slashing from 77% to 70% had begun previously to stimulate employment after World War II with the Employment Act of 1946 (Galbraith, 1991, p. 258), but Reagan pushed the measures to the limits. In fact, Reagan (and the Laffer curve) slashed the top tax rate from 70% to 50% and then to 28%, which *increased* the federal deficit more than 3 times from $900 billion to $3,000 billion (Wheen, 2004, p. 19),[6] resulting in "one of the most spectacular expansions of state indebtedness in world history" (Arrighi, 2010, p. 327). It seems the Laffer curve wasn't that funny, and something was very wrong with the math.

Wheen (2004) further noted that during Reagan's presidency, the United States "was transformed from the world's biggest creditor into the biggest debtor—and tripled its national debt for good measure" (p. 30), adding that "tax-cuts for the rich were central to the supply-side superstition" (p. 19). In comparison, the top tax rate prior to FDR's New Deal was only 24%, increasing to 63% and then 79% during his tenure before hitting 91% at the height of the Cold War in the mid-1950s (Krugman, 2009, p. 47).

There have been only three significant post-World War II increases in the federal debt as a percentage of GDP from beginning to end of a presidential tenure, as shown in Figure 9.1: Ronald Reagan (33.4–51.9%), George Bush Sr. (51.9–64.1%), and George Bush Jr. (58.0–70.2%). In the only other rise, Gerald Ford left the books only slightly more indebted (33.6–36.2%); thus, contrary to fiscal theory, Republicans increase and Democrats decrease indebtedness.[7]

Highlighting the seemingly fiscal recklessness of business-first politicians, President George Bush Jr. was as much a spender as a slasher, dealing a potent double whammy to the government coffers. Referring to George Bush Jr., his press spokesman and former George Bush Sr. speechwriter Tony Snow (2003) noted: "When it comes to federal spending, George W. Bush is the boy who can't say no. . . . The president doesn't seem to give a rip about spending restraint." During George Bush Jr.'s tenure, the national debt doubled, aided in part by preferential tax cuts for the rich, which cost the American taxpayers $2.74 trillion according to one calculation (Johnston, 2010).

Living beyond one's means is a problem, as any working person knows. With many more tools at its disposal (printing money, selling

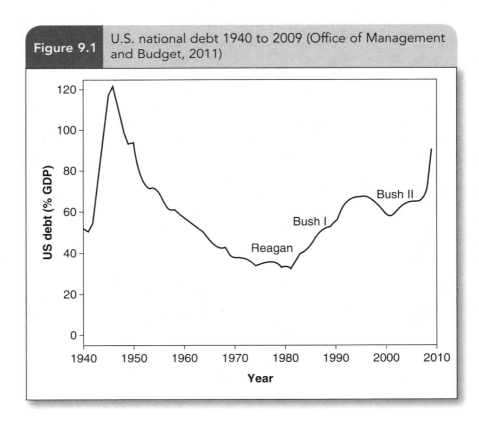

Figure 9.1 U.S. national debt 1940 to 2009 (Office of Management and Budget, 2011)

bonds, collecting taxes), a government living beyond its means is a problem only if the debt becomes unserviceable and grows so large that the tax take is used to pay off more interest than principle (i.e., essentially insolvent). What's more, growth becomes unsustainable when debt/GNP ratios become too high—for example, above 90%—or, worse in emerging economies, when the external debt/GNP exceeds 60% (Reinhart & Rogoff, 2010).

To quell the financial crisis upon his arrival in the White House, President Obama's first budget was the largest in American history, containing nearly $4 trillion in spending, raising taxes, and substantially increasing the national debt. The idea was to keep the damaged economy out of depression followed by a projected return to a more generally accepted annual deficit of 3% of GDP when the crisis blew over. Nonetheless, the national debt, which was $12.3 trillion or almost $40,000 per citizen at the start of his tenure in 2010, rose by $5,000 per person in his first year alone, while the total debt (national plus personal) was $39.9 trillion or $130,000 per citizen.[8]

Alas, we are at a point now where the debt load is becoming unsustainable (now more than 60% of GDP in the United States). What began in earnest with the Depression-era policies of John Maynard Keynes—who advocated spending to stimulate the economy and the new idea of an economy that could work at less than full employment, which "became, as ever, the new conventional wisdom" (Galbraith, 1958/1999, p. 139)—is not working. Although Keynes also advocated reducing the budget deficit when output increased, we seem to be caught in a permanent spending loop, an almost perverse form of the doubling game where one pays more but never gets ahead.[9] And things don't seem to be getting any better, especially when partisan politics make sensible reduction plans difficult, if not impossible.

The problem of increasing debt has become especially dangerous in Europe—as first seen in Greece, which required an emergency EU guarantee to cover its more than €20 billion in maturing bonds after repeatedly operating its budget at greater than the EU-stipulated 3% deficit as percentage of GDP (Germany and France had also exceeded the deficit limit, although to a lesser degree; Maddox, 2010). What's more, the Greek debt is likely worse than it appears because of derivatives trading that helped the government massage its accounts: "One deal created by Goldman Sachs helped obscure billions in debt from the budget overseers in Brussels" (Story, Thomas, & Schwartz, 2010). The problems may be even worse, given the systemic corruption within the country, as noted by its own finance minister (Mason, 2009, p, 205). In the end,

Greek bailout loans amounted to more than €130 billion. Furthermore, other peripheral countries became engulfed in the same sovereign debt crisis. To keep afloat, Ireland required an EU and International Monetary Fund bailout of €90 billion, as did Portugal to the tune of €78 billion.

In effect, all three countries were insolvent and needed to "restructure" or "refinance" their debts with reduced rates, longer repayment terms (which Ireland eventually secured), or bondholder write-downs (which Greece eventually secured), as well as implementing massive public spending cuts, including raising the retirement age at some point. The other alternative is to default, at great cost to international business and reputation—although, in the harsh reality of the European Central Bank's cost-benefit analysis, the cost of defaulting could far exceed that of imposing severe cutbacks (Kaletsky, 2011a).

Contagion to other EU countries (such as France, Spain, Italy), and, indeed, to the United States, has kept government leaders in crisis talks, hoping to avert a further meltdown to the world banking system. Furthermore, the real figures may be worse still, because major debt obligations are being placed in "off-balance-sheet entities," primarily pensions, "thereby masking the true position of the government debt burden" (Moyo, 2011, pp. 67–68, 82). As Moyo noted,

> This form of accounting trickery, by which the true value of the future pension obligation is set aside, usually as a mere footnote to the core balance sheet, enables governments to distort the figures they publish and conceal the true value of the debt burden that the public are carrying. (p. 82)

Clearly, no country can continuously increase debt as though no ultimate day of reckoning will come. To be sure, debt is fast becoming an all-consuming issue, seemingly tolerable when expressed as a percentage of GDP, as done on most government fact sheets. When expressed per capita or convolved to include a country's aging demographics, however, the numbers aren't so tolerable. The U.S. debt as a percentage of GDP (63%) ranks 29th out of 134 countries surveyed, but when expressed per capita (29%) ranks 10th (see Table 9.1). And U.S. debt is getting worse by the day because of an increased cost of repayments (a positive feedback adding more debt to the amount already owed).[10]

When demographics are factored in, countries with ageing populations will be especially vulnerable, and as the working population shrinks in coming years, the debt burden will grow even more. In the United States, the percentage of new retirees (i.e., those within a 5-year range from age 65 to 69) will increase from 4.1% in 2011 to as high as

Table 9.1	G8 and other select national debts (Central Intelligence Agency, 2011)				
Country	GDP (Trillion US$)	National Debt (Trillion US$)	National Debt (% of GDP)	National Debt per Capita (Thousand US$)	Percentage Population Over 50
Japan	4.31	8.61	199.7 (2)	68.1 (1)	43.5
Singapore	0.29	0.31	105.8 (9)	65.1 (2)	30.1
Iceland	0.01	0.01	126.3 (6)	48.0 (3)	30.2
Greece	0.32	0.45	142.7 (4)	42.2 (4)	38.8
Italy	1.77	2.11	119.1 (8)	34.6 (7)	39.5
Canada	1.33	1.12	84.0 (15)	32.8 (8)	36.7
Germany	2.94	2.45	83.4 (16)	30.1 (9)	36.1
USA	**14.66**	**9.22**	**62.9 (29)**	**29.4 (10)**	**41.0**
France	2.15	1.77	82.4 (18)	27.1 (12)	32.1
UK	2.17	1.65	76.1 (23)	26.4 (14)	34.9
India	4.06	2.05	50.6 (50)	1.7 (96)	16.7
Russia	2.22	0.20	9.0 (123)	1.4 (100)	34.1
China	10.09	1.64	16.3 (116)	1.2 (106)	35.5

7.1% in the following 10 years. In fact, a steady increase in new retirees may be expected for a number of years to come—as can be seen in Figure 9.2, which shows the distribution of ages in the United States, United Kingdom, China, and India—likely precipitating an increase in the retirement age in each country as the money runs out (which will also create an instant "brain boom" as fewer jobs become available and more undergraduates choose higher education). At the same time, the cost of Social Security benefits in the United States will rise in the next 20 years from 4.8% to 6% of GDP because of the ageing population (Krugman, 2010b).[11]

Not only is debt becoming bigger, but debt management is becoming more unbearable. The U.S. debt may have risen by $5,000 per person in 2010 but increases every day by more, causing renewed problems for Congress's built-in debt limits and future American solvency, as witnessed by Standard & Poors' downgrading of U.S. long-time credit worthiness to negative for the first time.[12] What's more, the future value of such overly extended investment planning is crippling the ability to pay back what is owed.

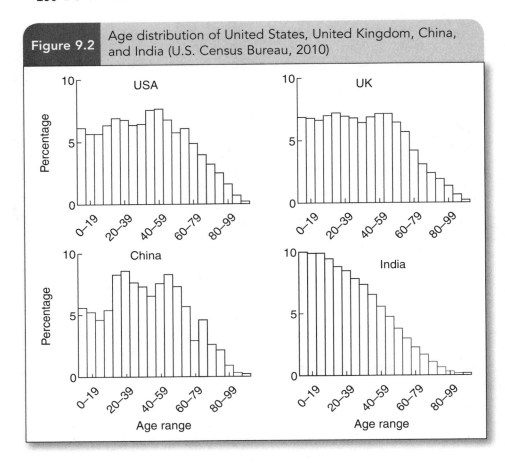

Figure 9.2 Age distribution of United States, United Kingdom, China, and India (U.S. Census Bureau, 2010)

When the weight of the debt pyramid becomes too great, wholesale relief will be required, if not wholesale restructuring of how taxes are collected and the basic idea of one's working life and who owes what to whom. As noted with personal debt, however, the system is a closed one, and so one must ask, to whom is all this debt owed? Certainly not to the regular citizen.

When asked about debt, most will cite bondholders as owners, i.e., the wealthy among us (including demonstrably undemocratic foreign governments), the interest payments for which are unevenly paid in taxes by regular citizens. It is interesting that debt relief on behalf of the poorest African nations, where tens of billions of dollars have been wiped off the slates, may soon be practiced closer to home. Such an amnesty, or "bondholder haircut," will help keep national coffers from an ever-worsening financial blight.

None of that is likely to happen anytime soon, however, and the very rich will continue to become richer at the expense of taxpaying citizens the world over. But the cracks are beginning to show, and the problems of excessive debt repayments are beginning to reveal both an inability to pay and an increased economic divide. What's more, uneven tax rates do little more than transfer money from the regular taxpayer to an already comfortable and finance-obsessed high-wealth group—all one would expect in any competitive game.

9.6. More Zero-Sum Losers and Winners: User Pay Versus a Shared Society

So what is the problem, then, simply put? Economics is an expression of collective common sense. It shouldn't be that hard. The main goal of government should be equitably distributing the wealth of a nation as well as providing common security. Why else would anyone want to belong to a society if not to share in its labors?

But balancing the public and the private is not easy. What if a taxpayer wants an income tax rebate because he doesn't have children or own a car? Should he pay for schools or roads? Few would disagree that schools and roads are in the public interest and that the cost should be shared by all, even if unused, but why not housing or health insurance? Furthermore, who decides what is in the public interest? ExxonMobil? Enron? The econometricians?

An excellent example of shared use is the local gym, where membership entitles one to use all the equipment, which would be idle most of the time if privately owned (Heath, 2009, pp. 86–87) and would no doubt be in the way, as in the bulky rowing machine in the living room. Libraries, parks, beaches, and all sorts of seemingly "uncommodified" places fit the bill, where we pay collectively for shared use. Heath called it "cross-subsidization," noting,

> Efficiency gains arising from the collective purchase (that is, the formation of an optimal sharing group) are sufficiently great that they outweigh the losses caused by the bundle of goods being less tailored to the needs of the individual consumer. (p. 87)

As such, taxes are nothing more than a "club fee" or "collective shopping," and shouldn't be begrudged since we would be paying for them individually anyway if they weren't imposed, and much less efficiently

(Heath, 2009, pp. 89, 90). How would it be if everyone were charged with providing their own library, park system, or police force to replace existing public services?

But not all such public services make sense. Food is a great example of a service not provided by the state, because there would be a disincentive for individuals to eat properly and such a system could easily be abused. But housing, less so, or social insurance and pensions that provide peace of mind about the future. Many shared services make sense and must be managed by a public trust if private industry is unwilling or incapable of fairly providing essential services and satisfying the public need.

However, many industries are economically subsidized and preferentially advantaged by tax breaks or pay no tax. Film companies are routinely supported at taxpayer expense, where subsidies can cover up to 40% of the film, as in a recent George Clooney film shot in Michigan (Cieply, 2011). Highlighting the absurdity of competing global subsidies, film companies are given preferential tax breaks to set up in foreign locales, where as a result, "the U.S. economy has lost some 47,000 jobs per year and an estimated $23 billion in economic benefits related to the production of theatrical length films alone since 2000" (Katz, 2006). Other companies think nothing of picking up lock, stock, and barrel to find the best breaks going, as we saw when Dell moved from Ireland to Poland in search of lower wages to better their bottom line.

Even more absurd is the economic trade-off between better subsidies and preferential wages, where many companies choose to operate elsewhere because of better tax rates—for example, in Ireland, which has the lowest corporate tax rate in the eurozone (12.5% compared with 40% in Germany and more than 30% in Britain), prompting one journalist to note that Ireland is in essence "a laundering operation for multinational industry in order to avoid tax" (O'Mahony, 2000). It is estimated that Google, which set up its European headquarters in Dublin, avoids paying hundreds of millions of dollars annually by being Irish registered (Watts, 2009).

And even though Ireland is seen as a tax haven to some who base their businesses there, others find even better tax-avoidance schemes elsewhere, such as the high-minded rock band U2, who after a lucrative artist subsidy was discontinued shuttled an estimated €30 million to the Netherlands, where they pay almost no tax on worldwide royalties (Coyle, 2010). Furthermore, when all allowances are accounted for, France may have a lower corporate tax rate than Ireland (Wighton, 2011). And, of course, Swiss banks and the Cayman Islands exist primarily to shelter money from the tax man, estimated by the Tax Justice Network to be worth more than $250 billion (Coyle, 2009).

What's more, disparate tax rates apply, permitted by various loopholes. In the United States, billion-dollar-per-year earners are able to pay tax at 15% instead of 35% by listing their earnings as capital gains (Krugman, 2009, p. 250). That's $6 billion less in total for the so-called "hedge fund loophole," which Krugman noted roughly equals health care for 3 million children (p. 250). Corporations can also list profits in lower-tax jurisdictions overseas, at an estimated cost of about $50 billion per year compared with the public returns (p. 258). Kocieniewski (2011) noted that in 2010 General Electric (the largest corporation in the United States) made profits of $5.1 billion in the United States yet paid no taxes: "In fact, G.E. claimed a tax benefit of $3.2 billion." In a collectively shared society, how is that possible? It seems that some of us are paying less into the collective pool than others.

But whose world is it? Does it belong to a select group, preferentially treated by business-oriented governments, or to the rest of us, who end up paying the bills, mostly in the form of tax (now at a premium to cover an enlarged debt)? Peter Vollmer, who is part of a millionaire's group advocating that the rich pay more taxes, noted that "the most decisive split is the one that exists between those who feel that wealth is a social responsibility and those who don't" (Marsh, 2009b). The Patriotic Millionaires for Fiscal Strength group in the United States, which includes ice cream kings Ben Cohen and Jerry Greenfield, echoes these sentiments and wants President Obama to stick to his promise to raise taxes: "In our nation's moment of need, we are eager to do our fair share" (Philp, 2011). Billionaire businessman Warren Buffett has made numerous pleas to the same effect. For some, how to spend the government coffers involves a more equitable arrangement than currently practiced.

Just as worrying are massive bonuses (some designed for simple tax avoidance), because such money paid to executives is then *not* paid to other employees or investors farther down the "trickle-down" spigot.[13] Smith (2009a) noted that bonuses in the U.K. financial services industry more than tripled to £16 billion from 2001 to 2008, with an estimated 4,000 city workers receiving at least £1 million. Furthermore, that some are ranked so highly by their obscene remuneration, incommensurate with ability, sends out the message that other jobs are much less valuable. Worse, the bonus culture itself is not geared to performance but to maximizing balance sheets, blamed by some as a start to the credit crisis (which we'll look at more in Chapter 11).

Some also question a banking bailout funded by taxpayers that leaves less in the pockets of regular citizens while freeing banks to continue business as usual, a so-called moral hazard[14] that socializes the costs but privatizes the benefits. Jackson (2009) noted that the U.S. banking bailout was financed with a "staggering $7 trillion of public money—more than the GDP of any country in the world except the U.S.—to secure risky assets, underwrite threatened savings and recapitalize failing banks" (p. 19). The world banking bailout has been pegged as high as $15 trillion (Mason, 2009, p. 53). Yet in 2010, Goldman Sachs[15] paid out almost $3 billion in bonuses after receiving a $6 billion government bailout (Jackson, 2009, p. 20). The Royal Bank of Scotland, which became almost wholly owned by the British government as a result of the crisis, posted operating losses of £3.6 billion in 2010 yet still paid £1.3 billion in bonuses (Griffiths, 2010).

As if to rub salt in the wounds, downturns can even benefit the well-off by penalizing others more, such as through income levies, a regressive solution to the problem of sharing the load. During the 1930s, many on fixed income did very well throughout the decade-long depression, and after the 2009 downturn, it was noted that the world's wealthiest had $2 trillion more to invest than before the crash (Jackson, 2011). It seems that the rich get richer in bad times as well as in good, although, as Frank and Cook (2010) noted, "a greater tax burden on the economy's biggest winners would not only help set our financial house in order but would also help steer our most talented citizens to more productive tasks" (p. 231).

Things may be changing, however, if President Obama's comments at the Pittsburgh G20 summit are anything to go by, where he addressed the needs of workers in relation to those of big business, as well as endorsing bottom-up solutions: "This is America. We don't disparage wealth. . . . But what gets people upset—and rightfully so—are executives being rewarded for failure. Especially when those rewards are subsidized by the U.S. taxpayers" ("President Obama's Remarks on Executive Pay," 2009). In 2010, he also pushed through legislation on much-needed Wall Street reform, which included financial early warning systems, creation of a new Bureau of Consumer Financial Protection, and the power to dissolve banks rather than bail them out (as would apply according to standard free-market principles).

Could it be that the balance is shifting because the economically well-off have helped themselves to too much of the cake and driven the economy into the ground? Could it be that too much wealth in the hands of too few people is counterproductive? Could it be that the tax take has been unfairly portioned? Is it time to recognize the failure of trickle-down thinking and the real cost of all this economic game playing?

9.7. The Hot 100: Are the Rich Getting Richer?

As a measure of success in pop music, best-selling charts have been going strong for more than 50 years, with the likes of Elvis Presley, who topped the American charts for 24 weeks in 1956 with four different No. 1s—"Heartbreak Hotel" (8 weeks), "Hound Dog" (5 weeks), "Don't Be Cruel" (6 weeks), and "Love Me Tender" (5 weeks)—and again in 1957 with two more—"All Shook Up" (8 weeks) and "Teddy Bear" (7 weeks)—and the Black Eyed Peas, who topped the charts for a record 26 straight weeks in 2009 with their back-to-back No. 1s, "Boom Boom Pow" (12 weeks) and "I Gotta Feeling" (14 weeks). Other notables such as The Beatles' "Hey Jude" (9 weeks, 1968), Debby Boone's "You Light Up My Life" (10 weeks, 1977), Whitney Houston's "I Will Always Love You" (13 weeks, 1993), and Mariah Carey and Boyz II Men's "One Sweet Day" (16 weeks, 1996) have all, in their time, been best-ever chart toppers.[16]

But can such data tell us anything about changing styles, differing attitudes, or even greater wealth disparity? From the numbers, one can certainly calculate how long a No. 1 stayed on top, which may reveal something about the music business since time at the top is a function of sales. From there, we may be able to comment on how competition and fairness in the marketplace change over time, trying not to get too bogged down in the subjective study of likes and dislikes or generational differences.

To start the analysis, the weeks at No. 1 on Billboard's Hot 100—which officially began on August 4, 1958—are plotted for every chart topper for 50 years as an amalgam of record sales and DJ plays (see top graph in Figure 9.3). That first week, Ricky Nelson's "Poor Little Fool" was No. 1 for 2 weeks before being knocked off by Domenico Modugno's "Volare." Before the year's end, Conway Twitty (2 weeks), The Kingston Trio (1 week), The Teddy Bears (3 weeks), and even The Chipmunks with David Seville (2 weeks) all had No. 1s. More than 2,500 weeks later, the list has seen out 10 presidents, four wars, and more than 1,000 different No. 1s.

Looking also at the calculated mean time at the top versus year (see bottom graph in Figure 9.3), some obvious differences are noticeable. First, the data is significantly more spread out since the 1990s, where a more recent No. 1 stays No. 1 longer. This can be seen in the number of No. 1s with longer than 10 weeks at the top—15 after 1990 compared

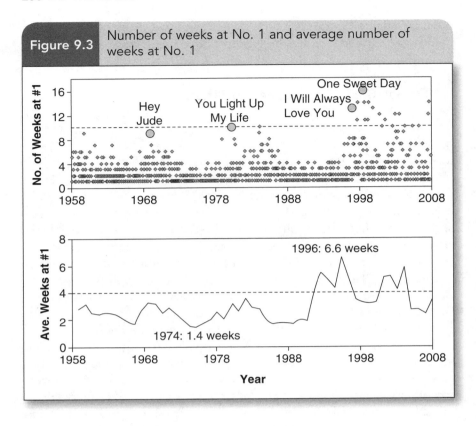

Figure 9.3 Number of weeks at No. 1 and average number of weeks at No. 1

with none prior to 1990—and by the higher average time at the top—all less than 4 weeks prior to 1990. In fact, prior to 1990, only five songs made it longer than 8 weeks at the top ("Theme From a Summer Place" by Percy Faith, "Hey Jude," "You Light Up My Life," "Endless Love" by Diana Ross and Lionel Richie, and "Physical" by Olivia Newton-John) compared with 24 after.[17]

Highlighting the lesser-performing No. 1s prior to the '90s, the years 1974 and 1975 saw 37 different No. 1s, with such immortal hits as "Hooked on a Feeling" by Blue Swede, "Can't Get Enough of Your Love Babe" by Barry White, and "Thank God I'm a Country Boy" by John Denver, all chart toppers for 1 week along with 51 others that ruled the roost in those 2 years. Furthermore, the average time at the top for each No. 1 went from a low of 1.4 weeks in 1974 to a high of 6.6 weeks in 1996, a year that saw only eight different No. 1s, including the longest-ever chart topper "One Sweet Day" (16 weeks). The average for weeks at the top per decade bears out the same conclusion: 2.0 in the 1970s and 3.6 in the 1990s.

One thought as to why this should occur is competition, measured here by diversity. Perhaps there was more competition in the '60s and '70s with lots of new pop songs and start-up bands, from the usual music industry stalwarts to any number of independent groups or one-hit wonders. One could also argue that songs today are better, although, without being too subjective about changing tastes, that seems unlikely as well as a much harder case to make. That the '60s and '70s fostered more diversity is beyond doubt, despite the perceived notion that modern independent bands abound today. The present notion of the independent music band seems to be a myth, at least when judged by sales.

More likely, the industry publicity and business machine has become more professional and better at promoting new stars, such that fewer choices are available to pop-music buyers today, typically teenagers, who are more susceptible to peer tastes and what's "in." Frank and Cook (2010) also noted that a "winner-take-all" market is at work, which favors winners through an iterative feedback loop (p. xvii; something we'll look at more in the next two chapters). In fact, they noted that "the top-selling 200 digital musical tracks on Amazon had a market share of 18.7 percent in 2008, up from only 14.5 percent in 2004" (p. xvii). Clearly, the music industry is very good at selling and reselling, an idea reinforced by the Hot 100 itself and other manufactured charts, such that one could ask whether the real No. 1 is the music business itself.

Interestingly, since the advent of TV talent shows such as *American Idol* and *X Factor*, winners who have made their way to No. 1 don't stick around for long (the first winner, Kelly Clarkson, is tops, maxing out her chart moment in 2002 at just 2 weeks), suggesting that prefab music isn't good enough, despite the slick industry machinery working to start such new stars off on their merry, chart-topping ways. But, at this point, we are wading into the grey area of accounting for tastes, a dubious practice at best, for which we could end up forever discussing the relative merits of Elvis Presley versus the Black Eyed Peas or Mariah Carey versus Whitney Houston, when all that can be said is that the Black Eyed Peas and Mariah had longer-lasting No. 1s than did Elvis and Whitney. As far as sales go, it little matters if one prefers "Time in a Bottle" (Jim Croce, 2 weeks, 1974) or "Money Maker" (Ludacris featuring Pharrell, 2 weeks, 2006).

The Hot 100 Countries?

Moving to a similar analysis of GNP per capita, however, we may be on surer ground. This time, instead of No. 1s (topped by Luxembourg or Switzerland over the past 50 years, followed by the usual high-GNP

Western nations), we look at the top performers as a percentage of total world GNP (University of California, 2003)—the "Hot 100" of countries, if you will. Not unlike the music Hot 100, the idea here is to see if wealth has become more, or less, concentrated in a few countries—i.e., are the rich countries becoming richer.[18]

Interestingly, it turns out that the rich countries aren't becoming richer, relatively speaking (i.e., by percentage of total world GDP), but certainly the poor countries are becoming poorer—much poorer. While total world GDP increased almost fourfold from 1960 to 2000 and average GNP per capita almost doubled over the same period ($3,099 to $6,280), the 10 poorest nations' share of the pot dropped from 0.84% to 0.21%, which is a fourfold decrease during a time when GNP quadrupled—a staggering statistic that dramatically highlights increasing relative world poverty. Apparently, the world is not flattening out with globalization, as some believe, but becoming much more unequal.

Some of this can be accounted for by inflation from 1960 to 2000, but there is no dismissing the relative sixteenfold decrease in wealth percentage in low-GNP countries while others have enjoyed a relative boom. It is even more staggering that just 14 countries account for more than 50% of total world GDP while comprising less than 15% of world population.[19]

To determine if the rich *within* a country are getting richer, however, we must look at individual wealth data to see how much more the top tier receives from the overall pot. Individual data is harder to obtain, but snapshots can be found in places such as the *Forbes* rich list, which states that the richest Americans have gotten richer every year but five since 1982 (Miller & Greenberg, 2009).

The distribution of American household incomes is also available through the U.S. Census Bureau, as shown in Figure 9.4, where the increasing wealth divide can clearly be seen. Note that the highest fifth now make more than half the income.

Figures quoted by economists tell a similar tale. Incomes of the top 1% of Americans more than doubled between 1977 and 1989, although the median income rose less than 7% (Krugman, 1992, p. 54). Since 1973, the top 0.01% incomes have increased by 7 times and the top 0.1% by 5 times, primarily as a result of paid compensation (stock options, bonuses, etc.; Krugman, 2009, pp. 129–130). CEO pay increased from 35 times that of an average manufacturing worker in 1974 to 400 times today (Frank & Cook, 2010, pp. xii, 67).

Heilbroner and Thurow (1998) noted, "In 1976 the top 1 percent of families in the nation owned 22 percent of all family wealth. In 1992 the top 1 percent owned 42 percent of all family wealth" (p. 192), a figure

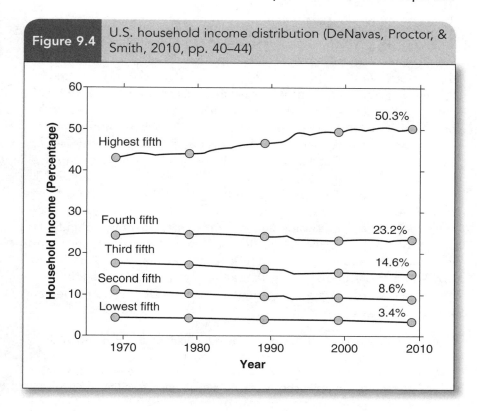

Figure 9.4 U.S. household income distribution (DeNavas, Proctor, & Smith, 2010, pp. 40–44)

comparable to that of the 1920s. They added that the United States had "the dubious distinction" (p. 193) of having the largest difference between rich and poor in the world (comparing the top decile to the bottom):

> In Finland . . . the total income of the top 10 percent was 2.7 times greater than the total income of the bottom 10 percent. In Norway it was slightly more—2.8 to 1. In the Netherlands it was 2.9 to 1; in Canada, 3.8 to 1. In the United States it was 5.7 to 1. (p. 193)

Today, the figures have become even more disparate. Davies, Sandström, Shorrocks, and Wolff (2008) reported that 2% of adults own more than half the world's wealth, while the bottom 50% own only 1%. Furthermore, 10% own more than 85% (Brown, 2006). Half of all American wealth is also inherited (Heilbroner & Thurow, 1998, p. 187).

Simple addition shows that the net worth of the *Forbes* Hot 100 tots up to a staggering $774.9 billion ($7.749 billion per capita!), which is equivalent to a country with a population of more than 123 million people earning the 2000-calculated average GNP of $6,280 ($774.9 billion/ $6,280 = 123,390,000). Others have noted similarly oddly weighted inequalities: Michael Portillo noted that America's top 74 citizens equal the bottom 19 million (Kaletsky, 2010c), and Atwood (2008) pointed out that the world's 25 million richest equal the 2 billion poorest (pp. 193–194), although the difference is likely worse given that 1.3 billion people live on less than $1 per day (Wheen, 2004, p. 238). Wheen noted that the world's poorest 10% lived on 72 cents a day in 1980, and the amount had risen a paltry 6 cents 20 years later, which he noted "hadn't even kept pace with inflation" (p. 261).

Heilbroner and Thurow (1998) further noted the "disturbing things that are going on in the American economy these days" (p. 7), especially with regard to a widening income gap: "To put it bluntly, the stream of incomes going to the richest families has been growing rapidly, while the stream going to the less than affluent has been falling" (p. 7). They noted how circumstances have dramatically changed since the formation of the state and how an attitude of making money any way possible now prevails:

> In the days of our Founding Fathers a hundredfold difference between one person's income and another's would have been regarded as a gross violation of ethical norms. By the mid-nineteenth century, the enlarged scope and growing personality of economic life steadily increased the acceptance of, even the admiration for, high rewards for economic success, as long as that success was attained by legal means. During the roaring twenties the sky was the limit, no questions asked. Something like this state of mind seems to have reappeared in our time: when the CEO of AT&T improved the profitability of his company by letting forty thousand employees go, his grateful stockholders rewarded him with a bonus of $15 million. (p. 192)

The Beatles may have sung "Can't Buy Me Love," but it seems that money can make a lot of "Daydream Believers," "Suspicious Minds," or, if one so desires, a rather large "Bridge Over Troubled Water."[20] In the words of another top band, although not reflected as such in the Billboard Hot 100, "Money, it's a gas."

9.8. The Fictional Commonwealth: You, Me, or Us

It is hard to convince people that excessive wealth is counterproductive to a prosperous society, which many think is solely an extension of their own ideals and goals. As Schumacher (1993) noted,

Now, one does not have to be a believer in total equality, whatever that may mean, to be able to see that the existence of inordinately rich people in any society today is a very great evil. Some inequalities of wealth and income are no doubt "natural" and functionally justifiable, and there are few people who do not spontaneously recognize this. But here again, as in all human affairs, it is a matter of scale. . . . They corrupt themselves by practicing greed, and they corrupt the rest of society by provoking envy. (p. 236)

It could be just a matter of scale, as in when is enough enough? Is one car enough? Two? Three? Ten? Is one house enough? Two? Three? Ten? A limit of 10 rental properties? Is $1 million a year enough to live on? What about $1 billion? Should there be limits to ownership, particularly if it is causing more harm than good? And is the measure of a human determined solely by wealth, as in one big children's board game? Wealth distribution is no doubt skewed, and a more equitable means of remuneration is needed to reset the balance.

If remuneration were directly related to training, skill, commitment, merit—all the things we think it should be connected to—one could argue for better pay for accountants than, say, nurses (and a trickle-down system). But if such attributes are not applicable, the argument fails. A bank manager should certainly make more than a bank teller—but how much more? Double, 5 times, 100 times, 3,000 times, as did the Barclays Bank president, who received £63 million (more than 3,000 times the pay of a front-line teller)? Even the U.K. business secretary branded that unacceptable: "He hasn't earned that money, he's taken £63 million not by building business or adding value or creating long-term economic strength, he has done so by deal-making and shuffling paper around" (Sylvester & Griffiths, 2010).

Prior to the $182 billion blowout at AIG—the largest bailout in American history—the head of the financial products division received $300 million in salary and bonuses over an 8-year period, primarily for engaging AIG in the ultimately disastrous world of mortgage-backed securities (Ng & Catan, 2010). But is anyone worth that much, especially when he or she ends up almost bankrupting the company? Or how about Wal-Mart CEO Lee Scott, who in 2005 as head of the largest corporation in the United States made more than 1,200 times as much as the regular nonsupervisory floor staff (Krugman, 2009, p. 139)?

Highlighting the reality of the zero-sum economic game, the starkest result of income inequality is that real incomes have been effectively reduced for all but a few, as we saw above. Krugman (2009) estimated that the reduction in the typical worker's income is about 35% since the 1970s

(p. 128). He noted that "only the top 1% has done better since the 1970s than it did in the generation after World War II" (p. 129) and that income gains for the richest 0.01% are 7 times those of 1973 (p. 129). At the same time, charitable giving for those on more than $500,000 per year decreased during the Reagan years by 65% yet increased by 62% for those making between $25,000 and $30,000, and those on less than $10,000 gave the most by percentage (Wheen, 2004, p. 29). These numbers by themselves refute the notion of a working trickle-down economic system.

Krugman (2009) also noted that a CEO today makes 300 times more than the average worker, compared with the 1970s, when he or she made only 30 times more (p. 136)—a particularly odious power law relation where lower-wage earners' incomes have increased little while higher-wage earners' incomes have increased much more. Over the same period, "schoolteachers have seen only modest gains" (p. 136). What's more, the average may have increased but only because the high end skews the numbers, another example of the mean not meaning very much. Krugman pointed out that just by walking into a bar, Bill Gates increases the *mean* but not the *median* income (p. 125).

The simplest understanding of the zero-sum game shows that we cannot all be billionaires and that wealth is relative, in the same way that a limited number of tourist dollars can't be spent in Mull *and* elsewhere. Bill Gates is rich because others are poor. It's not his fault; the system is designed to reward business success. But however hardworking or talented he may be or his wealth may suggest, he did not earn his fortune on his own. Even Gates himself recognized this imbalance, noting that "the most needy people get the least," and questioning, "Why do people benefit in inverse proportion to their needs?" (Charter, 2008).

Others argue that the system creates opportunists such as Gates on the backs of other less-business-minded creators (Barber, 2007, p. 71);[21] regardless, he could not have created or found his position without an already existing public structure, one created by years of public investment. As Schumacher (1993) observed, private wealth is made on the strength of public works:

> Private enterprise claims that its profits are being earned by its own efforts, and that a substantial part of them is taxed away by public authorities. This is not a correct reflection of the truth—generally speaking. The truth is that a large part of the costs of private enterprise has been borne by the public authorities—because they pay for the infrastructure—and that the profits of private enterprise therefore greatly overstate its achievement. (pp. 231–232)

However, the public keep paying the bills through regressive taxes, bailouts, and socialized costs. Heath (2009) noted that bankruptcy protection and limited liability can be thought of as "social programs for capitalists" (p. 42) that seek to protect the rights of lenders rather than borrowers and "leave 'society' to pay the bill for the improvident or foolish conduct of individuals" (p. 42). At the height of the banking bailout, trillions of dollars were funneled into private-sector relief, ultimately borne out by a greatly increased national debt—"socialism for the rich," as Stiglitz (2009) called it.

Barber (2007) noted that an educated public is required (even demanded) by business, although business pays little toward its expense or administration: "The American market system has become a paragon of how to socialize the costs but privatize the benefits of a supposedly private market economy" (p. 149). He cited an example of American pilots trained in the military at taxpayer expense who then move on to a welcoming commercial airline industry that has paid nothing for their costly professional development (p. 149).

But to whom does the world belong? We all stand on the shoulders of those who came before us so that we can see farther. Without the 1890 U.S. government census, there would be no IBM. Without the early 20th-century engineering pioneers of electronic amplification, there would be no radio, no Hollywood. Without Sputnik and the Apollo space program, there would be no personal computer. Without CERN and Tim Berners-Lee, there would be no Google or Sergey Brin or Larry Page or Facebook. The list is endless. Why, then, are such successes not shared and, worse, why are they turned into monopolistic profit-based ventures? Google is already in talks with the telecom giant Verizon to provide fast access to the Internet, available only on an exclusive user-pay basis. Coming soon, the two-speed Internet, a so-called doomsday scenario "that marks the beginning of the end of the Internet as you know it" (Foley, 2010).

As inequality increases and economies suffer, the questions that must be asked are, whose money is it, and why do some get such a heady tax-free view? It is the public kitty that ensures our collective wealth, without which none of us could survive in today's world, where taxes or development costs not paid by business don't end up benefiting the public either as infrastructure or shared services. And who pays for the public kitty?—not the untaxed wealthy. One must wonder why so few have so much (well beyond what they need) while so many have so little. Guns or butter? Tanks or levees? Bailouts or jobs? Ensuring a fair accounting is in everyone's best interest.

notes

1 **Early British national debt**: During the Seven Years War (1756–1763), the debt almost doubled from £74 million to £134 million (Ferguson, 2007, p. 34). As Ferguson noted, however, "the Seven Years War decided one thing irrevocably. India would be British, not French" (p. 35).

2 **Quantitative easing (QE)**: A new economic term to describe how a central bank buys debt using electronically printed money.

3 **What is a libertarian?**: A quiz by the Advocates for Self-Government aims to classify one as a libertarian, statist, liberal, conservative, or centrist based on five questions on "personal responsibility" and five questions on "economic responsibility" (see http://www.thead vocates.org/quiz). Accordingly, a libertarian is defined as one who advocates small government with no external regulations. Essentially, libertarians believe that the individual is supreme.

4 **Historical world powers**: The dates are not definitive and correspond to events marking roughly the beginning and end of each cycle, as noted by Arrighi and Braudel: 1450, the Age of Discovery begins; 1648, the Thirty Years War ends; 1628, the Mexican Silver Fleet is captured by the West Indies Company; 1784, the fourth Anglo-Dutch War ends; 1776, the United States Declaration of Independence is signed; 1917, World War I ends.

5 **Immoral imperialism?**: Ferguson (2007) wrote that Britain's empire "was paid for by British taxpayers, fought for by British soldiers, but benefiting only a tiny elite of fat-cat millionaires, the likes of Rhodes and Rothschild" (p. 283), noting that there were 72 separate military campaigns during the *Pax Britannica* of Queen Victoria's reign (pp. 255–256), with large populations wiped out across Africa (pp. 267, 280). But "most of the huge flows of money from Britain's vast stock of overseas investments flowed to a tiny elite of, at most, a few hundred thousand people. At the apex of this elite was indeed the Rothschild Bank" (p. 285). Ferguson further highlighted the mix of politics and money, noting that Disraeli, Randolph Churchill, and the Earl of Roseberry were all connected socially and financially with the Rothschilds (p. 285) and that Gladstone had invested in Turkish bonds while Disraeli had bought Suez Canal shares (p. 286). Given the height of its failures, including a British prime minister investing in foreign bonds and then overseeing the military occupation of Egypt, from which a "capital gain of more than 130 per cent" accrued (p. 287), one wonders if such imperialism was not just immoral but wanton exploitation.

6 **Republican debt increases**: In 2000 year dollars, the numbers are Ronald Reagan (33.2–50.5%), George Bush Sr. (50.5–60.6%), and George Bush Jr. (61.4–75%).

7 **Budget deficits**: The idea of a balanced budget was first discarded by Keynesian liberals to increase employment during the Great Depression in the 1930s (which substantially expanded and then contracted during and after World War II) and has since been a fixture of modern fiscal policy. Galbraith (1958/1999) noted that conservatives were first loath to abandon a balanced budget and a corresponding enlarged government (p. 138), an idea that did not extend beyond the times, as seen in the substantial increase in size of both debt and government during the Reagan, Bush Sr., and Bush Jr. presidencies.

8 **The U.S. national debt**: The debt was $12,292,059,103,725 at midnight on January 1, 2010, as seen on real estate developer Seymour Durst's national debt clock at Sixth Avenue in New York. Divided by a population of 310,232,863, this gave a debt of almost $40,000 per person (12.3 trillion / 310 million = $39,677). As of January 1, 2012, the debts had risen to $15.2 trillion ($48,700 per capita) and $56.4 trillion ($180,000 per capita). For continuously updated calculations of U.S. debt, federal spending, interest payments, and more, see http://www.usdebtclock.org.

9 **The world debt picture**: After the subprime crisis and near economic meltdown in 2009, the debt-to-GDP ratio is expected to rise to more than 100% by 2014 across the G7 nations (excluding Germany and Canada). In early 2010, the debt ratio in the G7 was already significant: Italy (117%), Japan (115%), Germany (76%), Britain (75%), France (73%), United States (67%), and Canada (31%).

10 **Debt figures**: The debt as expressed in Table 9.1 ($7.54 trillion) is from the *World Factbook* (Central Intelligence Agency, 2011) and thus differs from the debt clock number.

11 **Ratio of workers to retirees in the United States**: Martin (2000) noted that the ratio of workers to retirees was 3.25 to 1 in 2000 but is projected to fall to 2 to 1 by 2030 (p. 147).

12 **Rating agencies**: Rating agencies haven't been impartial since becoming for-profit companies in the 1990s. Furthermore, Kaletsky (2011b) noted that they "have had an almost unbroken record of sending false signals, especially in their analysis of AAA credits," not to mention guaranteeing bogus investments during the subprime buildup, which Mason (2009) likened to having an audience plant in a magician's trick (p. 93).

13 **Profits versus wages**: Former Australian prime minister, Paul John Keating, noted that the exceptional increase in productivity in the 1990s garnered not better wages but more profits (60% of which went to the top 1%), ultimately creating a watered-down workforce unable to pay its bills (Till, 2011, p. 227).

14 **Moral hazard**: A moral hazard exists when one operates knowing that any losses will be covered. Thus, a bank can lend recklessly (to high-risk customers while highly leveraged to increase profits) knowing that if their strategies fail the government will bail them out. As Mason (2009) noted after the mortgage lenders Fannie Mae and Freddie Mac and the investment bank Bear Sterns were bailed out, "The impression was developing that [the Fed chairman Hank] Paulson would, basically, save anybody" (p. 116).

15 **Goldman Sachs**: Goldman Sachs was a leading player in the original holding company doubling scam of 1929, contributing greatly to the stock market crash by buying its own stock to increase the value artificially, although it survived the carnage (Galbraith, 1992, pp. 86, 145). In the 2008 credit crunch, Goldman Sachs bailed out its own GEO fund when the deleveraging began, which may have stopped "a crazed rush to zero that could have put the entire financial system in peril" (Patterson, 2010, p. 240). For the most part, Goldman Sachs survived both disasters intact, although there has been much concern about its involvement in both.

16 **Music data**: All raw musical chart data is from Bronson (1992). Note that some No. 1s made an occasional reappearance at the top (although much less often than one might think) and are calculated as separate No. 1s.

17 **Soundscan technology**: In 1991, The Nielsen Company introduced better tracking technology to count unit sales in the music industry, using UPC bar codes and point-of-sale computers. Although such methods are more efficient, the effect on any changing buying pattern is thought to be minimal.

18 **Uncertainty in the method**: There is some uncertainty in this calculation because data was not available over the whole period (1960–2000) for every country.

19 **The GDP stakes**: The 14 countries, with GDP breakdown, are the United States (20.61%), Japan (6.02%), Germany (4.03%), United Kingdom (3.14%), France (3.03%), Italy (2.65%), Spain (1.98%), South Korea (1.89%), Canada (1.85%), Turkey (1.33%), Australia (1.20%), Taiwan (1.04%), Netherlands (0.96%), and Poland (0.94%), for an overall GDP tally of 50.67% for 1.02 billion (14.74%) people. All figures are calculated from the Central Intelligence Agency (2011) *World Factbook*.

20 **Money-themed songs?**: The Beatles' "Can't Buy Me Love" was top of the charts for 5 weeks in 1964, The Monkees' "Daydream Believers" for 4 weeks in 1967, Elvis Presley's "Suspicious Minds" for 1 week in 1969, and Simon and Garfunkel's "Bridge Over Troubled Water" for 6 weeks in 1970.

21 **Infrastructure for the masses**: Contracted to provide an operating system for the IBM PC, which he didn't have, Bill Gates bought DOS from its creator, who didn't share in Microsoft's great success. Microsoft similarly excluded browser maker Netscape from the computer market by bundling its own Explorer browser with other Microsoft software, for which Microsoft was charged with antimonopoly practices (although not before Netscape was significantly devalued). Andrew Carnegie bought up railway lines into Pittsburgh so that Carnegie Steel could exclude competitor access to production.

10

HOW TO GET ALONG

Be Nice, Clear Your Ice

Although mathematics is hard to avoid in today's world—especially in its many derived forms, such as engineering, economics, and statistics—not everything is about numbers. Morality, for example, seems outside the realm of mathematics, although in many ways even morality can be seen mathematically, with respect to strategy and maximizing one's chances for success. Should we let someone go ahead in line, look after only our own, attempt to win at all costs? These are some of the questions asked in this chapter, where we look at gas guzzlers, Olympic also-rans, feather-their-nest politicians, and aggressive advertising practices that serve to keep the world divided.

We also take a look at feedback loops and winner-take-all markets that turn too much of a good thing into a bad thing. We also gently lay the groundwork for better strategies of cooperation and community inclusiveness, rather than short-sighted and money-only exclusivity, which we will look at more in the next chapter. The non-zero-sum game is our ultimate goal, and the question we aim to ask in both chapters is, is what is good for me good for all?

As for the title of this chapter, it comes from an advertisement that encouraged people to clear their sidewalks after a snowfall to allow others to pass freely. In Toronto, one must clear the snow and ice from the sidewalk within 12 hours or face a fine, but this doesn't always happen. Since enforcing such a law is time-consuming and counterproductive, the government decided to try another tack and came up with the catchy slogan, "Be nice, clear your ice," hoping to appeal to one's better nature. The television advertisements featured local celebrities such as Wendel Clark, captain of the Maple Leafs hockey team; Lloyd Moseby, outfielder for the Blue Jays; and Ben Wicks, local newspaper cartoonist. Soon, the catch phrase beckoned as a popular reminder to all.

As I think of it 20 years on, the sentiment, "Be nice, clear your ice," embodies the simplest concept of civic-mindedness, community, and cooperation, suggesting even a transcendental understanding of one's place in the world. It is a simple statement of responsibility to one's neighbor, to one's self, and to a greater collective. This chapter, thus, takes a light-hearted look at some people who *don't* clear their ice, perhaps shedding some light on how we err on the side of self-interest rather than collective understanding or empathy. From there, we move on to a more formal introduction of the mathematics and theories of competition and cooperation, such as the prisoner's dilemma and the non-zero-sum game, which are both mathematical and moral expressions of inclusiveness.

10.1. Modern Etiquette, the Rise of the Road Hog, and Other Self-Interest Strategies

For many years, *The Times* etiquette writer Philip Howard kindly dispensed his how-tos of modern living, from the old standards—which knife goes where and what should one wear to the annual club function?—to the less familiar—can one use one's fingers when eating asparagus, thank a generous aunt by e-mail, or knock the stuffing out of a croquet ball in a backyard game among friends? As he recounted in one article, however, Howard (2006) got a little hot under the collar when he failed to negotiate a parking lot ticket machine with quick enough aplomb and was rudely told to step aside by an older man behind him (whom Howard referred to as the Commander) and his female companion (whom Howard took to be the Commander's daughter). In response to this affront, he temporarily lost the plot and told the impatient dolts behind him to do something he would never countenance in his role as an official manners man.

The point to the article was whether there are new manners and customs to learn with each newfangled contraption that comes along in our newfangled, contraption-filled world. And do morals change, or are good morals timeless? One might not think that the world revolves around such questions, but how we behave in a line, in rush-hour traffic, or in everyday interactions is an essential part of our society and presents a constant dilemma: whether or not to help when one sees another in need. As Howard routinely reminded us in his weekly column, custom and manners are society's grease and give both the elegant and less elegant a chance to behave acceptably without worry in the company of others.

Of course, good manners need never be reinvented, because the rules of decency apply as much today as in the days of hand signals and donkeys. Speaking loudly on a bus is no more acceptable now than before cell phones. Acting as though there are no other people in the world is selfish whatever the age. As for ticket machine queues in London or 7-Eleven store lines in Texas, one should never hurry or harangue another.

Are such interactions any different from the use of SUVs (sport-utility vehicles) or four-wheel drive vehicles on our roads? That they consume and pollute like tanks and contribute to a worrying increase in CO_2 emissions is an important concern, as is the damage they cause to the roads, but I suggest that the real dispute is still a case of old-fashioned manners. It is simply rude to drive such a big car (truck, tank, moon rover?) on the same road as less showy, more self-respecting cars. The SUV is a road hog and should be banned from the city—not to mention the military-like Hummers (which, oddly, receive a tax break for weighing more than 6,000 pounds). As Barber (2007) noted,

> The gloating Hummer owner may preen with macho pride, unaware or simply uncaring of the fact that he drives an ecological behemoth that squanders fossil fuel resources, pollutes the environment, and makes the United States more dependent than ever on foreign oil reserves—contributing quite inadvertently to the justification for Middle East military intervention he otherwise vehemently opposes. (p. 35)

Perhaps *Debrett's Etiquette and Modern Manners* said it best, and in a way that is most appropriate for selfish drivers of gas-guzzling SUVs who foist their ostentatious show-of-money lives on the world: "It is inconsiderate for a group of friends to walk abreast along a crowded pavement, thereby blocking the passage of others" (Howard, 2006).

But since it is also an issue of safety—which would you rather be driving in a head-on crash: an SUV or a mini?—and since only the more financially well-off can afford such safety, it is also an example of how ordinary concerns are excluded in our everyday world. Money and might is deemed right, where the only way to compete for your and your family's safety is to drive your own SUV, thus escalating the SUV arms race. With one in two new cars an SUV in the United States and one in four in Britain, the race has begun. Think globally and act locally? I don't think so. How insensitive. How rude. How inconsiderate.

Michael O'Leary, the CEO of the low-cost airline Ryanair, adroitly lampoons the do-good SUV drivers: "I smile at these environmental loons who drive their SUVs down to Sainsbury's on a Saturday morning. If you're concerned about the environment, stop driving. Aircraft account for 4 percent emissions in Europe, motor cars for 28 percent" (Billen, 2006). In fact, according to one study, the emissions breakdown is more like cars 41% and air passenger travel 7%[1]—but the point is well made.

Barber (2007) described it as "civic schizophrenia" (p. 128), a trade-off between first-order and second-order desires, where the infantile citizen wants something contrary to what is best for the community (pp. 221–222). Essentially, "toys for big boys," without any thought to others—as if "me first" is a strategy for better living. But we all know that the *way* one does something is as important as *what* one does. Don't we?

Barber (2007) further noted that first-order and second-order desires are where "the empire of impulse is allowed to trump the empire of the will" (p. 221), what can be called private versus public or me versus we (which we look at more below). In short, are we becoming a monoculture (essentially consumerist) instead of a balanced culture with shared, yet varied lifestyles? However hard that may be to obtain or maintain, what is culture if it cannot recognize the other?

Society is a simple example of rules that are designed for all but, in reality, can be achieved by only a few—whether larger and safer cars on our public streets, better policing and infrastructure in more affluent neighborhoods, or even the ability to pay a one-size-fits-all charge. Barber (2007) cited a seemingly trivial example: the rich having two cars—one with an even-numbered license plate and the other with an odd—to beat intercity restriction systems (p. 156), a clear example of a reverse democracy at work. As in our creeds for better living, the thought may be good but the practice less so.

So what can be done? A return to fair values and an understanding of manners or personal responsibility is a start. The bedrock of civility in society seems to have been chipped away so completely that no one questions bad manners anymore, from haranguing others in a line to hogging the road to the rich man thumbing his nose at us in his armor-plated SUV. Effective regulation also seems to be in order, especially restricting rising CO_2 emissions, which is increasingly important given the threat of global warming. Effective punishments also seem to be in short supply. In short, manners and civility help us ensure fair play.

But have we become stuck with the ways of a self-interested and less caring world? The manners man himself may have said it best in response to a reader who rather cheekily suggested that matters of etiquette are inconsequential and trivial:

The only "rule" of Etiket that matters more than a wren's feather is the Golden Rule: Treat others as you would be treated yourself. That is no triviality. It is the bedrock of society, civilisation and sane religion. (Howard, 2006)

How can we count up the costs of our actions without considering the moral in the equation? Right is right, no matter how hard it is to define the meaning of "right" or the collective understanding of responsibility, as we look at now in a few examples.

How to Win Friends and Become Infamous Against Your Better Judgment

The Olympics have been the backdrop for a number of events that could have brought lasting fame and glory to a few athletes if only they had seen past their self-interest. In three recent Olympics, judging errors provided the declared winners an opportunity to refuse their medals and thus act with humility and bravery, for which they would have been remembered for all time as heroes. None chose to do so and have been forgotten as holders of tarnished medals, highlighting how the path of self-interest is not necessarily the best or right choice.

The first case involved Kristen Babb-Sprague, an American synchronized swimmer in the 1992 Barcelona Summer Olympics. After another swimmer's performance in the preliminary round, a judge mistakenly entered an 8.7 into the electronic scorer (instead of 9.7). Although the judge declared the error immediately, she could not get the supervising referee to understand that she had made a mistake. She was not allowed to correct the electronic tally, and the result stood—a clear error against Sylvia Fréchette. And, thus, in the solo final the next day, Fréchette could not catch Babb-Sprague, who went on to win gold at Fréchette's expense. Had Babb-Sprague looked beyond winning a medal and spoken out against the mistake, however, she would be remembered today for true Olympian selflessness. Highlighting Babb-Sprague's failure to act, a year later, the International Olympic Committee awarded joint gold to Fréchette.

In the 2002 Winter Olympics in Salt Lake City, a figure-skating judging debacle led to a pair of Russian skaters winning gold over a Canadian pair who had clearly outperformed them. One judge confessed that she had been under "a certain pressure" to vote for the Russians. The situation is not exactly the same as the previous example in that no error was made, but the impartiality of the judging was clearly compromised (two judges were later suspended for 3 years and a Russian crime boss arrested on charges that he had fixed the event). But if the Russian skaters had stood up for all the world to see and said that they would not accept the medal no matter the difficulty or possible threat to their lives, we would hold them up today as models of fair play and honor, as great Olympians. Alas, no one remembers them. As for the hard-luck, second-place finishers, they were also belatedly awarded gold by the International Olympic Committee and feted by the likes of Larry King, Jay Leno, and Oprah (perhaps not such the great achievement it appears to be).

The last case involves the supremely talented American gymnast Paul Hamm at the 2004 Summer Olympics in Athens. A judging error was made in the men's all-around event, which incorrectly penalized Yang Tae Young of South Korea by 0.1 point. After Hamm's final performance, he was awarded the gold medal, having benefited from the difference. In this case, had the error been corrected prior to Hamm's performance, he might have won anyway, and some latitude is necessary in discussing the possible outcomes to this travesty. It is one of those hard-luck cases where no one is a winner. Nonetheless, the United States Olympic Committee suggested that Hamm voluntarily return the medal. Hamm refused and has been forgotten like all the others.

It is not easy to do the right thing, particularly when it contradicts the implied goal of competing. In all three cases, however, one would think the athletes' better selves would have found a way, at least, to offer to share the medals. Just think of the television deals and Wheaties boxes that would follow such a sacrifice. But we don't find a way. We typically think of ourselves first and foremost, whether for our own gain or for posterity.

The sporting world is rife with such examples, offering numerous chances to act with bravery instead of cowardice. Another apt example of choosing one's own goals at the expense of fairness comes from the world of international soccer. In the second leg of the 2010 World Cup playoff qualifier between France and Ireland, French team captain Thierry Henry handled the ball, which led to an extra-time goal. Despite the protests of the Irish team and seemingly irrefutable video evidence, the illegal goal stood and the French went on to win.[2] Worldwide condemnation followed for Henry and the French, who were labeled cheats for their underhanded means to victory. Henry was castigated and forced to apologize and call for a replay, which was not granted by the governing authority. Had Henry immediately owned up and refused to allow the goal, he would be recognized the world over as an exemplary sportsman. Instead, he is tarnished by allegations of foul play.

The "handball incident" also highlights a growing problem with ethics in general—that is, the refusal of a governing body or authority to accommodate exceptions to a one-size-fits-all policy. In this case, FIFA, the governing body of international football, refused to sanction a replay, despite incontrovertible evidence that the result was wrong and appeals from the likes of Arsène Wenger, the Arsenal manager and a French national. FIFA even refused to discuss the possibility of introducing video technology in future games, unconcerned that similar tragedies could occur. It was only a little ironic, then, when the French were victims of a clearly offside goal against Mexico in the ensuing World Cup group stages, which could have easily been corrected by video replay.

In other cases, airlines refuse to let passengers off a plane to get water or even to give out water during long delays, citing a policy beyond their means to change. Bus drivers refuse to let people off between official stops, citing safety regulations even in perfectly safe conditions. Junior players cannot join a community league because their entrance fee is a day late. Institutional inertia and malaise seem to be the institutions' masters.

To be sure, institutions are needed for organization, order, and maintenance, all the more so in fast-moving times, but they do not stand alone outside a collective moral basis. If they do, their only function becomes that of obtuse, legalistic, and at times inhuman systems that do not consider exceptions to their rules or the spread or variance in possible outcomes, contrary to their original purpose.

It is fascinating, then, when someone goes against type or party policy, seemingly for the sake of public betterment—for example, Nancy Reagan supporting stem-cell research, Dick Cheney advocating gay rights, or John McCain opposing torture. It is surprising how two high-ranking Republicans—one the wife of a president who had Alzheimer's, the other a vice president and father of a lesbian daughter who adopted a child—can espouse such anti-Republican-party ideals. But, of course, they have their own interests, where looking after the phylum one-ups party politics or a stated position (in McCain's case, he was tortured as a POW in Vietnam).

Happily, however, no one begrudges their efforts, and such behavior should be applauded regardless of the motive. Nancy Reagan, Dick Cheney, and John McCain should be thanked for drawing attention to their causes, although one wonders why it takes a family tragedy or personal circumstances to arouse our better selves. Surely, if the cause is worthy, it is worthy on its own.

The same can be said of one-issue candidates, who now and again enter and exit the political landscape: Ross Perot on business reform, Steve Forbes on flat tax, the American-born mother of nine Kathy Sinnott, who was elected to the European parliament on a campaign for better education for the disabled, including her severely autistic son. Another good example of a single-issue group is Mothers Against Drunk Driving (MADD), who courageously crusade for better laws against drunkards who get behind the wheel and kill children in resulting car accidents.[3] These causes are commendable, and it is natural to want justice for the loss of a child and to work to see that it doesn't happen to someone else.

After falling ill and requiring a new liver, Steve Jobs campaigned for a better organ transplant registry system and helped change some of the procedures. This is a good thing, someone advocating on another's behalf, although Jobs also highlighted the unfair side of the equation by moving to Memphis to receive a new liver, setting up a business there and buying a house to establish residency and then moving out after all was done—not at all what is intended in a registry system that hopes to match potential organ donors to needy recipients fairly.

The question here, however, is why such advocacy is not done by elected representatives, who *represent* on behalf of us *all*. To be sure, that representation has become skewed—where thousands will die on the road each year because the cost to government bent on preserving their mean voter base is too high, where important stem-cell research is held back because of poorly informed attitudes, where sanctions against so-called "too-big-to-fail" institutions are whittled down because of insider dealings. Doing the right thing is never easy, but at the very least it requires a basic understanding of one's place within a society, organization, or group, and not just of one's self—no matter who is calling the shots.

The Politician Who Came in From the Cold

We should be happy that promoting the welfare of others, whether motivated by tribal factors or self-interest, is not as blatantly self-interested as the feather-one's-nest or feather-one's-friend's-nest variety of promotion, typical of today's politicians. Michael Moore documented the backdoor connections of the Bushes and the House of Saud in *Fahrenheit 9/11* and of the White House and Wall Street in *Capitalism: A Love Story*. Who can count those costs? Amidst all Bill Clinton's White House shenanigans, his wife's brother was paid $300,000 for helping secure a presidential sentence commutation for cocaine trafficker Carlos Vignali, raising the question of why a president would want to pardon a drug trafficker in the first place. Billionaire Italian businessman-turned-prime-minister Silvio Berlusconi has been endlessly tainted by embezzlement charges and a host of domestic scandals (paying underage girls for sex, putting showgirls/girlfriends with no political experience on the election slate of his ruling People of Liberty party, obstructing justice, etc.).

The corridors of power are plagued with abuse. In Ireland, a government minister used a limousine to travel from one side of an airport to another at a cost of almost €500. In Thailand, a former premier abolished capital gains taxes so he could sell his business without having to pay capital gains of more than $1 billion. The indomitable Berlusconi doubled tax on a television service in competition with his own (Black, 2009), which also raises the question of how a prime minister can own a media company.

In the United States, one ex-governor tried to sell a senate seat (President Obama's vacant Illinois seat), another was mired in a prostitution scandal, and a lieutenant governor was on trial for misappropriation of funds. Numerous senators are also under investigation for sexual impropriety. In Britain, the ongoing scandals at Westminster take the cake for serving one's own interests and are an avowed embarrassment to politicians everywhere. During the Westminster expenses debacle, myriad trivial claims were made, including a kitchen refit (Tony Blair), a television subscription (Gordon Brown), and hundreds of sacks of gardening manure by one back-bencher, prompting renewed calls for reform (Riddell, 2009).

But, sadly, nothing seems to change. Even worse, the public rarely sees anyone go to jail. As President Obama expressed in response to two of his cabinet nominees failing to pay all their taxes, there should not be "two sets of rules: one for prominent people and one for ordinary folks who have to pay their taxes every day" (Couric, 2009). But one wonders, would Tom Daschle not only lose his nomination for Health and Human Services secretary but be penalized for failing to pay taxes on income in excess of six figures if there weren't already two sets of rules? Obama may have stated that "ordinary people are out there paying taxes every day" and that such actions send "the wrong signal" (Couric, 2009), but by extension, are the *extraordinary* people paying theirs? Who can count the costs of thousands of Americans who set up secret accounts solely to evade paying billions of dollars in taxes?

As if to turn the screw completely for avoiding accountability, the punishment seldom fits the crime. Fines related to the subprime meltdown were only a small fraction of holdings—for example, $75 million to settle charges brought by the Securities and Exchange Commission against Citigroup's $40 billion in total subprime holdings (less than 0.2%), a fine that was "roughly equal to one week's profits" (Rusche, 2010). Goldman Sachs paid only $550 million (or 2 weeks' profits) to settle charges that it sold subprime investments secretly designed to fail, while no senior executives were charged (Rusche, 2010). The investment bank is also being investigated for lying to Congress about its role in the financial crisis, where it claimed not to have held "big short" positions on the American housing market when it clearly did, as well as misleading clients about risk. But one wonders if any appropriate punishments will be handed out to redress the loss, not to mention how a bank can work for and against American interests at the same time.

Worse still, when Halliburton receives a government contract—untendered, at that, as in New Orleans (Barber, 2007, p. 158)—because of insider connections, another company doesn't. If financial managers reassure investors of the worthiness of their loans and company assets, yet cash in hundreds of millions of their own shares, then other shareholders are swindled (Rusche, 2010). If others receive backdated stock options to inflate their holdings, the regular investor is cheated. When even a president is in possession of illegally downloaded music—wantonly, it seems, as in the case of George W. Bush—the message becomes clear: The rules don't apply to all.

Why is this so? Perhaps Tony Blair was closest to the mark when he answered a student during a trip to China as British prime minister. The student asked why his wife made more money than he did, and Blair replied, "My wife is smarter than me, which is why she chose to go into law and not politics" (Sparrow, 2003), adding that he wished she made even more. Noting the number of rich participants at the World Economic Forum in Davos, he joked again that he had made the wrong career choice (Robinson, 2005).[4]

Democracy is full of leaders who show little concern for their constituents, the people each and every politician is meant to represent, but is it any wonder that governments fail in their public duty when their goals are so often directed toward a private duty? Nicolas Sarkozy may have summed up the sad reality of politics best: "The most important thing in democracy is to be reelected. Look at Berlusconi, he has been elected three times" (Bremner, 2009). It would be wise to remember, however, that all that glitters is not gold, even for those who only *represent* our dreams.

10.2. Needs and Wants as Advertised: How the Poor Stay Poor

In a self-interested society, individualism reigns and competition is king, ultimately resulting in greater inequality and economic disparity. Nowhere is this more evident than in the world of advertising, which peddles lifestyles instead of products and wants instead of needs, regardless of the consequences, as it attempts to create a globally branded inclusiveness. It is as though we are being coached to succeed at all costs, which, as we will see, is a race to the bottom in the mathematics of feedback loops.

Alas, today's media are paid from advertising and are unlikely to bite the hand that feeds them. The goal is to encourage spending, the numbers for which are staggering. It's not surprising that we judge a product by its image given the amount spent by today's advertisers. Heilbroner and Thurow (1998) noted that advertising expenditures increased in the United States from $50 million in 1867 to $500 million in 1950 and to $130 billion in 1998, "about a third as much as we spend on all primary and secondary education" (p. 166). Expenditures in 2007 were $290 billion ($630 billion worldwide; Coen, 2007). Is it any wonder we are more convinced by style than substance?

According to advertising giant Ogilvy & Mather, "Our job is to make advertising that sells, and the advertising that sells best is advertising that builds brands." Responsible for the likes of Huggies Happy Babies and the Barbie Fan Club, the ad agency knows the importance of name recognition, best served up as a catchy slogan in oft-repeated commercials. In reality, such slogans as "Don't leave home without it" and "Schweppervesence" add billions to the costs of everyday consumer goods.

Heilbroner and Thurow (1998) noted advertising's questionable purpose and do much to dispel its allure:

> Yet it is obvious that not all advertising serves a useful purpose. It is impossible to watch the raptures of "housewives" extolling different brands of soap, laxatives, or canned goods without thinking that perhaps the single most persuasive message of these minidramas is that grown-ups will say things they obviously don't really believe because they are paid money to do so. Is that perhaps the residual effect of advertising on our culture, including, not least, on our children? (p. 167)

A Christmas advertising campaign for the Irish post office illustrates how the capitalist ethos is focused on selling and not producing or even consuming. The ads, which are erected on mailboxes around the country in early November, come with the slick slogan, "The moment you send a card it's Christmas," coyly poking fun at the ever-earlier dates at which Christmas seems to begin each year, as if forgiving one for beating the rush. But the use of the word *send* spells out the true sentiment. To the marketer, Christmas is not when you write a card, it is not when a card is received, and it is certainly not the day itself. It is the moment the stamp enters the system and is canceled. In fact, it is irrelevant what happens to the product after it is sold and is purely a monetary experience, validated by the purchase. The marketer's wish is clear—that every day be Christmas.

A Canadian Christmas advert with the odd catchphrase, "Give like Santa but save like Scrooge," also betrays the underlying ethos of the capitalist mantra, i.e., to spend regardless. But why would anyone want to save like Scrooge? Scrooge was a sociopath who abused everyone around him and cared for no one but himself. Saving like Scrooge is deplorable and promotes individual solutions to collective problems. "Giving like Santa but saving like Scrooge" means buying gifts at inflated, pre-Christmas prices to boost seasonal sales in an annual consumerist joke played on the unwitting citizenry of today's materialist world.

But Christmas wasn't always the selling frenzy it is today. Boorstin (2000) noted that the previously little observed holiday became the "spectacular nationwide Festival of Consumption" (p. 158) we know today only after the American Civil War, when department stores such as Macy's and Woolworth's began encouraging greater Christmas giving to spur on greater sales. Christmas is an entirely made-up celebration, begun in a burgeoning, American consumerist society by eager "go-getters."

As if anointing the transformation from religious to consumerist celebration, F. W. Woolworth even instructed his store managers about the commercial importance of the Christmas season, saying, "This is our harvest time. Make it pay" (Boorstin, 2000, p. 159). Furthermore, to ensure full participation by the workforce, the Christmas bonus was created not to reward past service or to share the company profits, as one might think, but to keep employers from striking during the monthlong Christmas buying bonanza (p. 159).

Christmas consumerism had become so important that in 1939 President Roosevelt moved the date of Thanksgiving forward by 1 week to add to the buying season, which traditionally begins after Thanksgiving. As Boorstin (2000) noted,

> This trivial shift in the date of President Roosevelt's proclamation of a national Thanksgiving was significant mainly for what it revealed of the American Christmas; and for what it told of the transformation of this ancient festival into an American Festival of Consumption. (pp. 157–158)

Nonetheless, it's hard to resist the lure, sugarcoated by style instead of substance, where the consumer is told what to think. Is Superman gay (because of his red manties and hooker boots) or the second coming of Christ (because he comes from another world and saves people)?

It depends on the perception, scientifically cooked up with a dash of marketing, PR, and advertising—today's bywords for convincing yellow-sun mortals to spend money.

Is Coke better than Pepsi? McDonald's better than Burger King? Crest better than Colgate? Or is there, in fact, little or no difference?

> **Coke:** Carbonated water, sugar/glucose-fructose, caramel color, phos-phoric acid, natural flavors, caffeine
>
> **Pepsi:** Carbonated water, sugar/glucose-fructose, color, phosphoric and citric acids, caffeine, flavor

And yet soft-drink makers spend *billions* each year trying to convince a buying public that there is a difference, which only serves to add to the price and reduce competition because of restrictive advertising costs. The real point to advertising is not to say that Coke is better than Pepsi, or that McDonald's saves you more money than Burger King, or that you'll get more dates if you brush with Crest instead of Colgate, but to remind you that the product is *your* product and always has been.

As Ernest Dichter noted in 1961,

> But because our technological development has been so good and so fast, the fact is that almost all our products are uniformly good, so that there is in reality very little difference, in the same price category, between a prod-uct with one brand name and a product with another. What people actually spend their money on are the psychological differences, brand images permitting them to express their individuality. . . . We have reached . . . a psychoeconomic era. (Boorstin, 2000, p. 447)

Think back to the toothpaste your mother made you use as a child; chances are it is the same brand you use today. Why? Because it's *your* brand and always has been. But is there really a difference? Most wash products are nothing more than generic soap that has taken on an absurd meaning in an overly sanitized world of advertised dreams.

Of course, television is the main medium today, hardwired into our way of life as if to satisfy an addiction. As Boorstin (2000) noted,

> If the set was not on, Americans began to feel that they had missed what was "really happening." And just as it was axiomatic that it was better to be alive than to be dead, so it became axiomatic that it was better to be watching *something* than to be watching nothing at all. (p. 396)

Watching, that is, with regular commercial breaks every 15 minutes "to pay the bills."

Any image can be used to sell, from Da Vinci's *Mona Lisa* (toothpaste, film, computers, etc.) to Picasso's *Guernica*. However, not everyone will agree on the message. For example, in 1989, Germany's armed forces used Picasso's iconic *Guernica* in an ad to promote reunification and "provoke public debate" (Hensbergen, 2005, p. 315). When the German author Günter Grass acerbically noted that German planes had bombed the Basque town of Gernika in the first place and Gernika's mayor objected on "grounds of bad taste," the offending image was withdrawn (p. 316).

Nothing is deemed free from advertising, not when an object can be doubled up beyond its function to sell something to an uncritical audience. Even Picasso himself has been appropriated as the brand name of a car ("stylish and compact on the outside, huge on the inside"). One ad submitted in a 2011 Super Bowl contest portrayed Doritos and Pepsi as the body and blood of Christ, and although the commercial didn't win and thus didn't air on the day, its notoriety became an advertisement in itself, viewed on the Internet in any number of places. Despite its questionable taste, it exists as a testament to a prevailing culture that will use any image to sell.

Advertising and profit-only motivations have taken over all media. In today's constant consumer culture, companies pay fans to wear their gear en masse in a stadium and to wave on cue, as in one "ambush marketing" incident at the 2010 Football World Cup, where scantily clad models wearing orange dresses pretended to be Dutch fans when they were in fact advertising beer. Product placement is rampant, where even news is embedded into advertising, so-called "newsvertising." Elizabeth Hurley, Serena Williams, and Lily Allen mix their Twitter updates with unannounced product endorsements (Mostrous, 2011). More and more, the lines are blurring, creating tiers of access, markedly more commercial for the less well-off.

And what of the ever-increasing number of ads on the Internet, our latest great democratizing media outlet? The Internet is fast becoming the prevalent advertising medium, where users are subjected to thousands of advertising images per day in any number of new ways. Here, choices may appear to be multiplying, but there is no escaping the always-on sales pitch. In reality, choice has become an illusion—that is the price we pay in an autocratic market where profit is the only guide.

What is the true cost of all this unbridled selling? Foreseeing globalization and the movement of production to cheaper labor markets, Schumacher (1993) wrote that "the role of the poor is to be gap-fillers in the requirements of the rich" (p. 177)—slaves, essentially. With less choice, limited in access to infrastructure and technology, the poor are also limited in lifestyle—poor education, cheap consumables, lowest-common-denominator entertainment—and are constantly being ripped off by pay-as-you-go charges and debt. Saddling the poor with high-interest-rate repayments from credit cards and loans while encouraging them to buy and rebuy—you couldn't design a more stacked system.

Statistics can quantify the correlation between lack of wealth and performance, but one doesn't need the numbers to understand the obvious. Is it any surprise that reading skills of children whose parents are jobless are less than average (Hill, 2005), that fewer students from poorer backgrounds go to university (Schofield, 2006), that death rates among the poor are 3 times higher (Burke, Kenaghan, O'Donavan, & Quirke, 2004), or that life expectancy in some urban centers is lower than in Iraq (Gillan, 2006)? Is it any surprise that increased poverty is directly related to increases in cancer and stroke, a correlation unchanged since the late 19th century (Dorling, Mitchell, Shaw, Orford, & Smith, 2000)?

Wilkinson and Pickett (2009) argued that general health is a function of income and wealth, citing the greater likelihood of health problems the greater the divide between rich and poor, including depression, heart disease, and drug addiction. In numerous cases, they show that health and social problems are a function of one's zip code, alarmingly depicting how much greater mortality rates are for the poor (~90 per 10,000) compared with the rich (~50 per 10,000; pp. 12–13). The education background of the mother has also been found to contribute to the skills of the child. As Bennett (2010) noted, "Differences in educational attainment among preschool children are so stark that researchers believe that each extra £100 a month in household earnings when children are very young is worth a month of cognitive development." Should we also be concerned that childhood obesity figures are through the roof, that credit card companies target teens, and that alcohol advertising contributes to underage drinking?

Once again, we see that development is a function of basic needs and well-being a result of equality and access to resources. But when self-interest and excessive individualism rule, life becomes a game separated into winners and losers, exaggerated all the more by a world of self-interested sellers.

Feedback Loops and the Devil You Know

In today's interconnected communication world with speed-of-light transmission rates, a simple Twitter post about a skin rash can threaten to shut down a film festival because of a supposed infestation in the theatre, as occurred in Toronto at the start of its 2010 International Film Festival. The prime minister of Ireland was forced to apologize on national television for his hoarse comments after another Twitter post called into question his early-morning partying after a late-night party "think-in." A Florida pastor with a parish of fewer than 50 announced that he planned to burn the Koran on September 11 to protest Islamic culture, and the world came running with instant CNN, FOX, and network updates.

In such a viral world, misunderstandings, rumors, and sensationalist rants spread like wildfire through vast interconnected social networks, where everyone knows everyone else's business in minutes—the ultimate "global village," as described by media guru Marshall McLuhan. The same is true of any "feedback" loop or system, where a recursive relationship exists between what goes in and what comes out.

We are all familiar with feedback loops, from the screeching sound of a microphone held too close to a speaker, which amplifies the input beyond the limits of the speaker (and our ears), to the snowballing effect of exponential doubling. Division over a small number, excessive bank leverage, or the focusing power of a lens, which magnifies the output relative to the input, is similar.

Feedbacks loops also apply to the market, for example, VHS beating Betamax in video recording (despite being a poorer technology); Microsoft beating Apple in the operating system wars; advertising campaigns; even best-seller lists, where a small advantage leads to a dominant market share (as we saw in Chapter 7). Think of a movie that became a summer hit. Why did it succeed when others failed—word of mouth, advertising, poor competition? It isn't always down to merit. Regardless, one movie or another must become *the* summer hit, destined to be seen by moviegoers everywhere.

As noted by Frank and Cook (2010), "In all these processes, small differences at the early stages of competition can prove decisive" (p. 34). Such markets are defined as "winner-take-all" or "reward by relative performance" and are especially susceptible to feedback loops. They stated as an example that one cannot watch CBS, ABC, NBC, and FOX News at the same time (p. 29) and that there must be an ultimate winner in one's viewing preferences in the overall viewing sweeps, which then multiplies of its own accord. Wars are also susceptible to winner-take-all feedback (as we will see in the next chapter).

Such feedback systems can be seen in a simple probability game from evolutionary theory, which shows how one player (or side or gene or species) ultimately comes to dominate another when in competition (or

reaches "fixation," in evolutionary parlance), as cited by the British evolutionary biologist and geneticist John Maynard Smith. In a mathematical
study on ecology, evolution, and behavior, Sigmund (1995) gave the
computational rules: Put 50 red and 50 black marbles into a bag, double
each, then randomly select 100 from the 200 (pp. 81–84). The number of
red (or black) marbles is then observed over a series of trials (double the
100 to 200, randomly choose 100, etc.), showing the evolution of each
marble "species." As one might expect, one species will eventually dominate the other, despite the randomness of the selection process, and
sooner than later if the original balance is biased. Figure 10.1 shows the
red marble population for three initial conditions: red and black = 50 (top
graph), which takes more time to reach fixation; red = 20 and black = 80
(middle graph); and red = 80 and black = 20 (bottom graph), which both
reach fixation much quicker (two trials are shown for each). From the
results, we see that a little push at the start can make a huge difference,
highlighting the winner-take-all nature of a recursive feedback loop.

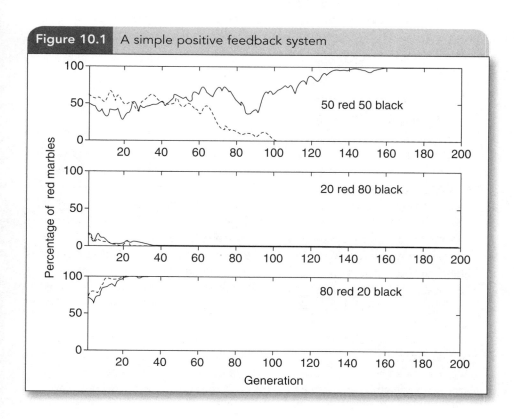

Figure 10.1 A simple positive feedback system

Capitalism, however, is meant to *reduce* prices with *increased* competition, where companies adapt to create viable alternatives with competitive prices. It would seem that in some cases, competition itself is the problem, where the irony is that the more successful a venture becomes, the more market share it realizes, which creates an advantage or monopoly position for some and, thus, *less* competition. In the process, we become losers in the race to pay more as our world is swamped with profit-only-minded junk that "result in inefficient patterns of consumption and investment" (Frank & Cook, 2010, p. 19). Without controls and intermediation, such success will always overheat the system.

The purpose of any market is choice, which comes not from a predetermined menu but from a wide spectrum of possibilities, which, as Barber (2007) noted, should also be free to be altered or improved (pp. 36, 139). In any winner-take-all market, however, the ultimate result is a reduction of choice, fueled by the feedback loop of commerce. It seems that in the marketplace, the devil we know is preferable to the devil we don't. How we get to know that devil is perhaps the most important feedback system of all, one that creates habits and in which one hopes there is more than enough choice to accommodate a multitude of identities. One hopes we can tell the difference.

10.3. No Need to Think: The Adult as Child and the Child as Consumer

Why is this happening? Barber (2007) called today's market thinking an "infantilist ethos" that induces childishness according to the demands of a consumer culture in a global market economy (p. 3), preferring easy over hard, simple over complex, and fast over slow (p. 83)—all the while "declaring war on variety" (p. 200). Jackson (2009) noted that children today "are brought up as a 'shopping generation'—hooked on brand, celebrity and status" (p. 152). From the principles of the Association of National Advertisers' (2011) *Marketers' Constitution*, which state that "marketing must become increasingly targeted, focused and personal" and "unencumbered by inappropriate legislation or regulation," one sees the scale of its determined effort.

According to Heidemarie Schwermer—a German teacher, psychotherapist, and author who set up her own skills-and-possession exchange shop and has lived without money for 15 years—consumerism is about "an attempt to fill an empty space inside. And that emptiness, and the fear of loss, is manipulated by the media or big companies" (Marsh, 2009a). An ever-greater plague of advertising has been visited upon our houses, but why is selling the purpose of markets anyway? Buying should be the purpose, based on need and not on manufactured wants, particularly important as we face limits to energy and resources.

This thinking hasn't gone without causing some consternation among economists. Galbraith (1958/1999) noted that the instinct for providing what is needed was once tied to the obvious and was even "maternal" (p. 116). Alas, something changed along the way in the rush to maximize profits. Frank and Cook (2010) noted that

> social critics have long identified advertising as perhaps the largest and most conspicuous example of pure social waste in a market economy. . . . Even the most enthusiastic proponent of advertising must concede that the private incentives to engage in it are larger than the social ones. (p. 141)

Galbraith (1958/1999) even goes so far as to question the whole value system of some for-profit production:

> Few economists in recent years have escaped some uneasiness over the kinds of goods which their value system is insisting they must maximize. They have wondered about the urgency of numerous products of great frivolity. They have been uneasy about the lengths to which it has been necessary to go with advertising and salesmanship to synthesize the desire for such goods. (p. 116)

In its lust to fulfill its own motives, modern advertising can be likened to a glass repairman breaking windows, a virus software company infecting files, a hotelier posting bad reviews about competitors on a tourist website, or, as Galbraith (1958/1999) noted, a doctor knocking over pedestrians to get patients (p. 129): "As a society becomes increasingly affluent, wants are increasingly created by the process by which they are satisfied. . . . Wants thus come to depend on output" (p. 129). In economic terms, "the marginal utility of present aggregate output, *ex* advertising and salesmanship, is zero" (p. 131). In fact, affluence is no longer required but, rather, the *need* for affluence, where the desire to be economically enhanced in an increasingly unequal society acts as a catalyst to buy.

Excessive buying is also related to a culture based on growth, where waste is encouraged and false values promoted. Barber (2007) cited a Stanford University study that reported "up to 8 percent of Americans, 23.6 million people, suffer from compulsive shopping disorder" (p. 239) and a 1998 Mintel marketing study that reported "almost one in four

Britons admits being addicted to shopping" (p. 241). Furthermore, the need to spend is built right into Irving Fisher's equation of exchange,[5] otherwise prices will decline (Galbraith, 1977c).

The worst commercial excess aims to hook children, essentially treating them as Peter Pans with pockets. Barber (2007) noted the dangers, especially since many children cannot distinguish between an ad and a program, whereas "adults can mute, filter or otherwise elude or ignore advertising" (p. 233).[6] Furthermore, "retailers do not draw the young to malls or theme parks or multiplexes to encourage them to socialize or hang out or cruise as they might 'naturally' do, but to put them to work shopping, to direct their play to commodities and for-pay entertainment, to turn the impulse to socialize into an impulse to consume" (p. 112). Naomi Klein called it "theft of cultural space," where the "tentacles of branding reach through every crevice of youth culture into psychological attitudes" (Brayfield, 2000).

The most damning indictment of a permanently switched-on, commercial world, however, is the dumbing down of critical analysis, which encourages a minimal understanding of products. Here, Barber (2007) noted that "for consumer capitalism to prevail you must make kids consumers or consumers kids. That is to say, smarten up the kids—'empower' them as spenders; and dumb down the grown-ups, disempower them as citizens" (p. 20). Such watered-down learning "delegitimizes adult public goods such as critical thinking and public citizenship (once the primary objectives of higher education) in favor of self-involved private choice and narcissistic public gain" and "is encouraged by a political ideology of privatization" (p. 15).

Despite the evidence that we don't need half the things we have, we buy without thinking, working more and more to afford what we don't need, unconcerned with the social or psychological ramifications of a consumer-first society or the environmental damages caused in the process. Barber (2007) pulled no punches in telling us that liberty is at stake:

There is considerable evidence to suggest that the ubiquity of consumerism, the pervasiveness of advertising and marketing, and the homogenization of culture and values around an infantilizing commerce together have created a cultural ethos which, although not totalitarian, robs liberty of its civic meaning and threatens pluralism's civic vitality. Combined with privatization and branding, this commercial homogenization has made us less free as citizens and less diverse as a society than traditional liberals conceive us to be or than traditional capitalist producers and consumers think we are. (p. 217)

Again, the issue is one of private versus public, the self-interested versus the collective. Where do we draw the line as private business encroaches further into the public space? What can truly be called public anymore? Do we still have a public agora where people come together and share ideas and thoughts and ruminate on shared problems and solutions, free from advertising or sponsorship? Or is it all one super-sized mall? "Buyer beware" should not be the ethos (slogan) of any society.

What end does an overweaned and overheated mass consumerism serve? Barber (2007) chillingly stated that "market philosophy is more than a threat to democracy, it is the source of capitalism's most troubling problems today" (p. 123), where "the modern tyrant is the enforcer of consumer capitalism's need to sell" (p. 125)[7]. Ayres called it a vicious circle, where consumption fuels growth and growth fuels consumption (Jackson, 2009, p. 76). Galbraith (1958/1999) suggested that the production of goods as the central problem of our lives is, in fact, a myth that creates unnecessary consumer demand and social imbalance (p. 209). Furthermore, "as we expand debt in the process of want creation, we come necessarily to depend on this expansion" (p. 148). Through an ever-expanding debt, the American consumer has become what Soros (2008) called—and not in a flattering way—"the motor of the world economy" (p. 98).

What started as an attempt to explain how a free-market economy works in an expanding industrial world, as practiced by Adam Smith and most economists since, has failed. Our markets are not free, and yet we are told that they are and led to believe that we are independent choosers in an open-market landscape, where natural laws about likes and dislikes provide a stable and self-correcting economy. In fact, consumerism for the sake of consumerism is rampant and democracy has become a slave to the market, fueled by its own feedback. Consumerism and buying for the sake of buying have become a constant, nicotine-spiked tobacco fix.

Yes, we've been hoodwinked, abused, and sold a bill of goods by the most manipulative and seductive power that society has ever known, one that readily remakes our every want into a need for the sole purpose of selling anything from mousetraps to mortgages, backed by numerical measures (some of which have questionable validity) to convince us that success depends on never-ending growth. But improving solely through buying is an illusion, which because of a flawed feedback loop in the design of capitalism serves only to expand inequality.

The public space, one that we share and engage in daily, where one is free from a constant nagging of commercial interests, must be protected. It is, after all, a shared world, where value matters more than price, where life matters more than lifestyle, and where freedom must be maintained. By saying "no" to the commercial powers that enslave us, we say "yes" to the future and "yes" to ourselves. By saying "no" to excessive consumerism, we seek to be more than gap-fillers in the lives of the rich.

Notes

1 **The complete oil breakdown**: Cars 41%, trucks 13%, chemical and plastic raw material 10%, air passenger travel 7%, factory heating processes 5%, home and office heating 5%, oil refineries 3%, road asphalt 3%, water freight 3%, agriculture 2%, electricity generation 2%, construction machinery 2%, military 2%, rail and air freight 2%, recreational vehicles 1% ("Oil: Where does a barrel go?" 2010).

2 **Irish sportsmanship?**: On the flip side, Ireland benefited from a controversial call against Armenia to advance from their 2012 European qualifying group, where video evidence suggested that an Irish striker had touched the ball prior to an Armenia red card. In this instance, none of the Irish called for a review.

3 **Road deaths**: In 2008, 13,846 (37%) road deaths in the United States involved alcohol, down from 26,173 (60%) in 1982. According to MADD, one person is injured every minute from an alcohol-related accident (http://www.madd.org/statistics/).

4 **Tony Blair's post-prime-ministerial jobs**: In short order, Tony Blair made up for his poor career choice as public servant and for making less than his wife: He earns more than £5 million a year as a privately paid Middle East "peace envoy," another £5 million a year for talks and contracts with JP Morgan and Zurich Financial Services, as well as signing a £5 million book deal 4 months after leaving office (Ungoed-Thomas, 2009).

5 **Fisher's equation of exchange**: Prices = Private Money times its Velocity plus Public Money times its Velocity, all divided by the total number of Transactions. If velocity of the public or private spending increases, so does price. This equation can also be used to see the impact of stimulus packages (Public Money) on price when the economy falters (e.g., with reduced Private Money).

6 **Children's advertising**: Some countries restrict underage access, such as Sweden and Norway, who have banned television advertising to children under 12, stating that children are unable to understand the purpose of advertising and have thus been declared "commercial-free zones" (Rowan, 2002).

7 **Supercapitalism and democracy**: Reich (2009) noted that democracy has become imperiled by "supercapitalism," a cutthroat global supply chain that circumvents hard-fought labor laws and considers shareholder returns at the expense of the public good, preferring consumer-investors to citizens; "Corporations now have little choice but to relentlessly pursue profits" (p. 50), the result of which has led to "the decline of democracy" (p. 222).

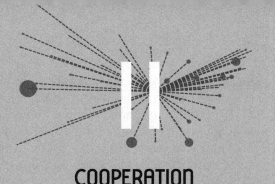

COOPERATION

The Art and Science of Agreement

I n this chapter, we take a more formal look at cooperative strategy, beginning with problem solving and continuing on to the art and science of agreement via the non-zero-sum game and the prisoner's dilemma as it applies to game playing and strategic thinking. Along the way, we pay particular attention to some of the economic strategies rooted in our money-based world, including the 2009 credit crisis, ever mindful of the mathematics and morality we've already seen and how money seems to have its own rules. In the end, we hope to see how such strategies can be used to curtail excess in the markets and in our lives.

II.I. Bottom-Up or Top-Down Learning?

Attending school to memorize facts and regurgitate them for exams is how many of us are taught but is not how we best learn. As the saying goes, genius is 1% inspiration and 99% perspiration, which identifies working as the best way to learn. And if we work toward self-directed ends, we teach ourselves and acquire the tools we need to solve problems. Most people ultimately teach themselves what they need to know using the resources available to them.

From time to time, obstacles appear along the way, all of which must be dealt with. The way one bypasses an obstacle is the key to solving problems. Perhaps the most tried-and-true method is that of trial and error. Faced with a locked door and a set of keys, most will turn through each key one by one in sequence until the right one fits, a perfect example of trial and error as a bona fide problem-solving method. It may take a while—the door opening on the first or last key—but the method will succeed, in time. Using the same method to find a phone number in a telephone book, however, would be hopeless, where bisecting the numbers by name is more fruitful (turn to the middle of the book; if high, turn to the middle of the first half; if low, turn to the middle of the second half, etc.). If efficiency is the stated goal (time, cost, ease of use), then *how* to solve a problem is perhaps the most important skill one can learn.

From the 14th century, we have Ockham's razor—also referred to as the law of succinctness—a maxim for solving problems: "Entities should not be multiplied beyond necessity." Ockham's razor is an early form of management-speak, such as KISS or "Keep It Simple Stupid," and states that the best choice between competing theories is that which has the fewest assumptions. Under such stricture, it is easy to see why, for example, the Copernican heliocentric solar system is a better model than the more cumbersome Ptolemaic geocentric model, with its myriad epicycles, deferents, and equants. The simple beats the complex every time. Similarly, multiplication in base 10 is much easier than in base 12 or base 60 or, indeed, using Roman numerals. Who would not choose ease of understanding and efficiency of calculation when dealing with constructed models? Who would not choose the shortest path from A to B?

Similar strategies such as the computer programmer's dictate, "simple, small, scalable"; the environmentalist's theme, "reuse, reduce, recycle"; and Winston Churchill's rules for writing, eloquently delivered according to his own mantra—"short words are best and old words when short are best of all"—all rely on simplicity and efficiency. Of course, simplicity and efficiency may not be as interesting as the road less traveled, but they will get you where you're going faster. And, of course, not all methods apply to all. What is elementary to Sherlock Holmes may not be elementary to Dr. Watson, or to you and me.

Numerous systems exist that suggest the "right" path, extolling the virtues of practiced learning—from the more logical methods of Edward de Bono to the not-so-logical ways of Tony Robbins and the like. In his 1967 book *The Use of Lateral Thinking*, de Bono championed the idea of "lateral" thinking—what today we would call "thinking outside the box"— followed by his "L-game" (an early, streamlined version of Tetris that uses seven differently shaped "tetrominoes") to practice strategic or combinatorial thinking, not unlike how one best packs a suitcase for travel or organizes the trunk of a car. On the other hand, Tony Robbins instructs us in *The Ultimate Relationship Program* DVD collection ($299 with free "Inner Strength" film of your choice) that "personal growth begins by asking yourself honest questions" and claims to provide "customized strategies for improvement based on your life goals." Perhaps $300 (sorry, $299) is a bargain for such wisdom.

One problem de Bono attempts to solve is that of ties in soccer, currently decided by a 30-minute extra period and then penalty kicks if still tied. An equitable solution is no simple matter and has included replaying the entire game (difficult to implement in scheduled tournament play), a sudden-death "golden" goal (discarded when games became dull, as teams played not to lose rather than to win), and penalty kicks, a solution that may solve scheduling problems and deter dull play but is brutal in the extreme. De Bono's solution awards the win to the team whose goalkeeper least touched the ball in regulation time.

Another solution is to remove players one by one—say, every 5 minutes—until a team scores (applicable in any sport for that matter). Regardless, such strategies, logical or otherwise, attempt one way or another to arrive at a solution that best serves everyone's needs, from players to fans (although FIFA, the sport's governing body, recently hinted at bringing back the golden goal, universally thought to destroy the flow in overtime games).

Another prescription about how problems are solved or how work gets done is the Peter Principle: "In a hierarchy every employee tends to rise to his level of incompetence," an idea first put forth by Laurence J. Peter in his 1970 book of the same name. Essentially, we all get promoted throughout our careers until we get a job we can no longer do well, and thus will no longer be promoted. Presumably, many have experience with this principle in action, which sadly suggests that organizations are top-heavy with deadwood.

More constructive is Parkinson's law and his many corollaries, which state that "work expands so as to fill the time available for completion," known intimately to every student who has ever left a paper for the last minute. First proposed in 1958 by C. Northcote Parkinson, Parkinson's law is a seemingly posh rewording of "Don't put off until tomorrow what you can put off today," an illogical turn on Benjamin Franklin's famed advice for success. Although Franklin also tells us in his *Advice to Young Tradesmen* that "time is money," unfortunately, money rarely expands to fill need.

Akin to Ockham's razor, common sense is essential to problem solving, although common sense itself can be wrong when based on false assumptions. Aristotle believed that heavier objects fall faster than lighter objects, an idea that seems perfectly reasonable at first thought. As he wrote in *Physics* (Book IV), "We see that bodies which have a greater impulse either of weight or of lightness, if they are alike in other respects, move faster over an equal space, and in the ratio which their magnitudes bear to each other." This idea was not wholly overturned for 2,000 years.

However, one has only to drop a large rock and a small pebble together to see how objects fall at the same rate, as Aristotle himself could have verified—although a feather and a hammer do not (because of air resistance), which may have led to the original misunderstanding. Perhaps if we lived in a vacuum (i.e., no air friction), where *all* objects regardless of their shape fall at the same rate (e.g., a feather and a hammer), we might have a better intuitive or "common" sense about falling objects. But we don't, and it took Galileo conjecturing that all objects fall at the same rate to set the record straight after an experiment believed to have taken place at the Leaning Tower of Pisa,[1] which was reenacted on the moon in 1971 by Apollo 15's commander David R. Scott using a 1.3-kg aluminum hammer and a 0.03-kg falcon feather (Williams, 2008a).

Aristotle also believed that a moving object when dropped falls backward and that an object spinning in a circle flies off in a curved path when let go. In fact, dropped objects move *forward* with the velocity in which they are moving, because the perpendicular motions forward and down are independent (think about how a plane leads the landing zone when dropping a parachutist or care package), and spinning objects fly off *tangential* to the radius of the circle they are orbiting, because the force is proportional to the acceleration, whose tangential direction is constantly changing. You can prove both of these facts by dropping a stone while walking and whirling a chestnut conker around and letting it go.

Most important, however, as shown by Galileo, problem solving can be couched in a more scientific way or method of thinking (Kuhn, 1957/1997, p. 95). Francis Bacon is credited with defining the "scientific method" in the early 17th century, although the phrase became fashionable only in the mid-20th century to help codify science. There is no one way, although all methods require repetition and verification, such that any assumption can be tested from a worked analysis of the data by experiment, observation, and conclusion to prove or disprove a hypothesis. To be sure, all of us use some form of scientific method to determine what makes sense.

But what methods are available to analyze data? Extrapolating from the specific to the general may well be the simplest form of analysis, which we all do, but may cause the greatest problems. Paulos (2001) called such inference "personalizing" data (p. 9), where a single instance of such and such is sufficient to constitute a fact, a trend, or a law. We may lament that toast always falls butter side down, which it doesn't (although it may well fall butter side down more often than not, say, because of an unbalanced spin rate); we just remember the bad results more than the good. When we see a bus arrive at a certain time on a number of days in succession, we may assume that it always arrives at the same time when, in fact, it doesn't. Television sports announcers routinely make the broadest generalizations from the slimmest of data, e.g., that so-and-so is good because he scored and so-and-so is not because he didn't on any given day (not to mention hedging their predictions on the outcome of a game at the same time).

Or as Veseth (2005) eloquently noted about Adam Smith's famous specialized pin-making factory, just because 10 men working together can make 240 times as many pins as 10 men working individually (48,000 versus 200 pins, because of division of labor: drawing out, straightening, cutting, pointing, grinding, etc.), it doesn't follow that all economic endeavors will so benefit and obtain, in Smith's words, "universal opulence" (pp. 41–45). In fact, according to Veseth, if most of us "perceive the complex patterns of global markets as simple images of hamburger stands and soda pop cans" (p. 39), we can blame the "Newtonian" rhetoric of Adam Smith and his "incredibly broad generalization about the division of labor" (p. 41) for the mistakes of capitalism, which among other things failed to account for social costs or competition from other nation-states in its thinking (p. 49).

Conversely, not all swans are white because one has never seen one that isn't; not all Americans speak Spanish, although in some parts of the United States it may seem so; nor do all politicians break their promises. In all such inferences, the sample is either too small or biased and we haven't yet recognized any underlying trend. In essence, we are drawing conclusions from incomplete information. As the Scottish philosopher David Hume noted, such inductive reasoning is why we believe the sun will rise tomorrow, the day after tomorrow, and every day thereafter, although such logic doesn't always follow, and in this case will not the very day the sun runs out.

Furthermore, one number is no luckier than another, even if it appears more often in a lottery, although many of us think *after the fact* that if a number scores us a win on a lottery, a horse, or a game, it is lucky, forgetting all the times when our lucky number wasn't so lucky. In a lottery, a person who wins may feel lucky or blessed, but it just isn't so. We confuse events that are unrelated as related because our subjective perception of the truth clouds our thinking.

Deductive reasoning, on the other hand, is more assured but depends on the original premises. "All books are full of words; the Bible is a book; the Bible is full of words" is correct provided that all books are full of words and that the Bible is a book. However, "Republicans favor the death penalty; the death penalty is evil; all Republicans are evil" fails on many grounds and is less assured, not least of which because not all Republicans favor the death penalty. Similarly, a light on in an office does not mean someone is there, nor does a light off mean no one is there. As Veseth (2005) noted, it would be crazy to conclude that globalization is democratic, because globalization requires decentralized networks "with no strict decision-making hierarchy" (p. 52) and networks are democratic "pushing power out to the grass roots" (p. 53), since there are lots of ways that networks exclude "those who are not part of the system (the digital divide)" (p. 53) and instead create an unequal world of haves and have-nots.

Incorrect assumptions can and often are made from such bad deductions, which can result in bigger mistakes as the wrong dominoes fall. As part of the Potsdam Declaration (which defined the terms for Japanese surrender) at the end of World War II in Europe, President Truman wanted to demand that the Japanese surrender unconditionally but was advised to demand only the surrender of their armed forces, which was then interpreted as a sign of weakness. The Japanese refused, which resulted in the dropping of atomic bombs on Hiroshima and Nagasaki and the loss of more than 100,000 lives.

Our own biased inferences play a large part in the interpretation of data, and, thus, it is imperative to verify all assumptions, lest we end up with "garbage in garbage out," or GIGO as the computer programmer would say. Even worse is to be right for the wrong reasons, which can be disastrous if such supposed acumen is then relied on for future direction and policymaking (e.g., Taleb's lucky fund managers, as we saw in Chapter 7).

Reasoning must be careful and cautious and is best conducted according to plan, not least to remember which trials have succeeded and which have failed. From the results, data is organized into tables and graphs to show dependency, whether a full-blown regression that determines the "goodness of fit" or our own determined analysis, from which we recognize patterns and make conclusions. From such inductive (bottom-up) or deductive (top-down) analysis, we begin to understand how things work—from planets orbiting the sun to organizational thinking, from the laws of physics to the uncertain workings of modern economics—and we begin to understand how best to proceed when given a choice, more confident of the results because of a studied and verified basis.

11.2. The Prisoner's Dilemma: You Scratch My Back, and Maybe I'll Scratch Yours

A *zero-sum* game is a game played by two people (or sides) with completely opposite results. For example, a chess game has one winner and one loser. In a baseball game (which by definition cannot end in a tie), the net score is +1 to the winner and −1 to the loser, with a *W* and *L* chalked up in the win and loss columns. If I make a bet and win $10, the person I bet loses $10, hence the name "zero-sum" ($10 − $10 = $0). Almost all two-player or two-team games are zero-sum games, with one winner and one loser. We've seen many examples of the zero-sum game in action, from winner-take-all markets to divergent trends in sports, music, and income—all based on some form of competitive game playing.

But what if it is distinctly in my advantage *not* to act in a wholly competitive way? What if pursuing what appears to be my gain in fact contributes to my loss? There are many real-world examples of *non-zero-sum games*, as they're called in game theory—including the complex interactions between people, such as trading, negotiations, or socializing—and are different from the single-minded, win/loss of zero-sum games (board games, sports, betting, etc.) that we have seen already.

Traffic is a good example. If I let someone in ahead of me, I lose my place on the road and may reach my destination later. But if I never let anyone in, I might not be let in down the road; thus, by letting someone in, I play by an unwritten rule of the road, where letting people in contributes to an overall shared ethos such that all of us can get ahead. Of course, if I am a mean-spirited person in a world of fair-minded drivers, I will benefit from being selfish as I whiz around hogging the road and cutting people off, oblivious to others' fair-mindedness. But if each selfish act lowers the general unselfishness on the road, other drivers may soon start to fend for themselves, ultimately resulting in an overall attitude of increased selfishness. Temperatures and tempers will rise, and traffic will snarl, all because of one person (me) who took no notice of his selfishness.

Similarly, in a crowded bus or commuter train, people not moving to the back (or middle in a multidoor bus or train) keeps others from getting on or off and may ultimately hinder the original self-interested commuters from getting on some future train as more and more people crowd around the doors and adopt a "why should I move?" attitude. The same thinking applies to multipronged lines, cafeterias, airport carousels, high-seas fishing, and even standing in an audience to get a better view, which forces everyone else behind to stand, resulting in no better view for anyone and much-diminished comfort for all.

Since the prevailing attitude is constantly changing because of various behaviors, it is interesting to look at how one (a driver, a commuter, society) reacts to changing attitudes, and how fast attitudes then change. Some change is linear (1 and then 2 and then 3 bad apples); some is exponential (2, 4, 8, . . . bad apples), such as a malignant tumor spreading out of control; and some is extremely complex because so many parts are interconnected and react at different rates, such as in diffusion (e.g., a perfume or foul smell transiting a room) or phase changes (e.g., a sudden transformation from individual to group behavior). For that matter, can a good apple make things better and turn one smile into many smiles?

How such acts are transmitted through the whole is important if we are to consider the effect of our actions on others and, ultimately, on ourselves. Is it better to act selfishly or unselfishly to attain one's goals? Is it better to try to win at all costs? At the outset of his groundbreaking book *The Evolution of Cooperation*, Robert Axelrod (1984) essentially asked the same question: "Under what conditions will cooperation emerge in a world of egoists without central authority?" (p. 3)—if you will, a mathematical version of a state of nature, as in William Golding's *The Lord of the Flies*.

Trading Up and Down: Good for Both?

The ways in which we agree and disagree is a fascinating subject that shows how something is valued and how we haggle over an as-yet-undetermined price (as we saw in Chapter 4). For example, when one barters, one exchanges one thing for another—say, a sports card collection for a bike. Both may be of equal value, yet each is deemed more valuable to the would-be new owners.

Any trade is meant to benefit both parties. When the Boston Red Sox sold Babe Ruth to the New York Yankees (for $125,000 and a $300,000 loan), they were holding off bankruptcy and ridding themselves of a seemingly eccentric player. But the Yankees believed Ruth was worth it, as he indeed turned out to be. Even Tom Sawyer's friends thought they were getting something in exchange for painting his Aunt Polly's fence, despite having to part with a kite in good repair, a dead rat, 12 marbles, and other "bartered" goods.

An interesting modern example of "trading up" involves the story of Kyle MacDonald, who in a series of improved trades swapped a paperclip for a pen for a doorknob for a barbecue for a generator for an entertainment center for a Ski-Doo for a trip for a van for a recording contract for a year's accommodation for an afternoon with Alice Cooper for a KISS snow globe for a credited movie part and, finally, for a house. MacDonald's inventiveness ingeniously highlights that price is whatever someone is willing to pay and trades don't always benefit both parties.

In essence, a trade is what is agreed on between two parties, typically to the betterment of both, though not always—as shown by Kyle MacDonald and Tom Sawyer, who got the better of their deals by dint of their own will and salesmanship. All such trades proceed from two concurrent valuations, without any need for an intermediary proxy or external authority, made all the easier today with the speed and ease of the Internet.

In a computer-simulated model of a trading society, called Sugarscape,[2] sugar is traded according to various rules, resulting in markets occurring naturally, with fluctuating prices and differing individual holdings over time (Ball, 2005, p. 438). Here, a perfectly flat distribution of wealth is unlikely, and inequality seems to be the norm, not least because commerce is not as great an interest to some, contrary to traditional economic thinking that assumes all of us make decisions for selfish reasons. The same applies to any "winner-take-all" market, i.e., markets that can be ranked (top 10 records, sports leagues, country competitiveness, etc.), where inequality naturally arises, as we have already seen. Some argue that an unequal distribution of wealth will always occur in any trading society, a power-law dependence between wealth and ownership, as in the Pareto principle (i.e., 20% own 80%).

Collaboration, on the other hand, is the ultimate *non*-zero-sum game, which is played to the benefit (or detriment) of both players. Two farmers (or more) agreeing to share the cost and use of a purchase, a coordinated discount organized by large numbers of buyers to avail themselves of much-reduced block sales,[3] someone lifting up another to open a locked window and then crawling through to let them both in, or even two people killing each other playing chicken on the road—results that neither party could have achieved alone. A non-zero-sum game may also have tiered rewards or penalties that differentiate the loss and gain, on which a number of models of human and genetic interaction are based.

Negotiation is a form of chicken, as is regularly seen in Washington, especially during the debt-ceiling talks between Republicans and Democrats in the summer of 2011. Both sides drew lines that they could not cross with regard to spending cuts and tax revenues prior to a possible government shutdown and government default. For a functioning government (or survival of a species), however, it is obviously best to

298 DO THE MATH!

cooperate (or swerve away from the oncoming car, in the chicken analogy), especially when the cost of losing is extremely high (death or loss
of face or election disaster) and, thus, it doesn't pay to dig in one's heals
too much (or to escalate the conflict, as we will see). As the August debt
deadline fast approached, talks intensified when neither party wanted to
be seen as failing to negotiate. In the end, a compromise was reached
that cut spending and raised the debt ceiling, rather than either side
adopting an unworkable "my-way-or-the-highway" strategy (although, in
essence, the problems were postponed to another time and another set
of negotiations).

To be sure, cooperation works, from marriages to government to corporations, not only because of specialization but because of directed
organization, implying that a society is better off if its resources and labor
are shared. Nonetheless, this does not always follow, and some are left
out, which leads to conflict and "group wars." As such, fairness is of
utmost importance to ensure against systematic abuse, including those
who control the medium (money or infrastructure).

In a study of traffic jams, it was found that many were caused by the
aberrant maneuver of a single driver (bad lane change, checking a phone
message, overreaction), which abruptly increased traffic density and thus
changed a "free-flow" or "synchronized" traffic system into a stationary
jam, highlighting how group behavior is changed by the individual (Ball,
2005, p. 217).[4] When everyone moves together, the flow is smooth, akin
to a coordinated school of fish or flock of birds that constantly monitor
their nearest neighbours for instant speed and directional feedback, but
all it takes is one bad apple to change the system. How information is
transmitted between the individuals thus becomes paramount to coordination and a beneficial emergent behavior (traffic can also break down
during high volume, i.e., the more successful).[5]

Cooperation is also needed in panic scenarios (essentially, high-speed
traffic)—especially so—as everyone rushes without thought to others (the
same applies as deadlines fast approach or in stock sell-offs during the
unwinding of a financial bubble). As hard as it is to do (or enforce), we
must consider others when rushing or we risk clogging the system, often
with disastrous results at obstacles or doorways (which is why emergency
exit doors open outward). It is true that the time to empty a room may
first decrease as people move faster, but only to a point, after which the
whole system gets bunged up. Below a certain threshold, a safe and
orderly flow exists, but above the threshold, cooperation is needed to
ensure mutual safety (Ball, 2005, p. 178). Indeed, it is in all our best interests—especially when different goals are evident—to cooperate.

The Mathematics of Cooperation

One non-zero-sum game proposed in 1950 at the RAND Corporation especially illustrates that what appears to be the best strategy may not be, and that cooperation leads to a better outcome and noncooperation to a worse outcome for all involved. The RAND Corporation was created in 1946 as an American think tank to advise the U.S. government on nuclear defense policy, not least how to rewire and maintain the command structure of a computer network after a nuclear strike.[6] Out of their work on systems of interaction came a concrete example of how cooperation can benefit those whose interests are interdependent—in particular, two thieves who are caught and whose punishments are intricately tied to how they cooperate while separated in police custody (their innocence or guilt was not included in the game, and the dilemma is constructed solely to partition reward and punishment).

Formalized by Albert W. Tucker, this new way of analyzing interactions is called the prisoner's dilemma, where the cost/benefit to the two prisoners is typically constructed in a 2-by-2 array, indicating the possible outcomes for Prisoner A and Prisoner B, as shown in Table 11.1. The game could well be called "to rat or not to rat."

If Prisoner A rats and Prisoner B keeps silent, then Prisoner A goes free and Prisoner B gets 10 years; conversely, if Prisoner B rats and Prisoner A keeps silent, then Prisoner B goes free and Prisoner A gets 10 years. If both talk, they both get 2 years, but if both keep their mouths shut, they both receive only a 6-month sentence. Such a scenario, where the trade-offs are tied to what you do *and* to what another party does, perfectly illustrates the benefit of cooperation. To *both* not rat is the best *joint* strategy, but honor among thieves not always being so honorable, saying nothing could get the silent partner 10 years if ratted out by the other, the worst single outcome. To be sure, it is tempting to rat, but if both rat, they still both get a hefty sentence.

There are many everyday examples of such dilemmas. People who crowd around airport luggage carousels pursue their own interests above others' to get their luggage as fast as they can, but in so doing, they

Table 11.1	Four possible prisoner's dilemma outcomes	
	B silent	*B talks*
A talks	A = 0, B = 10 years	A = B = 2 years
A silent	A = B = 6 months	B = 0, A = 10 years

make life more difficult for themselves and for others. If everyone were to take a few steps back, everyone could get a better perspective and pick out their luggage with ease as it passed.

Queuing is another example, although the problem of lineups at banks and supermarket checkouts has been solved by using a collective approach, as in a single feeder line. Not only is the individual better off adopting a strategy that seems contrary to his or her apparent best strategy, but the group is better off. One person can get a better place if he proceeds directly to an opening, but if everyone behaves the same, all are worse off; thus, it is in one's best interest to act to benefit the group.

Insurance is also a good example of a non-zero-sum sharing,[8] which limits the excessive (or minimal) occurrence to a few of its members to ensure that all receive in good and bad times (mathematically, it narrows the variance). Insurance minimizes individual loss and gain by pooling risk and is a forward-thinking idea to promote or secure balance in uncertain times, e.g., hurricane damage, decreased rainfall that reduces crop yield, market turmoil. Heath (2009) noted how a more socialized government arose in the farmlands of Saskatchewan, where mutual aid was (and is) essential to protect against crop failure and volatile prices, while a more conservative government arose next door in Alberta's less-capricious ranching and oil-laden lands (pp. 127–128).

The prisoner's dilemma can especially be seen in tournament groupings. For example, after two games in the round-robin stage of the 2010 Soccer World Cup, Uruguay and Mexico sat tied at the top and were scheduled to play each other in the final group match, where they both needed only a tie to advance.[7] But should they both play for a tie and ensure advancement, risking elimination if the other team scores and one of the third-place teams wins (e.g., the host nation South Africa)? In this case, Uruguay won 1 to 0. But much to Mexico's relief, they advanced anyway since their final goal difference was better than South Africa's, who did beat France 2 to 1 to tie Mexico in the standings. But had South Africa managed to win by more, the subtleties of the prisoner's dilemma would have provided the South African hosts their storybook result of advancing to the knockout stage, while raising questions about Mexico's strategy.

To be sure, it takes only one or two acting in their apparent best interest to clog (or destroy) the system, as in traffic or around a luggage carousel, known in game-playing theory as the "tragedy of the commons," where farmers grazing one cow each in a common pasture is fine, but grazing two each becomes a problem, and all the more so as the population of farmers increases or the availability of pastures decreases (Hardin, 1968). There are numerous examples, such as overfishing in understocked oceans, looting during a disaster and circumventing more orderly aid dispersal, even trade agreements where tit-for-tat subsidies between nations bung up the works, or corporate malfeasance that destroyed the likes of Enron. All (or most) of us must follow the rules and behave according to the plan—which, of course, is no easy matter—or the system breaks down.

In the airport carousel case, a better design—e.g., roping off an area by the carousel to exclude trolleys—and a sign or regular announcements suggesting cooperation would make life easier. For the bus or train, the driver need only herd passengers with regular reminders to move on. As for traffic, where the players (or agents) have more autonomy, better coordination and enforced laws with effective penalties are needed. Ball (2005) noted that "jam-inducing fluctuations . . . could be eliminated by driver-assistance systems" (p. 219), such as changing speed limits as traffic density changes. Furthermore, "some jams in heavy traffic can be smoothed away completely if just 20 per cent of the cars are equipped with automated systems which enable them to respond optimally to changes in traffic flow" (p. 219).

In economics and business, however, it is much harder for everyone to behave in a manner that first benefits all, and an organized structure—regulated and with penalties where needed—is essential to ensure cooperation, without which abuse and crime become common. When a building developer gets land rezoned illegally because of an under-the-table deal and the misdeed goes unpunished, the system fails. As seen in the Deep Horizon disaster in the Gulf of Mexico, in the absence of better regulation, penalties on the order of thousands of dollars (or worse, as was typical, only *threats* of penalties on the order of thousands of dollars) were meaningless to companies making profits in the billions of dollars. Without police forces to protect property and assets, it seems likely that life would be little more than organized chaos.

What's more, attitudes are reevaluated and a new morality prevails (not dissimilar to the so-called supply-and-demand price point). Somewhere down the line, another developer, perhaps fresh out of business school, learns to flout laws with impunity. Or a politician or banker or oil executive learns that he won't be punished for corrupt or careless actions. Every child learns as much by stealing a cookie and not being found out. In reality, all such changed attitudes are the result of poor cooperation.

Fool Me Once, Shame on You; Fool Me Twice, Shame on Me

An illustrative social example of the prisoner's dilemma is in the movie *An Affair to Remember*, where two lovers (Cary Grant and Deborah Kerr) agree to meet at an arranged time atop the Empire State Building, each trusting the other to show. Each asks the same question: The man says, if I go and she doesn't then I will look like a fool, but if I don't and she does then I am a fool; the woman questions the same in reverse. If they both go, hallelujah, and if they both don't, no one cares (other than not knowing if the other showed). But is the prospect of being made to look a fool better or worse than the possibility of love? That may well be the real dilemma.

In this case, the dilemma is a bit benign because one could argue that selfishly pursuing one's own goal is good here—who cares if you look the fool for a chance at love—and perhaps that's the difference between love and a longer prison sentence. It seems worth looking the fool for a chance at love. But if you look at both dilemmas, with regard to what the individuals agreed on prior to separating, the right thing to do is what you said you would. For criminals, shut up and don't rat out your partner (trust); for lovers, show up because you said you would and don't doubt the other (trust).

Trust is especially important in trading goods at market, such as ensuring fast delivery (without too much delay for inspection) and freedom from graft or pilfering (or large-scale corruption), which partly explains why modern countries with established supply chains and rules of law succeed and continue to succeed while developing countries fail. Beattie (2010) noted that improving infrastructure in developing countries helps, but improved practices are also needed, all of which rely on established trust (p. 209).

So, we could say that the prisoner's dilemma, the lover's dilemma, and the trader's dilemma are all based on the same requirement: *trust*. New love is the ultimate reward/punishment prisoner's dilemma as it looks for affirmation that love given will be love returned, where the whole is thought greater than the sum of the parts. For crowds, the right course is that which benefits the crowd, not one's self, i.e., cooperative, social behavior that trusts that others will not block me in or will let me off when my turn comes, similar to Kant's categorical imperative: "Act only according to that maxim whereby you can at the same time will that it should become a universal law" (which basically asks the question, "What if everyone did as I did?"). This, in turn, can be construed as Jesus' commandment to his followers to do unto others as they would want done unto them, an almost transcendental understanding of how individual behavior affects all.

It seems that the Christian prescription understands the collective need, and the need to look first at the log in one's own eye before taking odds with another's speck when solving problems—and to make amends when we do not (i.e., apologize, forgive). Here, the nonobvious route (i.e., given that most of us find it hard to see our own fault in an argument or apologize) gives the greatest reward. Certainly, if problem solving is the collective goal, it seems that cooperation is essential, as is trusting that the other will concur. All it takes is trust and looking at our own failings or contributions to a problem to gain important insight into a solution—whether in traffic, trading, or love.

11.3. The Iterated Prisoner's Dilemma: "No More Mister Nice Guy" Rules, Again and Again

Continuing with the prisoner's dilemma as it applies to more real-world problems, we begin to gain insight into the value of cooperation, and under what circumstances cooperation fails. There are many examples of transactions between two people whose future behavior is dictated by their past behavior, the so-called *iterated* prisoner's dilemma. Paulos (2001) cited a famous example of two women drug traffickers who exchange drugs for money in brown-paper bags left beside each other on a street corner (pp. 138–139). If the exchange goes according to plan, everyone is happy, but if one decides to dupe the other, it is unlikely the pair will ever agree to a future exchange. As Paulos noted,

> Political scientist Robert Axelrod has studied the iterated prisoner's dilemma situation wherein our two women drug traffickers (or businessmen or spouses or superpowers or whatever) meet again and again to make their transaction. Here there is a very compelling reason to cooperate with and not try to double-cross the other party: you're probably going to have to do business with him or her again. (p. 141)

Sigmund (1995) noted that "it is the *expectation* of further dealings which makes cooperation so alluring" (p. 184). Axelrod (1984) listed many such examples, including the number of times one will invite another for dinner without being invited in return, people exchanging favors, a journalist trading publicity for a story, and businesses colluding on prices (pp. 4–5). In most cases, exchanges between people occur on an ongoing basis, as in a working or spousal relationship, where compromise is continually agreed (as opposed to the one-off drug deal gone bad)—hence the name *iterated* prisoner's dilemma.

As such, how we behaved in the past dictates our dealings with others over time. As in the lover's dilemma above, I am less likely to show again if the other didn't the first time, as in the adage, "Fool me once, shame on you; fool me twice, shame on me." In *An Affair to Remember*, Cary Grant was so miffed at Deborah Kerr for standing him up at the top of the Empire State Building that when they finally met, he behaved guardedly, even to the point of churlishness (not knowing that she had been hit by a car while running to make their meeting).

Axelrod (1984) analyzed numerous examples of the iterated prisoner's dilemma—including business and political strategies, the nuclear arms race, and trench warfare in World War I—to help model human behavior over time, where a one-off decision about whether to rat (defect) or not to rat (cooperate) in the prisoner's dilemma is applied more generally to any transaction, the outcome of which can affect the next transaction. Honor is no longer assumed of one's accomplice, and, thus, acting greedily or kindly affects how one is treated in future dealings.

According to Axelrod (1984), an iterated prisoner's dilemma existed in the trenches during World War I—"a live and let live system" (p. 74)—where mutual restraint was exercised on both sides of the front during miserable weather, for an hour of private time in the morning, and in extensive fraternization at Christmas (p. 78), including a soccer match and the singing of "O Tannenbaum." He further noted, "One thing the soldiers in the trenches had going for them was a fairly clear understanding of the role of reciprocity in the maintenance of the cooperation" (p. 87). In other wars, enemies have foraged for fuel together (Peninsular War), bartered with each other under the protection of white flags (Crimean War), and imposed truces to negotiate surrender or swap tobacco and other provisions (American Civil War) before continuing with normal hostilities (Wade, 2010).

Here, the dilemma is a matter of life or death: in this case, don't do unto me or I'll do unto you. Interestingly, such mutual cooperation typically occurs in times of need, e.g., when rations are delivered, as reported by a soldier on the front line in 1915:

> It would be child's play to shell the road behind the enemy's trenches, crowded as it must be with ration wagons and water carts, into a blood-stained wilderness . . . but on the whole there is silence. After all, if you prevent your enemy from drawing his rations, his remedy is simple; he will prevent you from drawing yours. (Axelrod, 1984, p. 79)

Even in war, it seems, a means to cooperate is essential for survival. Even in war, there are good and bad rules for maintaining order.

Cooperative models have also been used to explain evolutionary change, where the success of transmitting new genetic information from one generation to the next depends on better adaptation to new environments (Sigmund, 1995, pp. 181–206). Cooperation is often rewarded, akin to family cliques that come to dominate a species. As Sigmund (2005) noted, "The payoff is rigged: you have a vested interest in your partner" (p. 190). The ability to recognize one's own is literally a matter of life or death. However, randomness also applies (as always in evolution)—where genetic information can mutate and transmit mistakes—and, as such, forgiveness is needed to avoid the possibility of endlessly damaging recriminations (e.g., a feud or the end of a species). As Sigmund noted, "It is in one's own interest to develop a high sense of gratitude and a limited dose of tolerance" (p. 194).

Furthermore, interspecies cooperation, as in sexual reproduction, is especially advantageous because sex "reshuffles" the genes to hasten the removal of deleterious mutations (as well as to fight off parasitic disease). Genes don't always work for the benefit of the collective, but here group selection wins out. Although there are short-term goals to asexuality, they are bad in the long term because of an eventual buildup of bad genes.

What's more, species that produce variety among their offspring stand a better chance of adapting to a new environment and thus surviving (Sigmund, 1995, pp. 140–147). As Sigmund noted (after Williams), buying 100 lottery tickets will increase your chance of winning, but not if they are all the same numbers: "Multiple copies of the same genome would be as wasteful as multiple purchase of the same lottery ticket" (p. 147).[9]

Evolutionarily speaking, stability (including adapting to change) is most important; thus, cooperation is preferred. It isn't always about who escalates or retreats (whether in evolution, games of chicken, house invasions, arms races, etc.) but about how cooperative behavior is achieved. What's more, such strategies (for the species anyway) are more robust if they are learned rather than hardwired, which itself is, paradoxically, a hardwired genetic trait (Sigmund, 1995, p. 177).

Modeling Human Behavior

To learn more about ongoing human behavior, Axelrod (1984) created an organized "computer tournament" where the iterated prisoner's dilemma was run as a game using different programmed strategies against each other (pp. 27–54). In total, 63 entrants submitted strategies, and it was found that the more altruistic strategies fared better than the meaner strategies (p. 36). But not all were blindly altruistic. The best (and simplest) was a "tit-for-tat" strategy, which, as its name suggests, returns like for like—a surefire way to arrive at cooperation or, rather, to prevent uncooperative behavior from continuing in ongoing dealings (p. 31).[10] He also identified four essential characteristics for getting along: *non-envious, nice, reciprocating,* and *not too clever* (pp. 23, 110).

In Axelrod's parlance, *non-envious* means not comparing oneself to others who score better such that you defect only to score points, which ultimately leads to mutual and possibly endless (echoing) recriminations: "A better standard of comparison is how well you are doing relative to how well someone else could be doing in your shoes" (p. 111), as in the criteria for scoring in duplicate bridge. It seems counterproductive to try to keep up with the Joneses, or to covet your neighbor's belongings, when you can properly compare yourself only to yourself. Simply put, how well one does should be measured against one's own life.

As the name implies, *nice* is a cooperative strategy, defined as not being the first to defect, as we saw above for cross-purpose traffic, commuter trains, and airport carousels. "Not being nice may look promising at first, but in the long run it can destroy the very environment it needs for its own success" (Axelrod, 1984, p. 52). Of the 63 programs entered in the tournament, 14 of the top 15 were "nice" and "the overall correlation between whether a rule was nice and its tournament score was a substantial .58" (pp. 44, 195–196). Such a tournament is only a model for human behavior, but it may be appealing to many that being nice has its rewards.

Reciprocating returns action in kind—cooperation for cooperation, defection for defection (an eye for an eye, a smile for a smile, aka the law of equivalency or retribution)—ensuring that one is not continuously taken advantage of but is also forgiving of another's indiscretions. The winning tit-for-tat strategy "is nice, forgiving, and retaliatory. It is never first to defect; it forgives an isolated defection after a single response; but it is always incited by a defection no matter how good the interaction has been so far" (Axelrod, 1984, p. 46). This strategy is tough but effective and seemingly the best and most robust means for cooperating with others in a more aggressive, occasionally misfiring world. A nice but reciprocating strategy is happy to cooperate but is nobody's patsy.

Not too clever relates to being transparent in one's dealings, allowing others to benefit from your cooperative manner (Axelrod, 1984, pp. 120–123). Not knowing the strategy of another (a stranger or friend) can be as dangerous as not letting another know one's own intentions, as in countless spy stories where agents question the validity of each other's intentions or an uncertainty about a new neighbor or new kid on the block. The nuclear arms race or proliferation of weapons of mass destruction can also suffer from overly secretive maneuvering. The same applies to a misfiring love relationship, when one doubts one's partner's vows and acts foolishly (see any number of Shakespeare's plays).[11]

As Gowa (1986) noted, in its simplest form, a tit-for-tat strategy "turns conventional wisdom on its head," preferring Hammurabi's Code to the Golden Rule (p. 167). However, as Axelrod (1984) further noted about his simulated behavioral world, a strategy of being a bit more forgiving than the simple tit-for-tat model (what he called "tit-for-two-tats") would have won the tournament if it had been submitted, and "the implication of this is striking since it suggests that even expert strategists do not give sufficient weight to the importance of forgiveness" (p. 39). How one achieves such cooperative behavior, once again, becomes an issue of trust, which is built up (or destroyed) over time.

Nonetheless, "if the main danger is from strategies that are good at exploiting easygoing rules, then an excess of forgiveness is costly" (Axelrod, 1984, p. 120), which, in the absence of a greater punishment, the one-tit-for-one-tat strategy specifically guards against. The main problem of repeated transactions between models in Axelrod's computer world, therefore—as it often seems in life—is that the good are taken advantage of when inadequately protected, or when punishment or at least the potential for punishment is absent. One has to be nice, *but* reciprocating and a tit-for-tat strategy (or the threat thereof) are required for protection.[12]

The Real World

But how can good behavior be enforced in the real world, where unsociable, illegal, and immoral actions are routinely left unpunished and in fact encouraged? Short of convincing everyone that we need to cooperate for all to succeed, upping the penalties by the appropriate authorities (or fairly applying existing penalties) would protect against the less virtuous, especially if following like with like doesn't seem right. Clearly, if lawbreaking goes unpunished—whether for top earners, government officials, or the common thief—it's harder for the average person to abide by the rules.

Shared resources are particularly subject to exploitation and will become unusable if all seek to advance only their own gain, as we have seen in a number of "commons" problems, from airport carousels to financial obsession. As Sigmund (1995) noted, cooperative strategies do poorly "in a population of inveterate defectors" (p. 189), whereas "kinship facilitates cooperation" (p. 190). In the absence of representative authority, all-out clique wars or Mafia-style living appears to be the most viable option to ensure success for one's own, but leaves many less well-off. An increase in selfishness may in fact be a direct result of lax regulations.

One might think that random changes of strategy can help keep one's opponents on their toes, a seemingly well-practiced modus operandi in today's world. Bluffing in poker is a good way to create uncertainty about one's holdings or a series of holdings over the course of a session. If people are unsure about another's actions, they may better treat him or her as if dealing with a stranger. As Sigmund (1995) noted, "By judiciously mixing their strategies, players can always maximize their minimal payoff, or, what amounts to the same, they can minimize their opponent's maximal payoff" (p. 163). But in poker, players begin to form ideas about who's who based on their play over the long run, from the out-and-out chancers to the more rapacious randomly bluffing sharps. Highlighting the value of hiding one's holdings, Norwegian poker player Annette Obrestad even won a tournament without once looking at her cards!

However, a player who bluffs too much exposes the ploy and encourages others to counter with more aggressive strategies, which can work for or against that player in further games. The boy who cried wolf is an example of someone who bluffed too much and was eventually ignored. A particularly good strategy is to bluff until you get caught. If you are someone who plays only with good cards, it is essential to bluff and get caught now and again, showing your cards when you do to make a point of being found out as someone who bluffs and is not to be trusted, thus increasing the chance of a bigger pot down the line with a nonbluffing, winning hand (provided you haven't gone broke in the process).

In baseball, even a hard-throwing pitcher will be caught out (or caught up to) eventually if he doesn't mix his pitches. The same is true in tennis and soccer, where the pace and location of a serve or penalty kick must be varied to keep an opponent off balance, lest he guess your intentions. In any game or sport, the point of keeping your opponent guessing is to mix your strategy lest you be figured out. Sigmund (1995) even suggested flipping a coin to ensure maximum randomness (p. 171).

But poker and sports are zero-sum games, where deception (or not being transparent) is encouraged, and, as noted, random strategies fare badly because their intentions are so unclear by design, not to mention promoting a lack of feeling or joyfulness in a world full of random strangers with unconnected lives. Being unclear may work in poker, but in life random actions can be wholly misleading. In Axelrod's (1984) tournament, the one wholly random strategy finished second to last (p. 193).

To couch the dilemma in more human terms, we can consider oppos-ing strategies as erring toward one strategy or the other, particularly if, like most, we tend to act in one way (presumably good) but occasionally "go against the grain," whether intentionally or otherwise. Is it better to err on the side of not ratting or ratting, cooperating or defecting, cau-tion or bluffing, freedom or security, openness or secrecy, with or with-out clear communicated intentions? Or is it better to be predictably good, holding oneself open to being taken advantage of on occasion, or hard-nosed so you can never be burned? We all know the diligent worker who gets ploughed under with more and more work, while the do-nothing slacker is tolerated—where the old refrain, "If you want something done give it to the busy person," hardly seems fair. Is a com-bination of strategies thus required—sometimes good, sometimes bad (leaning one way or the other)—so that others will know they can't take you for granted?

Or perhaps in between the reactionary, eye-for-an-eye strategy and the more authoritative, heavy-punishment strategy lies a middle ground where I don't do wrong because by so doing I invite wrongdoing on myself. I cannot protect myself other than by a deep understanding that it is wrong to wrong another because in a fundamental way I am wrong-ing myself; thus, I must police myself, which in essence is at the core of the premise of cooperative behavior—trust and responsibility.

Imagine a world without external authority, where we ensure our own right actions: no speed limits, no security guards, no checkout sales staff. Such *Star Trek*-like future worlds seem unlikely, although in Eastern cul-ture accepting responsibility when caught is paramount, as is shame over one's wrongdoing, which in effect reduces wrongdoing. We may not be ready for Utopian living in an often harrowingly cutthroat Western world, but we seem to be missing a greater sense of responsibility to others with our prevalent "winner-take-all" and "don't-get-caught" attitudes.

The social contract asks us to be good so that we can expect others to be good and can go about our merry ways, socializing, working, making money, and doing whatever other things we do, without fear of being taken advantage of. But somehow this doesn't happen, and the world gets divided into givers and takers, where the more honest are taken advantage of or abused. It seems that without proper laws to protect the honest citizen, all that remains is to return like with like, hoping not to incite endless recriminations that come with mistrust, as seen by the num-ber of ongoing conflicts throughout the world and in our own lives.

In a world of only a few cheaters (i.e., defectors), a tit-for-tat strategy loses out and requires like-minded others to succeed (i.e., cliques).

Sigmund (1995) noted that for the prisoner's dilemma payout, a tit-for-tat strategy wins if more than 6% are similarly inclined (p. 189). However, in such a world, misunderstandings become highly problematic and can be thought of mathematically as "noise" in the system (i.e., a misfiring or mistaken defection). Noise can be especially costly in human interactions, as it sets off an immediate defection, often taking years if not decades to overcome. Again, the solution is to be clear about your intentions and to give others the benefit of the doubt when required—though, as Sigmund adroitly noted, "not always!" (p. 194).

It seems that nice guys can do well, but only in a trusting world with rules can we *all* do well—in that egalitarian, equal-opportunity world we are all told exists. Furthermore, it seems that altruistic actions (or even honest living) can be preserved only from the outside, when organized protection and regulation exist free from bias and persuasion. Left to their own devices, those in power become all the more powerful.

Is It All Money Madness?

Interestingly, the incentive for being nice may be evolutionarily hard-wired into our genetic makeup, along with the knowledge that we are better off as a species when we cooperate (parasites notwithstanding). Ball (2005) noted,

> People do not, in the absence of a higher authority, necessarily seek to exploit one another in the way Hobbes envisaged. But neither do they desist from it because of a "reason" instilled in them by God. The "reason" can come from nature alone: from the inexorable mathematics of interaction coupled to the winnowing effect of natural selection. (p. 532)

Being nice also has its rewards in promoting harmony in the workplace and more efficient production. Pryce-Jones (2011) noted that leaders and bosses get more from their underlings by being nice, which saves money in the long run by reducing sick leave and turnover, and instills better focus: "[Likability] is a key element for high performance and happiness at work."

This seems, however, not to be the case where money is concerned, particularly when seen as an unfair feedback loop that increases wealth for the already wealthy. As for ensuring a fair playing field in a world of much inequality, the dilemma is all the greater. Echoing the conditions of a non-zero-sum game, former United Nations Secretary-General Kofi Annan (2009) elegantly wrote, "No one's stability, security and prosperity can be guaranteed unless we strive to tackle the gross inequality of wealth, opportunity and influence in our world." Former United Nations High Commissioner for Human Rights Mary Robinson noted, "At a time of unparalleled prosperity for some, 54 countries are poorer now than they were a decade ago" (Jackson, 2009, p. xv). It seems that business ethics—at home and in an ever-more-globalized world—are contrary to those of society, fostering a feeling of suspicion or doubt in our most basic transactions.

Today's society has numerous tiers of wealth and advantage, with groups of varying degrees of self-reliance. The rich are artificially separated, as if a human species unto themselves, and the poor are seen as dependent or willfully incapable. How do we have such polarized groups, especially given that we preach a language of equality, freedom, and brotherhood for all? Cooperation is meant to create societies for everyone's betterment.

We have already seen the improbable figures. In the United States, 1% own 33% of the wealth (Davies, Sandström, Shorrocks, & Wolff, 2008); in the United Kingdom, 1% own 70% (Marsh, 2009b). Furthermore, 5% make more than 20% of the income, with the 0.1% in the highest income bracket receiving 7% of the income at an average of $3.5 million per year (Krugman, 2009, p. 259). As a result of the credit crisis, 25 million people depended on emergency food relief (Schama, 2010). More than 40 million Americans live below the poverty line (Morello, 2010). Globally, the richest 2% own half the world's wealth (Brown, 2006).

But such disparity is wrong, not just because it is exclusive but because it doesn't allow for fair transactions or promote equal opportunity (thus setting off a never-ending series of misfiring recriminations). The misguided solution has been to segregate—even to criminalize—those who lose out in such a mismatched game, thus ignoring the real problem. If all the poor were abusers and criminals, one could argue that the more industrious must be protected, but we know this is not true. In fact, if time at work were used as a measure, one could easily argue that the rich are the wasters and abusers.

Krugman noted the effects of inequality on democracy: "The ugliness of our politics is closely tied to the inequality of income. . . . The people who have the most influence are not interested in having good public services, because they just don't use them" (Free Exchange, 2007). What is worse, as though mired in Victorian-era thinking, we are led to think that the poor are a parasitic group—begging, borrowing, and stealing—when it is the wealthy who routinely live off the work of others, propped up by a neo-Christian, social-Darwinist order designed to protect them and their selfish ways. If equality is used as a measure of a society's success, then ours is not very advanced.

Narayana Murthy, chairman of the multibillion-dollar Infosys and considered the Indian Bill Gates, understands how excessive wealth can detract from society:

> When rich people get rich, they cut themselves off from the context that has earned them these riches—the context of the common man. They forget they are part of society. I believe that unless we are in touch with reality and the common people, we will not be in a position to add value to a society. (Hell, 2005)

Kenneth Clark (1969) noted that "great wealth is destructive," and he might well have equated those who choose wealth to the exclusion of other pursuits (*wealthmongers*?) with the barbarians, who once obliterated civilization in Western Europe. Warren Buffett, the world's second-richest man (before donating three-quarters of his $31 billion fortune to the Melinda and Bill Gates Foundation), described the United States as "a great meritocracy," yet disparaged inherited wealth:

> I cannot think of anything that's more counter to that than dynastic wealth. My kids have had all kinds of advantages, let alone being given billions of claims checks on other people. The idea of passing those from generation to generation so that your descendants can command resources just because they came from the womb flies in the face of a meritocratic society. (Lister, 2006)

Additionally, he told *Fortune* magazine, "It's neither right nor rational to be flooding [my kids] with money. Dynastic mega-wealth would further tilt the playing field that we ought to be trying instead to level" (Loomis, 2006). Even Gates cursed his own wealth when it was revealed that he was the richest man in the world: "I wish I wasn't. . . . There's nothing good that comes out of that" (Burkeman, 2006).

The prisoner's dilemma shows us that we must *all* cooperate, lest we all suffer. The religious version—"Do unto others as you would have them do unto you"—is an equivalent expression for success, which if applied keeps the bad dominoes from falling, be they rudeness, a general attitude of selfishness, crime, or hoarding wealth. No wonder it is held up as the fundamental tenet of Christianity and considered an ideal strategy for problem solving. Nonetheless, the bad eggs are ruining things for others with their obsessions and me-first, money-driven attitudes, unduly privileged by unfair laws and protected from their own wrongdoing by nonexistent punishments.

Winning at All Costs: How to Become Less Successful

The either/or battle of the individual and the collective is not unlike how we choose to err, whether on the side of compassion or security, safety or money, or for or against a particular position (which we looked at in Chapter 4). Competition—as it is said to exist in a free market—globalization, and sports today fit the same bill—i.e., some competitive ventures spur us on not to excellence and innovation but to escalating costs and the limits of illegality, contrary to intentions. Fairness, again, is the issue, where the dark side of the dilemma is to be *too* good.

Anyone who has ever played a simple schoolyard game recognizes the need for fairness. When the teams are picked, the sides are divided to ensure a fair match. But it wouldn't do to stack one team with all the best players, and so a simple control is introduced to enhance enjoyment for all. Without such a control, the game would quickly fizzle out as the unfortunate sad-sack team cried foul.

The same cannot be said of more organized sports, however, which are designed to maximize the chance of winning, not competiveness. Professional teams go to great lengths to pick the best players and win by as much as possible. Much rests on the prestige of being the winner in any number of games. But in the rush to be better than the rest, have we forgotten the point—to compete within a fair set of rules?

As we have seen, professional sports teams are becoming increasingly trapped in ongoing wage crises as they attempt to better their opponents, where the costs to owners (and fans, who ultimately pay the price) are not only becoming prohibitive but result in a reduction of competitive teams, since only the most well-healed can pay and play—as seen around the world in leagues dominated by only a few "elite" teams. If competition is the stated goal, regulation is essential, as in salary caps and reverse-order player drafts to ensure a basic fairness so that the game is then decided on the field.

Of course, there are always ways around league regulations, and vigilance is essential to maintain the highest levels of fairness. College football in the United States is notorious for cutting corners to get an edge, even going so far as to offer side payments and prostitutes to prospective high school recruits (Frank & Cook, 2010, p. 136). The National Hockey League reworked its own salary-cap structure to allow a 15-year, $100 million contract to forward Ilya Kovalchuk, seriously undermining league fairness. And, as we have already seen, world soccer is collapsing in a heap of debt as it persists in the unregulated, free-for-all betting up of player salaries.

Competition will always suffer if not regulated. Frank and Cook (2010) cited dueling, which ultimately became regulated (prescribed distance, small guns, one shot) to minimize the amount of deaths, since unregulated dueling served no one's purpose (presumably to settle grievances) and only increased the chances of mutual death (pp. 167–168). As in most competitive games and market economies, dueling was characteristic of "winner-take-all markets," as we saw in Chapter 10, which "almost invariably result in mutually offsetting, and hence socially wasteful, patterns of investment" (p. 168).

What's more, the ever-increasing cost of unchecked competition becomes a "positional arms race" (Frank & Cook, 2010, p. 127), where "rewards that depend on relative performance will lead to excessive investment" (p. 145), a hallmark of the iterated prisoner's dilemma, or in this case the iterated competitor's dilemma. Regulations or "positional arms control agreements" are needed to combat spiraling competition in school starting age, SAT tests, hockey helmets, "blue laws," campaign finance limits, and taking more than one spouse (pp. 178–184, 209). All are regulated to prevent unlimited competition deleterious to all, although, as Moyo (2011) noted with regard to fair competition on the global economic stage, such a notion is "nice in theory, but only works when everyone plays by the same rules" (p. 152).

Axelrod (1984) was fundamentally aware of the problems of excessive competition and the environment of mutual power from his prisoner's dilemma computer tournament: "Even expert strategists from political science, sociology, economics, psychology, and mathematics made the systematic error of being too competitive for their own good, not being forgiving enough" (p. 40). The same wasteful behavior is seen in intraspecies competition, where "suboptimal" results occur when a trait that was originally successful for a few "naturally selected" winners is universally adopted to much less success—for example, obtrusive, though sexually attractive peacock feathers, which Heath (2009) noted "is a biological example of what economists call a 'race to the bottom'" (p. 30).

There are many examples of escalating tit-for-tat exchanges that do nothing to improve or increase competitiveness and instead increase only the cost. One of the more extreme examples is the madness of the Cold War and its escalating nuclear arms race, which one can argue was not won by the United States because of the superiority of the American way of life or the politics of the right but, rather, lost by the Soviet Union because they bankrupted themselves trying to keep up with the American Joneses. Election spending (especially so-called super PACS) also increases the costs without expanding the debate, which seriously undermines the democratic process.

War has been particularly susceptible to winner-take-all thinking, as improved technology makes killing easier yet doesn't necessarily decide the outcome any sooner. Stalemates appear despite the advances, increasing only the killing efficiency on both sides. In the U.S. Civil War (considered the first modern war, having taken place in the midst of the Industrial Revolution), better bullets, mass-produced rifles with faster reloading times, and advanced infrastructure turned the bitter conflict into a war of attrition lasting more than 4 years with more than half a million casualties.

Auto racing is similar, requiring a limit on the amount of fuel cars can take on for safety reasons (as well as other such restrictions). If the rule didn't exist, one team would add more fuel, with the others following suit to keep up, making the sport more dangerous without changing its competiveness. In reality, such unrestricted competition serves only to shift the baseline or "norm" at considerable extra cost to everyone.

As we have seen, the same failed thinking applies to modern economics via the supposed free market, which puts economic performance first, forcing out lower-end players and ultimately restricting competition, contrary to its stated goal. We have also seen that in such cases, the difference grows, with the rich getting richer relative to the poor not because of any inherent ability but because a small return becomes a bigger return. Better access, more efficient means, and better representation produce an unfair feedback loop detrimental to the rest.

So do we compete at all costs, using money as the arbiter of success, and watch the world grow more unequal? Or do we cooperate and seek to explore new ways to succeed? Frank and Cook (2010) noted that "citizens of the world at large, for example, might fare better if we spent more dollars on food and health care, and fewer dollars on improving the picture clarity of HDTV" (p. 146), regardless of the apparent positional costs to individuals or nations who do not play the game. Heilbroner and Thurow (1998) similarly wondered, "Is it better for a million consumers to buy cheap cameras, or for one hundred thousand steel workers to have higher incomes?" (p. 210).

The choice is ours. The great advances of our world can occur without preferring the business side of the human equation and its emphasis on the speed of change and excessive profits. The totality of humankind must be included in any measure of excellence and advancement. In fact, without ensuring fairness for all, we limit our chances for future success.

Notes

1 **Galileo's drop test**: Kuhn (1957/1997) noted that Galileo likely didn't perform the famous experiment, but rather one of his critics who supported Aristotle did, and had set out to refute Galileo (p. 95). In this case, the heavier object hit the ground first (no doubt because of shape and air resistance).

2 **Sugarscape**: Sugarscape simulates a trading society and was developed by medical doctor Joshua Epstein and computer modeler Robert Axtell, both professors associated with the Brookings Institution (Epstein & Axtell, 1996). John Conway's Game of Life also simulates cellular evolution and interacting "cultures" (Gardner, 1970).

3 **Bulk buying**: Bulk buying has been around for years, including discount bulk food marts, low-cost department stores, or the Internet coupon company Groupon. It is important to distinguish, however, between one-off products discounted because of bottom-up consumer-organized demand bulk buying and a top-down seller's release of undersold supply.

4 **Other traffic jam causes**: Mathematician Daniel Bernoulli noted the relationship between pressure and speed in his analysis of fluid flow, as witnessed by anyone whose water has fizzled out because of a crimp in the backyard hose.

5 **Emergent group behavior**: For fascinating models on how group behavior emerges from a collection of individuals, see "Microsimulation of Road Traffic Flow" (www .traffic-simulation.de) or "Panic: A Quantitative Analysis" (http://angel.elte.hu/~panic/).

6 **Internet origins**: The seeds of the Internet were planted by ARPA (Advanced Research Projects Agency) when a nodal, noncentral structure was formulated as the best way to keep a system running after one or more of its components was knocked out, thus creating the necessary means for a bottom-up, self-organizing, robust network.

7 **Tournament groups**: The top two teams advance; 3 points for a win, 1 for a tie, and 0 for a loss (a points system that often creates prisoner's dilemma-type scenarios).

8 **Insurance and non-zero-sum games**: Insurance is non-zero-sum by benefit, although technically zero-sum in its accounting if operated not for profit.

9 **Genetic exploitation**: Sigmund (1995) wryly noted that the move from asexual reproduction (where the female of the species is exploited by parasites) to sexual reproduction (where she is exploited by males) can be construed as a "switch from bad to worse" (p. 153).

10 **The golden mean strategy**: Tit for tat did not win any matches but was the best *overall* strategy.

11 **Deceptive evolutionary practice**: In contrast, being deceptive in higher organisms (discouraging cooperation) is essential to keep parasites (e.g., viruses and microbial diseases) from breaking through the immune system and for sexual reproduction as opposed to asexual reproduction, which aids parasitic attacks (Sigmund, 1995, pp. 148–153).

12 **Advanced behavioral simulations**: In a further tournament, Axelrod's (1984) results showed that clusters of good working together can modify the meaner strategies, where a number of tit-for-tat programs beat up on the less-cooperative ones (p. 65). Such behavior, however, requires even more astute and regulated cooperation.

THE ROOTS OF ECONOMIC DISASTER

Making Money From Money

I have argued that greed is destroying our lives, not in any naïve, Utopian way that suggests the abolition of private property or in a moralist sense, as in the love of money is the root of all evil. Money has become an overly complex system, and the rules involved in its transactions are neither fair nor transparent—essentially making it a game for insiders and, worse, one even they don't fully understand. In short, money has become too great an arbiter in our lives. What's more, given that the near collapse of an entire world economy came from within the financial sector and not from resource shortages, labor problems, or security issues, as many believed would happen one day, should we also question whether there is something fundamentally wrong with making money from money?

Heilbroner and Thurow (1998) noted that half of the 3,000 big businesses (defined as having assets greater than $250 million) in the United States are in finance (p. 46).[1] Soros (2008) observed that the financial industry has been growing in the United States since 1972 and that finance made up 14% of stock-market capitalization in 1990, growing to 23% by 2006 (p. 125). As can be seen in Figure 12.1, the financial sector has been growing at the expense of manufacturing (shown here from 1987). Given the corresponding social costs of economic inequality and lost jobs, could it be that too much of our economy has to do with money?

Arrighi (2010) noted that the shift from an expanding materials-based economy to an ultimately unsustainable financial one is marked by the change from noncompetitive regulated spheres of loosely cooperating ventures to the cutthroat competitive jockeying of organizations, where "the losses of one organization become the condition of the profits of

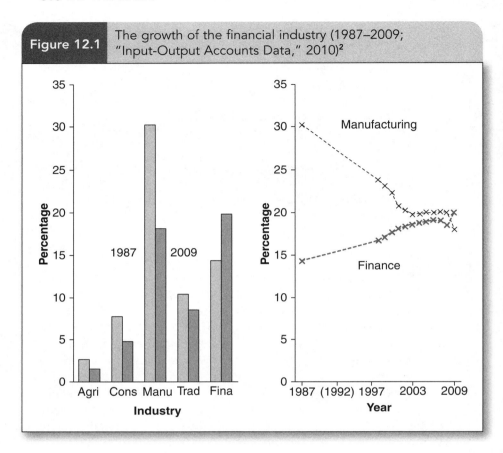

Figure 12.1 The growth of the financial industry (1987–2009; "Input-Output Accounts Data," 2010)[2]

another" (p. 233)—the very definition of a zero-sum game or "even a negative-sum game" (p. 233)—which in turn is marked by a fight for resources, whether because of more players or limits to supplies.

The state of the stock market today bears this out, where in short order, Wall Street has changed from an organizing force for raising capital to little more than a casino. More than 70% of all trades are now high-frequency trades,[3] where punters bet on the short-term rise and fall of a stock, often within minutes, facilitated by ultra-fast computers and encouraged by investment banks that take a percentage of every bet. During the 2008 meltdown, some brokers even had a vested interest in customers playing and losing, such as Goldman Sachs selling subprimes while at the same time betting against them, and Merrill Lynch and Citigroup withholding bad stock ratings from their clients (Mason, 2009, pp. xi, 72–73).

Money itself has become the new material, with the U.S. banking sector composing 20% of GDP (Moyo, 2011, p. 22) and corporate profits from the financial industry at more than 40% (Ferguson, 2010). As Soros (2008) noted, "The financial industry was allowed to get far too profitable and far too big" (p. 146). The transformation from makers and users to lenders and borrowers to gamblers is almost complete, marked by the extraordinary rise in a new lending class of financial merchants beyond the traditional rentier class.

This is doubly troubling since religious leaders, philosophers, and politicians through the ages have told us that living a good, balanced life nourishes spirit and body, and that straying from a good, middle way leads to excess, depression, insecurity, and an uncertain future. The world is littered with those who can't manage, from economic down-and-outs and impoverished working-class families to the now more common, stressed-out middle class. But is this because of injudicious choices, as in a lingering social Darwinist view that uses money to define success? And can any of us say that we are in charge of our economic lives as perhaps we once were in the pioneering days of yore, when one could make it solely by the sweat of one's brow?

Our economies have morphed from primitive "might-is-right" transactions to trading economies of value, bartering and swapping goods and services, to money as a currency of transaction to an increasingly intricate state-controlled proxy system, where money has become the exchange rate of our saved money, the interest rate of our borrowed money, and the rising or falling value of our invested money. Have we forgotten that none of us can take it with us when we leave?

To be sure, making money from money has been around forever, but never has everyday usury been so complex. A quick survey of the simplest so-called financial "instruments" includes subprime mortgages, arbitrage, and derivatives, as well as the more complex collateralized debt obligations (CDOs), credit default swaps (CDSs), and other asset-backed derivatives. But "subprime mortgage" is just a fancy name for legalized loan-sharking, sold to the poorest of borrowers who can't afford it. Taking advantage of price differentials is the same as betting for and against the outcome but is called arbitrage in the business world and is the basis of hedge funds, another fancy name for gambling. Derivatives relate to an underlying investment, shadowing the ups and downs of an asset (or even the *perceived* ups and downs). Few understand the arcane world of the more complex derivatives, such as CDOs (sliced and diced and repackaged to suggest a higher intrinsic value and minimized risk) and CDSs (offer insurance against a failing derivative and are traded to offset excess risk).

Summers (2010) noted that by 2007, the volume of derivatives trading was 100 times as large as in 2001. Zakaria (2009a) noted that

> derivatives need to be better controlled. To call banks casinos, as is often done, is actually unfair to casinos, which are required to hold certain levels of capital because they must be able to cash in a customer's chips. Banks have not been required to do that for their key derivatives contract, credit default swaps.

Gary Gensler, a one-time Goldman Sachs partner turned head of the American Commodity Futures Trading Commission, called for better regulation all round: "I disagree with anyone who says derivatives did not play a part in the crisis. Like San Francisco after the earthquake, we had a calamity, and now we need building codes" (Bowley, 2010). Summers (2010) noted that a "tough-minded" financial reform bill is essential to keep the democratic process afloat, particularly when lobbyists were spending $1 million per member of Congress to derail its implementation. Soros called for a ban on CDSs and likened them to "buying life insurance and then giving someone a license to shoot the insured person" (Hosking & Jagger, 2009).

To listen to the news today, one can't help but wonder what it all means. But like it or not, we have to deal with today's financial institutions and the mess brought on by underregulated financial institutions, overleveraged banks, toxic subprime loans, and the poorly advised and injudicious governments that brought us here. Like it or not, we have to try to understand the jargon of the latest and greatest financial crisis.

The Financial Mess: Where to Start?

So where does one start? Who is responsible may be the first question, but the answer depends on who one asks. According to Soros (2008), who has been engaged in the stock market for more than 50 years, the blame should be laid squarely at the feet of the Reagan-appointed, former chairman of the Federal Reserve, Alan Greenspan (p. 121). Others wouldn't disagree, including John Casti of the Santa Fe Institute, Nouriel Roubini of New York University, Nobel economist Paul Krugman, and former U.S. Secretary of Labor Robert Reich.

Greenspan certainly played it fast and loose with low interest rates post-2001 and discouraged the regulation of CDSs, but Soros (2008) faulted Greenspan's political leanings:

I was impressed by [Greenspan's] forward-looking, dynamic approach, which stood in sharp contrast to the static, rearview-mirror assessment of European central bankers. He can be faulted, however, for allowing his Ayn Rand-inspired political views to intrude into his conduct as chairman of the Federal Reserve more than would have been appropriate. He supported the Bush tax cuts for the top 1 percent of the population and argued that the budget deficit should be reduced by cutting social services and discretionary spending. And keeping federal funds at 1 percent longer than necessary could have had something to do with the 2004 elections. Responsibility for the real estate bubble can be justly laid at his feet. (p. 121)

In his defense, Greenspan admitted that he wasn't "flawless" and couldn't say he "made no mistakes" (Crossley-Holland, 2009), stating to the House Committee on Oversight and Government Reform that the economic orthodoxy he had previously believed existed was wrong: "Those of us who have looked to the self-interest of lending institutions to protect shareholders' equity, myself included, are in a state of shocked disbelief" (Andrews, 2008).

Soros (2008) also noted that a super-bubble was created prior to the housing bubble, resulting from the reestablishing of laissez-faire economics, which he thought had been forever laid to rest, during the Reagan and Thatcher era (p. 94). "Under its influence the financial authorities lost control of financial markets" (p. 94), where "ever more sophisticated financial instruments were invented, and new ways to keep assets off balance sheets were found. That was when the super-bubble really took off" (p. 117). The French, in particular former Finance Minister Christine Lagarde, have also gone on record to blame the unregulated laissez-faire markets of the past 20 years, citing the freewheeling excess and high degree of complexity, leverage, and greed, particularly in the American and British systems. Perhaps, then, Reagan and Thatcher are to blame?

Soros (2008) also noted that the international banking crisis led to greater bank freedoms in the United States, especially the merging of investment banking and commercial banking as well as the selling of debt down the line by banks that didn't want to keep loans on the books and sold them off to other unregulated investors (p. 117). In this free-for-all period, aggressive banking practices created a market where homes could be bought with no money down and no questions asked, during which "Americans added more household mortgage debt in the last 6 years than in the prior life of the mortgage market" (p. 85). In everyday terms, that means living beyond one's means.

Perhaps, then, easy credit was the start of what brought the world economy to the brink? Or did the dominoes of doom fall prior to the housing bubble, the super-bubble, and Reagan and Thatcher? Do we have to go back prior to the emergency G20 meetings in London, Pittsburgh, Toronto, etc.; prior to the bank bailouts; prior to AIG, Lehman Brothers, Bear Sterns; prior to the accounting irregularities at Enron, the 2000 dot-com bubble, the '98 Long-Term Capital Management bailout, the '97 East Asian emerging market crisis, the '86 savings and loan crisis, the '82 international banking crisis, the 1970s real estate investment trusts, the 1960s conglomerate boom, the 1920s Great Depression, the 1907 banker's panic? How far until we get to the real problem?

Ferguson (2008) stated that since 1870 we have had "148 crises in which a country experienced a cumulative decline in GDP of at least 10 per cent" and "87 crises in which consumption suffered a fall of comparable magnitude." Galbraith (1977a) noted that the time between crashes is about 20 years. Perhaps, then, modern capitalism is the real culprit. It seems that crisis is the de facto norm to capitalism.

Arrighi (2010) reiterated that capitalism itself is to blame because of its tendency to expand financially at the expense of manufacturing in reaction to competitive pressures. He noted that this process has reoccurred in historical capitalism since the 14th century by successive powers of accumulating capital, including Genoese-backed Spain, Holland, Britain, and the United States, which all ultimately declined after hitching their hopes to a small financial elite. Referring to the dynamics of global crisis, he noted, "Faster than under any previous regime, the *belle époque* of the US regime, the Reagan era, has come and gone, having deepened rather than solved the contradictions that underlay the preceding signal crisis" (p. 309).

Furthermore, the Keynesian solution to worldwide depression may have unleashed a massive 20th-century material expansion of world capitalism, but the modern corporation undermined its workings, making nonsense of a system that purported to be in the interests of fair competition and free trade. As evidence of an unsustainable financial new world order, Arrighi (2010) cited two-tier oil pricing, which marked down U.S. costs by 40%; the emergence of floating currencies, which focused corporate capital on speculative day-to-day shifts in exchange rates; and offshore money markets more concerned in petrodollars and eurodollars than in trade and production (pp. 220–224). In effect, capital itself may be to blame or, rather, the use of capital to fuel its own growth, a classic doubling game that fails to recognize its own failing existence.

Others, however, held different views about the most recent financial brinksmanship. The governor of the Chinese central bank, Zhou Xiaochuan, found George W. Bush's Fed chairman appointee, Henry Paulson, culpable. In a meeting with the former Goldman Sachs chairman and treasury secretary under Bush, Zhou said that "overconsumption and a high reliance on credit is the cause of the US financial crisis" (Preston, 2008). In his speech to Paulson about ways to right the ship, Zhou noted that "the US should take the initiative to adjust its policies, raise its savings ratio appropriately and reduce its trade and fiscal deficits" (Preston, 2008). However, as leader of a country with a $2 trillion foreign exchange reserve, Zhou may have been a little more than worried about a global collapse started by American excess.

Kaletsky (2010a) would not disagree:

> Had it not been for the inexplicable policy blunders of Henry Paulson in mid-2008, above all the decisions to wipe out shareholders in Fannie Mae and to bankrupt Lehman Brothers, but also his refusal to counteract the speculation that drove the oil price to $150 in mid-2008, the world would probably have suffered nothing more than a mild recession.

Others have said that the financial meltdown was caused simply by financial organizations making bad loans, which they then sold to other financial organizations that were selling the same bad loans back to them to off-load risk levels, playing the small percentages against each other, until the whole house of cards came tumbling down. From the outside, that looks like plain old greed.

One can point to Lehman Brothers, the largest ever underwriter of real estate loans in the United States, which was leveraged to the hilt at 44 to 1 (loans to capital; Smith, 2009a) or the mortgage lender Ameriquest, the Wal-Mart of borrowing, selling to anyone and everyone via its unchecked "stated income" guarantee—not to mention its "teaser-rate" mortgages, which started buyers off low and sharply increased in time. As one area manager put it, qualification amounted to "if you could chew gum and walk at the same time" (Smith, 2009b).

However, when house prices didn't keep rising, as in any failed doubling game, the subprime mortgage market—which borrowed at 2% or 3% from Wall Street to lend at 8% (Smith, 2009b) and guaranteed its risk by unregulated CDSs—was found out as the biggest of shams. Linking housing to the vagaries of an unfettered free market may well be the biggest folly of growth-obsessed capitalism, which marks everything as a commodity.

Moyo (2011) agreed that unfettered risk by profit-hungry banks was to blame but was made possible by a risk-averse government, which encouraged banks to lend freely to anyone with no downside (p. 39). Government-backed guarantees for the banking sector preferentially allowed lenders to "embark on unrestrained and reckless risk-taking" (p. 38), which led to the creation of a new category of ultimately disastrous subprime mortgages. "Had the government-guaranteed safety nets not been there in the first place, Freddie Mac and Fannie Mae bondholders would have been much more vigilant" (pp. 39–40). Add that to a pumped-up housing market that, in effect, was artificially fueling economic growth and lenders lost all connection to lendees (p. 44). As Soros (2008) put it, "Our current system has broken down because the originators of mortgages have not retained any part of the credit risk" (p. 173). In essence, the regulation was there, but the regulation was wrong-minded.

As Heath (2009) noted, "The solution, therefore, lies in cleaning up the system, in order to eliminate the moral hazards" (p. 145)—i.e., a return to a fair-minded world that respects industry and creativity and healthy competition, not a world that applauds cheats, rewards single-mindedness, and champions insiders.

The Capitalist Overhaul: A Lube and Tire Alignment or a New Engine?

Few would argue that the system needs an overhaul, but one wonders how that can happen if governments won't reign in the big boys playing games with the economy. With the introduction of CDSs—worth more than $60 trillion (Smith, 2009b) before the cracks started to show—the financial system didn't stand a chance as greed got greedier, not to mention trumped-up trading profits for a financial industry more interested in its bottom line than in serving its clients.

As Patterson (2010) noted, the high-stakes race for profits had transformed a once-staid banking industry "into hot-rod hedge funds fuelled by leverage, derivatives and young traders willing to risk it all to make their fortunes" (p. 177). They were given free rein to make money any way possible, trading over the counter supposedly correlated sliced-up debt products meant to securitize risk (CDOs and CMOs) and arbitraged insurance that played long and short positions at the same time (CDSs), all without regulation. In fact, regulation was discouraged.

Mason (2009) noted that the 2008 meltdown can be traced to a single moment when the Gramm-Leach-Blily Act replaced the Depression-era Glass-Steagall Act and the 1956 Bank Holding Company Act (p. 56). In one scrawl of Bill Clinton's pen, investment banks could once again invest ordinary people's savings at ever-greater margins. The date was November 11, 1999. A year later, Clinton also signed into law the Commodities Futures Modernization Act, effectively excluding derivatives from being classified as gambling (p. 57). For the "predatory lending" and "insatiable demand for high-risk loans" that followed, Mason puts the blame squarely on Wall Street (pp. 90–91).

In effect, the stock market was turned into the biggest casino going, in which many of the best mathematical minds of a generation gathered to ply their correlations and trend analysis formulae to create the best black box moneymaking machine ever and, in so doing, reap billions to assert their talent or curb their demons—all on the backs of credit or, in some cases, the perception of credit. Patterson (2010) noted that the quant labs cooked up a nightmare of "complexity built upon complexity" (p. 191) and that the CDO models were self-reinforcing such that "when the slightest bit of volatility hit in early 2007, the whole edifice fell apart" (p. 195). When the whole tangled mess went viral and the market unraveled, "this fizzing concoction would play a critical role in the credit meltdown of 2007 and 2008" (p. 191).

Krugman (2009) believes that the "wrong-headedness of conservative economic philosophy" (p. xiv) and the resultant rise in political partisanship is to blame for the climate of deregulation, which pushed the system past its limits and expanded inequality beyond the perceived norms of the prior, long-standing FDR New Deal era that had, among other things, created the American middle class and with it a period of prolonged prosperity and general equality (pp. 3–6). In the process, income inequality has risen as high today as it was in the 1920s (pp. 4–5).

Beattie (2010), however, believes it was not lack of regulation that failed to stop banks from doing what they did but the failure to apply or enforce the regulations. "National regulators and national policymakers had lots of tools to stop fantasy financial assets being created, given ludicrously unrealistic prices, and sold on. In a whole string of countries, they chose not to use them" (p. 287). In other words, the system would have worked if everyone had cared to let it or not dared to override its supposed self-restraining mechanisms.

Nonetheless, the balance is out of whack because governments relaxed essential measures put in place to stop what is known to happen whenever the market is run as a free-for-all—for example, the Securities and Exchange Commission in April 2004 loosening the amount of money a bank must keep in reserve, which freed the banks to expand their subprime lending spree, or Alan Greenspan nixing the idea of controls on derivatives. We've seen the same whenever government errs on the side of Wall Street versus Main Street, self-interest versus community, on back to the Federalist views of Alexander Hamilton versus the anti-Federalist views of Thomas Jefferson. As far back as that and farther.

Galbraith (1958/1999), however, may have been closest to the mark when he warned of increased debt and unsustainable practices as far back as 1958 in *The Affluent Society*, as though predicting the whole mess: "It would be entirely permissible to foresee the gravest results from the way consumer demand is now sustained by the relentless increase in consumer debt" (pp. 150–151).

Figure 12.2	The main financial players leading up to the crisis: from left to right, President Ronald Reagan (1980–1988), President Bill Clinton (1992–2000), Federal Reserve Chairman Alan Greenspan (August 11, 1987–January 31, 2006), and Secretary of the Treasury Henry Paulson (July 3, 2006–January 20, 2009)

SOURCES: http://en.wikipedia.org/wiki/File:Official_Portrait_of_President_Reagan_1981.jpg; http://en.wikipedia.org/wiki/File:Bill_Clinton.jpg; http://en.wikipedia.org/wiki/File:Alan_Greenspan_color_photo_portrait.jpg; http://en.wikipedia.org/wiki/File:Henry_Paulson_official_Treasury_photo,_2006.jpg

A New World Thinking

Whatever the causes, our situation is a lot more complicated than George Bailey's, who so devotedly and stoically stood by his fellow man in the perennial Christmas favorite *It's a Wonderful Life*, encouraging all to share in the fruits of community labor. If that was a Norman Rockwell painting, today's unfettered free-for-all is a Jackson Pollock mess, where one thinks one sees a pattern and is enamored by its beauty but, in reality, hasn't got a clue. In the words of Soros (2008): "Clearly an unleashed and unhinged financial industry is wreaking havoc with the economy" (p. 144).

But we are not out of the woods yet, and the ripples of this crisis will be felt for years to come. To try to make ground against the swell of greed and resultant tide of negativity, however, perhaps one solution stands out: the separation of speculative and protective investment, *again*, which will promote a more stable banking sector where no one bank can get too big (contrary to the basic tenets of capitalism). Better asset rating (no more dog assets), appropriate leverage (say, 12% or an 8-to-1 capital adequacy ratio), real regulation (with real punishments), and a simpler tax code (currently running to 14,000 pages in the United States) to remove the need for creative tax-avoidance schemes are also needed. In return, the banks will have to operate with lower profits while the regular taxpayer makes more.

How can this be done? Better qualification of assets and cash flow within a derivative (a structured CDO has AAA, AA, and B assets bundled together) would keep the badly qualified B assets from bringing down the whole structure (a direct cause of the subprime meltdown). A knock-on effect of properly rated assets would eliminate the need for riskier CDS derivatives, basically an insurance against failure—if the asset is bona fide, no need to insure it and then sell it on. The Basel III banking accord is attempting to ensure that derivatives are related to the cash flow attributable to the asset.

Leverage (capital adequacy ratio, or CAR) must also be maintained within the range of normal operating conditions, such that a bank has adequate reserves to suffer any downturns in its investments. At 100 to 1, the profits are greater, but so are the losses, and they can be catastrophic when a few randomly bad days pile up. What's more, if executive remuneration is tied to the adherence of proper leverage, managers will take more interest in how an investment is doing. To ensure compliance, if the CAR tots up to 12% or less (say, 10% on equity and 14% on project finance), Mr. Big gets his bonus at the end of the year. If not, the accounts will immediately show the extent of a manager's recklessness or a bank's incompetence.

Regulation is always the hard part. Nonetheless, regulation is needed to prevent one from doing what one shouldn't. The 2010 Dodd-Frank Wall Street Reform and Consumer Protection Act attempted to make some sense of the mess but, at more than 2,000 pages, is only a stop-gap awaiting a new round of abuse and loopholes. What's more, although Dodd-Frank attempted to put some form of regulation into a market that hasn't had proper regulation for years, it failed to invest sufficient regulatory power in one single authority rather than the current regulatory quagmire that exists today (the Fed, regional Feds, Federal Deposit Insurance Corporation, Securities and Exchange Commission).

How regulation is implemented and policed is a function of what kind of financial sector a country wants and needs and is similar to obedience training school for dogs. Without proper rules, the dogs will do as they please. What's more, not everyone is playing by the same rules.

However, if 40% of Congress is composed of lawyers and 48% millionaires,[4] it shouldn't be a surprise that lots of well-crafted laws exist for the rich. Furthermore, moral hazards still exist in government guarantees. As a former inspector charged with policing the Troubled Asset Relief Program noted, "The credit agencies remain committed to granting the largest institutions enhanced ratings based on the assumption they will be bailed out once again" (Frean, 2011).

Some other suggested bandages include a tax or commission on assets held for less than a specified time (to curb high-frequency trading, which has nothing to do with investing in a company), taxes on derivatives (which are essentially high-end games of chance), and removal of tax-avoidance schemes (which serve only to fuel speculation and expand inequality). Other more radical changes will be needed if the endless bandaging doesn't work, such as limits to income and ownership, benefits applied to all citizens (why is only one class protected?), standardized unions for all, and much more government intervention in basic economic affairs.

To those who believe that private industry, private regulations, and private markets know best how to serve innovation and growth, how can the market know best if price is its only measure and is so often manipulated? When price is the only measure, greed becomes greedier, especially when viewed only as ups and downs in a stock index or good and bad numbers in a spreadsheet.

Capitalism is not a license to act only for profit—its first great proponent, Adam Smith, said as much—although as Baker (2006) noted, "The incentives to manipulate financial accounts for the goal of personal enrichment are too numerous in financial markets." Somehow, we have to find a way to promote investment in people instead of profits; otherwise, money will always be used as the arbiter of success.

One can certainly wonder how in the midst of the American mortgage market meltdown one person was able to make $1.25 billion in 1 day (and $4 billion over the course of the crisis) while so many others suffered. It is shocking that markets are being played like a children's board game—pushed ever more beyond safe operating limits, like an economic Chernobyl. There has to be a better way.

How to Move Beyond Bubble Economics: Countercyclical Feedback Systems

Many argue that our now increasingly globalized economy is unable to keep up with the speed at which financial systems are changing—whether because of differing national laws, unregulated financial practices, or the still-prevalent unequal living and labor standards throughout the world—which simply postpones asking the hard questions about whether greed and the demand for more should be at the root of economic practice. Zakaria (2009a) wrote that "no system—capitalism, socialism, whatever—can work without a sense of ethics and values at its core." What we should be asking is, why do we want a system that is detached from all that makes us human?

Zakaria (2009b, p. xv; 2009a) further wrote that the world economy can be likened to a race car, faster and more complex than any before—alas, one that has crashed. But why do we need to live at breakneck speed with ever-increasing financial growth? Excessive speed is not good, and neither is excessive growth. If an economy continues growing, it must by definition reach an end—like a balloon stretched beyond its limits—and, as we have seen, the more we grow, the faster the end comes.

Economist Nouriel Roubini—the so-called Dr. Doom, who prefers to call himself Dr. Realist—said it best when he noted that the goal of a regulatory body (the Fed) is to take away the punch bowl after the party gets going (Smith, 2009b), a corrective or *negative* feedback system (as first noted by Fed chairman W. M. Martin Jr. in the 1930s). But neither the Fed nor the American government did remove the punch bowl, and in fact they encouraged the worst of drunken excess. As Roubini noted, they were "adding vodka and whiskey and even more toxic stuff to the punch bowl and making everyone drunk with irrational exuberance" (Smith, 2009b), a nod to Fed chairman Alan Greenspan's famous quip after the Dow Jones reached a record height in 1996. To be sure, when something on the scale of the world economy fails, it is a failure of governance, management, and strategic thinking, and certainly not the fault of everyday working people.

In times of prosperity, the economy is expanded (fueled in part by easy credit and increased debt) when it is least needed, whereas in more frugal times, spending is decreased, once again the opposite of what is needed. It is one thing to stimulate the economy in bad times, as intended by Keynesian policies, but quite another to do so when not needed. Stimulating an economy in good times is exactly the wrong thing to do, made worse because of the viral feedback nature of growth, which expands a system beyond its manageable limits and ultimately results in a busted boom.

What's more, what will happen to the world economy when China and India are unable to sustain double-digit growth rates and start to shrink? If economy is meant to be a way for all to succeed instead of a reflection of individual needs, planning for endless growth is a mistaken strategy. One may see temporary gain, but it can never last. Indeed, in the midst of one crisis, the next has already begun.

12.1. The Future in a Nutshell

The American Dream, as personified not by the likes of Andrew Carnegie, Bill Gates, or Mark Zuckerberg but by the vast unnamed citizenry, is long gone. We simply cannot all be rich, and it is time to recognize that cooperative strategies and not growth-based competition are needed. Our futures depend on it.

There is, of course, much to do. Kaletsky (2009) believes that "we are where astronomy was when Copernicus realized that the earth revolves around the sun"—pretty depressing, to be sure, when you think about how wrong and complicated the old geocentric system was. With the likes of CDSs, which Soros calls a "toxic market" (Hosking & Jagger, 2009), we really are in the economic Dark Ages.

But, sadly, Kaletsky (2010b) also noted that despite being in the worst recession since the Great Depression, things are not likely to improve:

> Not only have the banks escaped any wide-ranging regulation, but politics, at least in Britain and America, has reverted to the language of the Thatcher-Reagan period. Genuine jobs can only be created by the private sector. There must be minimal interference with market forces, whether in managing trade and exchange rates, subsidising science or culture or in shifting incentives to encourage a transition from fossil fuels.

Jackson (2009) asked "whether economic growth is still a legitimate goal for rich countries, when huge disparities in income and well-being persist across the globe and when the global economy is constrained by finite ecological limits" (p. 17), further noting that "progress towards sustainability remains painfully slow" (p. 184). Of course, the challenges are great and seemingly more pressing as we begin to reach the limits of our resources, which were once thought inexhaustible. One hopes we have not reached the limits of our imagination or inventiveness.

Axelrod (1984) noted that "mutual cooperation can emerge in a world of egoists without central control by starting with a cluster of individuals who rely on reciprocity" (p. 69). In many ways, Axelrod showed that biblical prescriptions work, such as "do unto others as you would have them do unto you." He also showed how forgiveness helps deescalate conflict and that cooperation turns one-off, zero-sum interactions where people care little about meeting each other again into ongoing, non-zero-sum interactions, which serve to stabilize mutually beneficial futures—in short, good community relations. As Axelrod noted, "This continuing interaction is what makes it possible for cooperation based on reciprocity to be stable" (p. 125).

Trust and reciprocity are also at the core of the microfinancing miracle of Graneem Bank, started amid the poverty of Bangladesh in response to uncaring banks whose usurious practices give little thought to community. Faceless interactions create faceless communities. As Axelrod (1984) noted, "The principle is always the same: frequent interactions help promote stable cooperation" (p. 130). In reality, without people there can be no profits, a maxim that seems to have been forgotten by today's profit-first megabanks.

As a start, George Soros created the Institute for New Economic Thinking with $50 million of his own money, hoping to help right a failing economic system. As noted by the director of the institute, "The economic crisis and the failure of economists to predict it and protect society illustrate that economics as a profession needs to re-earn its reputation and regain its mantle of expertise" (Frean, 2010a). Furthermore, he hopes "to steer the discipline away from the champions of the free market and deregulation who . . . share the blame for the global economic crisis" (Frean, 2010a).

After the failure of Enron, one Canadian bank charged with helping Enron commit accounting fraud spent $50 million on corporate-governance initiatives (Haskayne, 2007, p. 163) and began ethics training for its 50,000 employees. The CEO of another company called for "vigorous enforcement of existing laws—heavy prison sentences and the forfeiting of embezzled funds," adding, "I am convinced that will do more for chilling the occurrence of future Enrons than writing a thousand laws" (p. 152).

But will the coming revolution remove the individual from the center and recognize the collective? Will we begin to realize the ideal of shared living? Alan Greenspan stated that future crises will happen again unless we change human nature, but given that our organized governance has instructed us all how to behave since the beginning of civic society, blaming human nature seems an unfair cop-out. Some of us just aren't behaving as we should and, what's worse, are doing so with impunity.

A New Beginning or More of the Same?

Unfortunately, if Gordon Brown's meeting with a newly elected Barack Obama on a snowy March 2009 day in Washington is any indication, circumstances don't seem to be getting any better. In the midst of an almost global economic meltdown, Obama seemed to suggest that Brown had inherited a financial mess, saying that he had "taken the helm of the British economy at a very difficult time," presumably noting how he had inherited an American-sized mess from his predecessor George W. Bush. Although Brown had become prime minister only 7 months prior to their meeting as leaders, he had been Chancellor of the Exchequer for the previous decade and had been given unprecedented control over the British economy. If anyone could be charged with overseeing a culture of greed and lax regulation and be considered responsible for the "very difficult time," it was the chancellor-turned-prime-minister James Gordon Brown. It certainly wasn't the taxpayers, who were duped out of billions in savings and saddled with repaying a toxic debt run up by an underregulated and highly suspect banking system.

More worrying, however, was Brown's (2009) address the following day to the joint houses of Congress (only the fifth British prime minister to do so), which was sprinkled with more of the same platitudes, even drawing on his religious upbringing to endow his rhetoric:

> My father was a minister of the church and I have learned again what I was taught by him: that wealth must help more than the wealthy, good fortune must serve more than the fortunate, and riches must enrich not just some of us but all. And these enduring values are the values we need for these times.

Wonderful words, especially the telling use of the word *again*, but one wonders where he was during the 10 years when he was charged with putting a stop to exactly what he was saying had gone amok. As chancellor, he routinely swept away safeguards in an effort to appease the financial community, loosening standard regulation in 2006, 1 year prior to the failure of the first British bank, and championing "not just a light touch but a limited touch" system (Brown, 2005). It does no good to close a door after the horse has bolted.

President Obama's Treasury Secretary Tim Geithner furthered the argument, noting that Brown's so-called "light touch" approach "was designed consciously to pull financial activity away from New York and Frankfurt and Paris to London. That was a deeply costly strategy for financial regulation" (Fleming, 2011b). As in any competitive game, what started as a light touch in Britain became a light touch everywhere, as banks and economies competed with one another in the most dangerous of "winner-take-all" games.

But scariest of all was that Brown (2009) advocated business as usual, with the same mindless, counterproductive growth strategies used in the past. In his 38-minute speech to Congress, he called for doubled growth as a panacea for all ills:

Over the next two decades literally billions of people in other continents will move from being simply producers of their goods to being consumers of our goods and in this way the world economy will double in size. Twice as many opportunities for business, twice as much prosperity, and the biggest expansion of middle-class incomes and jobs the world has ever seen.

Egads! Shades of the hare, shades of Schumacher, shades of failing pyramid scams. If the ongoing crises of capitalism have shown us anything, it is that the economy should grow but only in measured steps that reward shared industriousness designed to benefit all—not as an endless doubling game that smacks of system tinkering and spreadsheet decision making.

Of course, in Brown's case, electoral defeat got in the way of his great plans, but they will, nonetheless, be resurrected by the next Great Promiser—although, in the wake of more ongoing financial crises, growth projections in Britain were revised to a more sobering 2% on the backs of massive public service cuts and corporate tax breaks. As reported prior to one G20 meeting, "The strategy also involves reducing the corporate tax rate from 40 percent to attract more investment from foreign countries" (Selb, 2010), as is planned in other countries, including the United States. In essence, such plans undercut others in a tit-for-tat round of global one-upmanship, serving only to increase the social divide and reduce government coffers.

As we've seen in the zero-sum game of modern economics, such breaks are meaningless in the context of foreign investment and instead circumvent tax (and thus decrease public infrastructure while increasing national debt). If everyone is doing the same, as in a coordinated G20 strategy, it is nothing more than a carnival game of hide the peanut. Worse, the usual doublespeak calls for reduced growth today but will not do so for long.

President Obama has promised a $1 trillion tax increase on those earning more than $250,000, has started to investigate ending tax havens for American companies abroad, and has announced more spending on science, particularly in non-fossil-fuel industries. He has even curtailed some of Wall Street's excess, but there is still much to do. The famous Bush tax cuts for the rich remain in place, without any apparent will to repeal them, and continue to benefit the wealthy and hamstring the less well-off. At the same time, the repayment of an ever-increasing and ever-rolling-over national debt is widening the inequality gap, giving more to those with money and taking more from those with much less.

If the new money men succeed, we may well end up living in a fairer world one day—that is, if the next round of reforms is more than the usual bag of tricks and sleights of hand designed not only to maintain but to promote inequality.

12.2. A Final Word on Myths and Fables: A Final Word on Numbers

The world is a mixture of overspenders and long-term savers, a mixture of risk-taking entrepreneurs and conservative investors looking for sure things. Is it better to be like the hare, hopping from extreme to extreme, or like the tortoise, cautiously weighing things? No thought for tomorrow or the golden mean? If using too much creates excessive consumption, pollution, waste, and an unhealthy divide between rich and poor, is it worth it? Is there a limit to our resources, and will our own consumption become catastrophic? Are there alternatives?

Imagine two people in a life raft. They have a limited water supply. Do they ration? Do they reduce their consumption, knowing that help may not be imminent? How does one balance the needs of today with the unknown needs of tomorrow? One certainly can't imagine the drifters doubling their intake every day with no thought of tomorrow. More interesting, do they divide the water in half or drink it based on need?

Why are we so enamored with growth when our economies stagger, cause excessive pollution and waste, and do not create full employment or equal opportunity? We are not in frontier times, endlessly expanding with vast resources. Modern economics has hit the wall, and so we must change from a system that questions if we *can* do it to one that asks *how* we do it (and *should* we do it).

The increasing encroachment of private interests into what has historically been the public domain and the subsequent failure of democratic governments, religious institutions, or businesses to uphold the necessary caveats against excess, exploitation, and abuse have helped create a world of myths, fantasies, and bad thinking based on me-first attitudes, specious inductive generalizations, and quickie solutions. It is an age of unreason and stupidity, where everyone is talking and few are listening.

Some argue that it is not society's place to protect people from themselves, lest freedoms be taken away, but since society builds economic systems (which include taxes), makes national and international law (which must be abided by), and regulates almost every facet of modern living to the benefit of many and the detriment of many more, great imbalances occur in favor of those who control society and the administration of money and power. A small percentage of people own most of the wealth, and yet billions live on practically nothing. The haves (hares?) are destroying the world.

Heath (2009) argued that management and employees in government-run companies (or even in some top-heavy and massive corporations) have less incentive to be industrious or profitable and will fare worse than the private firms, a standard concern about "controlled" economies (p. 198). Here, money is deemed the most efficient measure. But, surely, money is not the only incentive, the only measure, or the only variable in the work-to-live equation. Surely, we can find the common goals in all our quests. Money suggests an absolute certainty in its use, goals, and ideals, but it should not be. Money is the means by which we trade on our industriousness with one another. What's more, an economy run to benefit only the controllers is not a *free*-market economy, as we have been led to believe, but a *stacked*-market economy.

Marx and Engels (1848/1992) analyzed history in terms of class conflict, the happy bourgeoisie versus the downtrodden proletariat. They called for the workers of the world to unite and create a classless society, where resources are shared, not least of which by workers doing the grunt work, turning crop into fiber, fiber into textile, textile into garment, resource into usable commodity. That call is no less valid today, where, despite the commonplace luxuries of the modern worker—cars, televisions, personal computers—a new oppressed, debt-ridden, stressed-out wage earner exists in a world of rich, poor, and poorer—a world that is becoming even more uncertain for many. The call to change has never been more important.

Of course, communism in its totalitarian guise failed miserably to live up to its promise and was understandably given short shrift by a world weary of government oppression. Krugman (2009) made the case that racism has kept socialism from taking shape in the United States (pp. 129, 179), arguing that since minority blacks in general are poorer, there is an institutional resistance to avoid bettering their cause with socially minded practices such as welfare, benefits, and health insurance, noting that Ronald Reagan began his gubernatorial career in part on a promise to repeal the Californian fair housing act in favor of discrimination (p. 86). Krugman added, "Broadly, the higher the black fraction of a state's population, the lower its social spending per person" (p. 180). As Martin Luther King, Jr. wryly noted, mass unemployment in the black community is considered a social problem, whereas mass unemployment in the white community is called a depression (Carson, 2006, p. 350).

It can also be argued that socialist agendas were mollified by New Deal reforms that created social security, health care, and jobs or that Marx's atheism was too antithetical to a religious United States. Or, as was perhaps best described by Albert Camus in *The Rebel*, it can be argued that Soviet communism failed because the ultimate goal of freedom cannot be won by enslaving, a reworking of the end not justifying the means.

But the idea that an equitably shared wealth is bad or unmanageable because it has been corrupted by past powermongers and demagogues is mistaken, especially given that the economies of today's richest nations are mixed and have taken equally from Marxism (progressive income tax, public infrastructure, free education, and child labor abolition; Galbraith, 1991, p. 137) and from capitalism (free enterprise, creative entrepreneurism, organized investment, and protection from government control). As Martin Luther King, Jr. commented,

> Moreover, Marx had revealed the danger of the profit motive as the sole basis of an economic system: capitalism is always in danger of inspiring men to be more concerned about making a living than making a life. We are prone to judge success by the index of our salaries or the size of our automobiles, rather than by the quality of our service and relationship to humanity. Thus capitalism can lead to a practical materialism that is as pernicious as the materialism taught by communism. (Carson, 2006, p. 21)

To say that modern Western economies are not planned is false. To say that a government does not dole out its tax treasury foremost to preserve order and the status quo of an unequal state is naïve. To say that governments do not preferentially treat groups of people (classes) is wrong. To say that electors rather than business control our government representatives—a disproportionately university-educated, lawyer class of millionaires (40% and 48% of the U.S. Congress)—is a lie. Justice is meant to be blind to wealth and status. As Barber (2007) noted, "Equal access to power and security is among democracy's chief virtues" (p. 155).

Whether in a family, a clan, a tribe, a culture, a society, a nation, or a world, sharing is either right or it is not. Assets can be based on labor and fair-mindedness and not on money and sheltered investment. And governments, which, after all, represent the wishes of a collective citizenry, can redress the economic imbalances that exist—not just because it is right but because it is essential to keep our resource-based system from failing.

Following in the spirit of other political treatises of its time—including the Franco-Swiss philosopher Jean Jacques Rousseau's 1762 *Social Contract* and English statesman Thomas More's 1516 *Utopia*—*The Communist Manifesto* contemplated a better world for all, equitable and free from corruption by those who would abuse. They are all guides to better living. In the *Social Contract*, Rousseau asks what to do about the private versus the common. In *Utopia*, More's practical suggestions run the gamut from lockless doors to viewing potential mates au natural to spot flaws. To be sure, not all suggestions from the past make sense.

Religious books, such as the Bible and the Koran, also have value today, with their tenets of fairness and inclusiveness. Who has not heard of "do unto others as you would have them do unto you" or that usury is forbidden in Islamic banking?[5] We need such wisdom in our time, where the case has never been clearer: We must hold back the abuser, the deceitful double-talker, the downright crooked, whether banks, credit card companies, corrupt politicians, Christmas sales, false advertising, or a government-backed financial elite that is spreading inequality throughout the world.

So, how do we stop the madness? If the world is becoming more unequal, not for any meritocratic reasons, clearly we are being abused. And given that government regulates so much of our lives, such abuse is encouraged. So, we must look elsewhere—to one another—to recognize that the individual is supported by the collective. We must band together again. There is no going back.

We must act responsibly, whether reducing waste to sustain a collective existence or conducting ourselves more authentically in personal relations. We must work together to understand our differences, as in today's parable of the non-zero-sum, where to not trust a partner is to not trust oneself. The main purpose in life is not to increase personal wealth but to increase mutual well-being.

We must continue to debate the value of the modern corporation (the de facto superpower) and of the state (the de jure regulator of excess and abuse), which greatly shape our lives. Martin Luther King, Jr. commented after the violent riots of 1965, in which more than 30 people died in the South Los Angeles neighborhood of Watts, an area with the highest population density in the United States:

> When all is finally entered into the annals of sociology; when philosophers, politicians, and preachers have all had their say, we must return to the fact that a person participates in this society primarily as an economic entity. At rock bottom we are neither poets, athletes, nor artists; our existence is centered in

the fact that we are consumers, because we must first eat and have shelter to live. This is a difficult confession for a preacher to make, and it is a phenomenon against which I will continue to rebel, but it remains a fact that "consumption" of goods and services is the raison d'être of the vast majority of Americans. When persons are for some reason or other excluded from the consumer circle, there is discontent and unrest. (Carson, 2006, p. 295)

How one goes about reconciling discontent and unrest is not easy, but we must stop blind consumerism, wasteful consumption, organized misinformation, the excessive growth of the financial class, and the insecure selfishness of our petty dickering, for we are all one and the same. Since we all come from one common ancestor, we are in fact fighting ourselves.

What's more, numbers and mathematics have become increasingly more important in today's world, from how we play games to how we buy and sell, from how we grow to how we err to how we judge to how we reward and punish. We can't vote without understanding the importance of numerical weights, we can't debate entitlements versus tax revenues without knowing the size of our debts and the financial solvency of our countries, and we can't understand the value of our shared society without understanding our own involvement, our own footprint. Mathematics is integrated into everything we do and is paramount to understanding the modern debate. Fortunately, with increased numeracy comes increased understanding of the many challenges we face in the ongoing human dialectic.

There are many sides to a story, which means only that the truth is hard to find. But the truth must be sought and learned, for there is no better freedom. Life isn't about left and right but about decency, for which we all have the capacity. It is about treating others how we would like to be treated—the basic contract we have with our spouses, our children, our neighbors. How we treat one another is our first currency, and we simply cannot improve at the expense of others. We must look to shared solutions.

Notes

1 **American corporate wealth**: In 2010, the top 100 companies' total revenues were more than $6 trillion with profits of more than $200 billion (including losses of more than $90 billion by Freddie Mac and Fannie Mae). The top 500 companies' total revenues were $9.8 trillion with profits of $391 billion ("Fortune 500," 2010). Heilbroner and Thurow (1998) noted that the biggest 100 firms account for half of all sales (p. 46).

2 **The growth of the financial industry**: All data is from the U.S. Department of Commerce Bureau of Economic Analysis. Data was available for 1998–2009 and the 1987 benchmark year. Note that the graph is extrapolated backwards to 1987 and that only five industry categories (agriculture, construction, manufacturing, trade, and finance) are shown.

3 **Day trading**: Day or high-frequency trading is the buying and selling of a stock in 1 day, not for investment purposes, but to bet on the short-term rise or fall of a stock. Moyo (2011) noted that "high-frequency trading accounts for as much as 73 per cent of US daily equity volume, this figure being up from 30 per cent in 2005" (p. 119). The so-called 2010 "flash crash" that saw the largest 1-day Dow Jones loss (9%) has also been blamed on high-frequency trading.

4 **Number of lawyers and millionaires in government and the general population**: In the U.S. House of Representatives, 36% are lawyers (162 out of 441), and in the Senate, 54% are lawyers (54 out of 100; "List of Lawyers in the 111th Congress," 2009). There are 1,116,967 lawyers in the United States (Ellis-Christensen, 2011), or about 1 in 300 of the general population. Thus, there are about 120 times as many lawyers in Congress as in the general population. In the U.S. House of Representatives and Senate, 48% are millionaires (320 out of 669; Center for Responsive Politics, 2009). There are 10,541,000 millionaire households in the United States (Deloitte, 2011), or about 1 in 30 in the general population. Thus, there are more than 15 times as many millionaires in Congress as in the general population.

5 **Financial instruments in Islam**: Financial instruments exist to charge interest that complies with Sharia law (called *sukuk*) and account for more than $1 trillion in such Sharia-compliant investments.

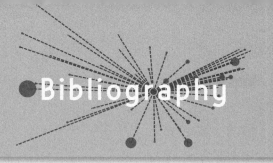

Bibliography

Ali, S. (2010). *Dubai: Gilded cage.* New Haven: Yale University Press.

American Gaming Association. (2011). How many pathological gamblers are there? Retrieved from http://www.americangaming.org/industry-resources/faq/how-many-pathological-gamblers-are-there

Andrews, E. L. (2008, October 23). Greenspan concedes error on regulation. *The New York Times.*

Annan, K. A. (2009, January 26). 'No country, no matter how powerful . . . can control the forces of globalisation on its own.' *The Times.*

Arrighi, G. (2010). *The long twentieth century: Money, power, and the origins of our times.* London: Verso.

Association of National Advertisers. (2011). *Marketers' constitution.* Retrieved from http://www.ana.net/constitution/show

Atwood, M. (2008). *Payback: Debt and the shadow side of wealth.* Toronto: O. W. Toad.

Ayres, C. (2006, May 5). Idol rich: Rude Cowell wins $36m. *The Times.*

Axelrod, R. (1984). *The evolution of cooperation.* London: Penguin Books.

Bagli, C. V. (2010, December 4). After 30 years, a rebirth is complete. *The New York Times.*

Bailey, E. (2008, January 17). Ten common tax audit flags: What does the IRS look for before auditing an individual? Retrieved from http://personalbudgeting.suite101.com/article.cfm/ten_common_tax_audit_flags

Baker, G. (2006, June 27). Giving generously, the American way. *The Times.*

Ball, P. (2005). *Critical mass: How one thing leads to another.* London: Arrow Books.

Barber, B. (2007). *Consumed: How markets corrupt children, infantilize adults, and swallow citizens whole.* New York: W. W. Norton.

Bayley, S. (2010, January 11). Britain should be a workshop, not a casino. *The Times.*

Beattie, A. (2010). *False economy: A surprising economic history of the world.* London: Penguin Books.

Bell Burnell, J. (2010, November 12). *Will the world end in 2012?* Dublin: Trinity College Dublin.

Bennett, R. (2010, January 27). Gap between rich and poor at its widest since the war. *The Times.*

Bennetto, J. (2005, January 6). A £180bn credit card industry accused over hidden costs. *The Independent.*

Biggs, B. (2009, January 12). The affinity Ponzi scheme: It's incredible that Madoff could have sucked so many rich and very sophisticated people into his Ponzi scheme. *Newsweek.*

Billen, A. (2006, June 20). When I stuff BA, I'll quit: No frills, no chilled champagne, but Michael O'Leary lives the life of Ryan. *The Times.*

Black, C. (2009, August 1). Sex-addict senators, kidney-running rabbis, corrupt mayors. How did the U.S.A. become . . . the sleaziest country in the world? *The National Post.*

Boeke, K. (1957). *Cosmic view: The universe in 40 jumps.* New York: John Day.

Bone, J., Reid, T., & Spence, M. (2009, June 20). Cricket billionaire faces 250-year sentence after pyramid scheme arrest. *The Times.*

Boorstin, D. J. (2000). *The Americans: The democratic experience.* London: Phoenix Press.

Bowley, G. (2010, March 12). Former foe of regulation emerges as one of its champions. *International Herald Tribune.*

Bradsher, K. (2010, September 10). U.S. union challenges Chinese subsidies. *International Herald Tribune.*

Brayfield, C. (2000, November 10). The anti-brand leader. *The Times.*

Bremner, C. (2009, April 17). Sarky Sarko: French leader sneers at "weak" Obama. *The Times.*

Bronson, F. (1992). *The billboard book of number one hits* (3rd ed.). New York: Billboard.

Brown, D. (2006, December 6). Richest tenth own 85% of world's assets. *The Times.*

Brown, G. (2005, November 28). Gordon Brown's speech to the CBI. *The Financial Times.*

Brown, G. (2009, March 4). Gordon Brown's speech to US Congress. *The Guardian.*

Burke, S., Kenaghan, C., O'Donavan, D., & Quirke, B. (2004). *Health in Ireland: An unequal state* (Youth edition). Dublin: Public Health Alliance Ireland. Retrieved from http://www.phaii.org/uploads/publications/Health%20in%20Ireland%20-%20An%20unequal%20State_Youth%20Version.pdf

Burkeman, O. (2006, May 5). Bill Gates: I don't want to be the world's richest man. *The Guardian.*

Burns, R. B., & Burns, R. A. (2008). *Business research methods and statistics using SPSS.* Los Angeles: Sage.

Cable News Network. (2009a). President: California. *Election Center 2008.* Retrieved from http://edition.cnn.com/ELECTION/2008/results/county/#val=CAP00p3

Cable News Network. (2009b). President: New York. *Election Center 2008.* Retrieved from http://edition.cnn.com/ELECTION/2008/results/county/#NYP00map

Cain, F. (2008, March 19). Geologist finds a meteorite crater in Google Earth. *Universe Today.* Retrieved from http://www.universetoday.com/13263/geologist-finds-a-meteorite-crater-in-google-earth/

Carlson, S. (1985, December 5). A double-blind test of astrology. *Nature, 318,* 419–425.

Carson, C. (Ed.). (2006). *The autobiography of Martin Luther King, Jr.* London: Abacus.

Carter, J. (1977). President's proposed energy policy. *Vital Speeches of the Day, 43*(14), 418–420.

Casey, M. (2006, July 16). Ireland's wealth just isn't working. *The Sunday Times.*

Casti, J. (2009, April 22). *Zeitgeist: Socionomics and the science of surprise.* Dublin: University College Dublin.

Center for Responsive Politics. (2011). Net worth, 2009. OpenSecrets.org. Retrieved from http://www.opensecrets.org/pfds/overview.php?type=W&year=2009

Central Intelligence Agency. (2011). *The world factbook.* Retrieved from https://www.cia.gov/library/publications/the-world-factbook/

Charter, D. (2008, January 25). Gates uses forum to call on businesses to find radical ways of reducing world poverty. *The Times.*

Cieply, M. (2011, January 21). In a budget bind, U.S. states consider cutting film subsidies. *The International Herald Tribune.*

Clark, K. (Presenter/Writer). (1969, April 27). The smile of reason [Television series episode]. In M. Gill (Director), *Civilisation* (Ep. 10). London: BBC2.

Coen, R. (2007, June 26). *Insider's report: Robert Coen presentation on advertising expenditures June 2007.* Retrieved from http://graphics8.nytimes.com/packages/pdf/business/20070627_ADCO.pdf

Cohan, W. D. (2009, December). Endless summers. *Vanity Fair.*

Cole, G. H. A., & Woolfson, M. M. (2002). *Planetary science: The science of planets around stars.* Bristol: Institute of Physics Publishing.

CoreLogic. (2011, September 13). Retrieved from http://www.corelogic.com/about-us/researchtrends/asset_upload_file591_13850.pdf

Couric, K. (Presenter). (2009, February 3). *CBS evening news with Katie Couric* [Television broadcast]. New York: CBS.

Coyle, C. (2009, March 1). Tax protesters rattle U2's gig with hypocrisy jibes. *The Sunday Times.*

Coyle, C. (2010, July 18). All the tax U2 can leave behind: Cap on artists' tax-free income has sent €30m abroad. *The Sunday Times.*

Crossley-Holland, D. (2009, September 15). Dr Greenspan's defence: 'It really wasn't my fault.' *The Spectator.*

Davidoff, S. M. (2011, April 27). The declining worth of a good name. *International Herald Tribune.*

Davies, J. B., Sandström, S., Shorrocks, A., & Wolff, E. N. (2008, February). *The world distribution of household wealth* (Discussion Paper No. 2008/03). Helsinki: United Nations University, World Institute for Development Economics Research.

Death Penalty Information Center. (2011, October). Number of executions by state and region since 1976. Retrieved from www.deathpenaltyinfo.org/number-executions-state-and-region-1976

Debates and proceedings. (1995, March 14). *The Legislative Assembly of Manitoba, sixth session of the Thirty-Fifth Legislature.* Retrieved from http://www.gov.mb.ca/legislature/hansard/6th-35th/vol18/h018_3.html

Deloitte. (2011). The next decade in global wealth among millionaire households. Retrieved from http://www.deloitte.com/us/globalwealth

DeNavas-Walt, C., Proctor, B. D., & Smith, J. C. (2010, September). *Income, poverty, and health insurance coverage in the United States: 2009*. Washington, DC: U.S. Census Bureau. Retrieved from http://www.census.gov/prod/2010pubs/p60-238.pdf

Dorling, D., Mitchell, R., Shaw, M., Orford, S., & Smith, G. D. (2000, December 23–30). The ghost of Christmas past: Health effects of poverty in London in 1896 and 1991. *British Medical Journal, 321,* 1547–1551.

Dove, A. (1968, July 15). Taking the Chitling Test. *Newsweek.*

Ducker, J. (2009, June 12). Ferguson in fight to halt exodus after cashing in on Ronaldo. *The Times.*

Election 2004 (California). (2004a). *USA Today.* Retrieved from http://www.usatoday.com/news/politicselections/vote2004/PresidentialByCounty.aspx?oi=P&rti=G&sp=CA&tf=l

Election 2004 (New York). (2004b). *USA Today.* Retrieved from http://www.usatoday.com/news/politicselections/vote2004/PresidentialByCounty.aspx?oi=P&rti=G&sp=NY&tf=l

Elliott, S. (2009a, March 22). Madison Avenue's chief seer. *The New York Times.*

Elliott, S. (2009b, April 21). Among advertisers, aging U.S. baby boomers looking better than ever. *International Herald Tribune.*

Ellis-Christensen, T. (2011, September 16). What percent of the US population do lawyers comprise? WiseGEEK. Retrieved from www.wisegeek.com/what-percent-of-the-us-population-do-lawyers-comprise.htm

Epstein, J., & Axtell, R. (1996). *Growing artificial societies: Social science from the bottom up.* Washington, DC: Brookings Institution Press; Cambridge: MIT Press.

Ferguson, C. (2010, October 3). Larry Summers and the subversion of economics. *The Chronicle of Higher Education.*

Ferguson, N. (2007). *Empire.* London: Penguin Books.

Ferguson, N. (2008). *The ascent of money: A financial history of the world.* London: Penguin Press. (Excerpted in *The Times,* October 31)

Ferris, M. (1999, June 24). Tone's lessons for today's Orange Order. *An Phoblacht Republican News.*

Fisk, R. (2006). *The great war for civilisation: The conquest of the Middle East.* London: Harper Perennial.

Fleming, G. (2005, April 10). *The most important forecast of the century: Will our climate change?* Presented at the Irish Science Teachers' Annual Conference, Carlow, Ireland.

Fleming, S. (2011a, February 14). Emerging nations' rampant growth to widen the growing energy gap: Oil production to plateau 'by end of decade.' *The Times.*

Fleming, S. (2011b, February 23). Geithner applauds tough stand on deficit. *The Times.*

Foley, S. (2010, August 6). Has Google finally sold its soul to big business? *Evening Herald.*

Ford Pinto. (2006, October 24). *The Engineer.* Retrieved from http://www.engineering.com/Library/ArticlesPage/tabid/85/articleType/ArticleView/articleId/166/categoryId/7/Ford-Pinto.aspx

Forston, D. (2010, August 8). BP audit: Errors on disaster rig. *The Sunday Times*.

Fortune 500: Our annual ranking of America's largest corporations. (2010). CNN Money. Retrieved from http://money.cnn.com/magazines/fortune/fortune500/2010/full_list/

Frank, R. H., & Cook, P. J. (2010). *The winner-take-all society: Why the few at the top get so much more than the rest of us*. New York: Virgin Books.

Frean, A. (2010a, April 5). Soros backs Oxford to refresh economics. *The Times*.

Frean, A. (2010b, June 4). BP is the target of jokes and curses as Obama fights back. *The Times*.

Frean, A. (2011, July 18). The Wall Street reform act that's big, but not too big to fail. *The Times*.

Free Exchange. (2007, August 21). Krugman on inequality and democracy. *The Economist*. Retrieved from http://www.economist.com/blogs/freeexchange/2007/08/krugman_on_inequality_and_demo

Friedman, T. L. (1999, March 28). A manifesto for the fast world. *The New York Times Magazine*.

Galbraith, J. K. (Presenter/Writer). (1977a). The manner and morals of high capitalism [Television series episode]. In M. Jackson (Director), *The age of uncertainty* (Ep. 2). London: BBC.

Galbraith, J. K. (Presenter/Writer). (1977b). The prophets and the promise of classical capitalism [Television series episode]. In M. Jackson (Director), *The age of uncertainty* (Ep. 1). London: BBC.

Galbraith, J. K. (Presenter/Writer). (1977c). The rise and fall of money [Television series episode]. In M. Jackson (Director), *The age of uncertainty* (Ep. 6). London: BBC.

Galbraith, J. K. (1991). *A history of economics: The past as the present*. London: Penguin Books.

Galbraith, J. K. (1992). *The great crash 1929: The classic study of that disaster*. London: Penguin Books.

Galbraith, J. K. (1999). *The affluent society* (40th ed.). London: Penguin Books. (Original work published 1958)

Gambling boom boosts British. (2009, April 26). *The Sunday Times*.

Gardner, M. (1970, October). Mathematical games: The fantastic combinations of John Conway's new solitaire game "life." *Scientific American, 223*, 120–123.

Gibney, A. (Director), Klot, P., & Motamed, S. (Producers). (2004). *Enron: The smartest guys in the room* [Documentary]. United States: Magnolia Pictures.

Gigerenzer, G. (2002). *Reckoning with risk: Learning to live with uncertainty*. London: Penguin Books.

Gillan, A. (2006, January 21). In Iraq, life expectancy is 67. Minutes from Glasgow city centre, it's 54. *The Guardian*.

Giorgini, J. D., Ostro, S. J., Benner, L. A. M., Chodas, P. W., Chesley, S. R., Hudson, R. S., et al. (2002, April 5). Asteroid 1950 DA's encounter with earth in 2880: Physical limits of collision probability prediction. *Science, 296*, 132–136.

Gladwell, M. (2008). *Outliers: The story of success*. New York: Little, Brown.

G.19 consumer credit. (2011, October 7). Federal Reserve Statistical Release. Retrieved from http://www.federalreserve.gov/releases/g19/Current/

Goddard, J. (2010, July 24). Alarms 'were silenced' before BP oil well blast. *The Times.*

Goldacre, B. (2009). *Bad science.* London: Fourth Estate.

Gowa, J. (1986). Review: Anarchy, egoism, and third images: *The Evolution of Cooperation* and international relations. *International Organization, 40*(1), 167–186.

Griffiths, K. (2010, February 26). Bank loses £3.6bn—but still finds £1.3bn to pay its staff bonuses. *The Times.*

Grossman, J. (2010, October 20). *The Erdös number project.* Retrieved from http://www.oakland.edu/enp

Hald, A. (2003). *A history of probability and statistics and their applications before 1750.* Hoboken: John Wiley.

Hardin, G. (1968, December 13). The tragedy of the commons: The population problem has no technical solution; it requires a fundamental extension in morality. *Science, 162,* 1243–1248.

Harris, J. M., Hirst, J. L., & Mossinghoff, M. J. (2008). *Combinatorics and graph theory.* New York: Springer.

Harris, N. (2010, March 28). Yankees on top in global pay review, Premier League in the shade. *Sporting Intelligence.* Retrieved from http://www.sportingintelligence.com/2010/03/28/yankees-on-top-in-global-pay-review-premier-league-in-the-shade-280301/

Haskayne, D. (2007). *Northern tigers: Building ethical Canadian corporate champions.* Toronto: Key Porter Books.

Heath, J. (2009). *Filthy lucre: Economics for people who hate capitalism.* Toronto: HarperCollins.

Heatley, C. (2006, July 19). Road deaths hit 400 in weekend of carnage. *The Irish Examiner.*

Heberling, M. (2002, Spring). State lotteries: Advocating a social ill for the social good. *The Independent Review, 6,* 597–606.

Heckman, J. (2006a, June 7). *The economics of child development.* Dublin: University College Dublin.

Heckman, J. (2006b, June 9). *The technology and neuroscience of skill formation.* Dublin: University College Dublin.

Heilbroner, R., & Thurow, L. (1998). *Economics explained: Everything you need to know about how the economy works and where it's going.* New York: Simon & Schuster.

Hell, I. (2005, January 12). 'The power of money is to give it away': Infosys' yogi has a mantra for our times. *The Independent on Sunday.*

Hensbergen, G., van (2005). *Guernica: The biography of a twentieth-century icon.* London: Bloomsbury.

Higgins, B. (Director). (2003). Mayday [Television series episode]. In A. Barro & G. Salzman (Producers), *Unlocking disaster* (Series 1, Ep. 1). Toronto: Discovery Channel.

Hill. A. (2005, February 6). Children of rich parents are better at reading. *The Observer.*

Historical CPI-U data from 1913 to the present. (2011). *InflationData.com.* Retrieved from http://inflationdata.com/inflation/Consumer_Price_Index/HistoricalCPI.aspx

Historical crude oil prices [Table]. (2011). *InflationData.com*. Retrieved from http://inflationdata.com/inflation/Inflation_Rate/Historical_Oil_Prices_Table.asp

Hobbs, E. (2010). *Energise: How to survive and prosper in the age of scarcity*. Dublin: Penguin Ireland.

Hoefer, M., Rytina, N., & Baker, B. (2010, January). Estimates of the unauthorized immigrant population residing in the United States: January 2009. *Population Estimates*. Washington, DC: Office of Immigration Statistics, Department of Homeland Security. Retrieved from http://www.dhs.gov/xlibrary/assets/statistics/publications/ois_ill_pe_2009.pdf

Hosking, P., & Jagger, S. (2009, December 9). 'Wake up, gentlemen', world's top bankers warned by former Fed chairman Volcker. *The Times*.

Howard, P. (2006, April 17). Ps and Qs and ATMs. *The Times*.

Hubbert, M. K. (1956, March). *Nuclear energy and the fossil fuels*. Presented at the spring meeting of the Southern District Division of Production, American Petroleum Institute, San Antonio, TX. Archived, Box 85, Folder 4, M. King Hubbert Collection, University of Wyoming American Heritage Center.

Huff, D. (1993). *How to lie with statistics*. New York: W. W. Norton. (Original work published 1954)

Hughes, M. (2009, December 31). "Debt-free" Chelsea aiming to pay way despite £44m losses. *The Times*.

Huxley, T. H. (1995). On a piece of chalk. In J. Carey (Ed.), *The Faber book of science* (pp. 139–147). London: Faber & Faber. (Original work published 1868)

Input-output accounts data. (2010, May 25). U.S. Department of Commerce, Bureau of Economic Analysis. Retrieved from http://www.bea.gov/industry/io_annual.htm

Jackson, E. (2011, June 24). Folio: Concentration of wealth. *The Globe and Mail*.

Jackson, T. (2009). *Prosperity without growth: Economics for a finite planet*. London: Earthscan.

Jacobs, J. (1985). *Cities and the wealth of nations: Principles of economic life*. New York: Vintage Books.

Jarvis, C. (2000). The rise and fall of Albania's pyramid scams. *Finance and Development, 37*(1), 46–49. Retrieved from http://www.imf.org/external/pubs/ft/fandd/2000/03/jarvis.htm

Johnston, D. C. (2010, September 24). So how did the Bush tax cuts work out for the economy? *Tax.com: The Tax Daily for the Citizen Taxpayer*. Retrieved from http://www.tax.com/taxcom/taxblog.nsf/Permalink/CHAS-89LPZ9?OpenDocument

Kaletsky, A. (2009, February 5). Economists are the forgotten guilty men. *The Times*.

Kaletsky, A. (2010a, January 11). Disappointment ahead for UK—and big test for euro. *The Times*.

Kaletsky, A. (2010b, September 15). We still haven't learnt the lessons of Lehman. *The Times*.

Kaletsky, A. (2010c, November 10). The pay gap is putting democracy in danger. *The Times*.

Kaletsky, A. (2011a, February 23). Ireland is small enough to hold us to ransom. *The Times*.

Kaletsky, A. (2011b, April 20). US recovery is more creditworthy than ours. *The Times*.

Kampen, P., van, Wemyss, T., & Smith, D. (2009, January 29). *Physics education workshop*. Dublin: University College Dublin.

Katz, S. (2006, July 30). U.S. runaway major feature film production continues to grow as more countries introduce federal tax incentives, continuing study shows [News release]. Encino, CA: Center for Entertainment Industry Data and Research. Retrieved from http://www.ceidr.org/CEIDR_News_3.pdf

Kay, O. (2010, April 2). Revealed: The £150m cost of debts owed by Premier League clubs. *The Times*.

Keeley, G. (2006, May 10). Gullible globe. *The Times*.

Kocieniewski, D. (2011, March 24). G.E.'s strategies let it avoid taxes altogether. *The New York Times*.

Kosmin, B. A., Mayer, E., & Keysar, A. (2001). *American religious identification survey*. New York: Graduate Center of the City University of New York. Retrieved from http://www.gc.cuny.edu/CUNY_GC/media/CUNY-Graduate-Center/PDF/ARIS/ARIS-PDF-version.pdf?ext=.pdf

Krugman, P. (1992, March 23). Disparity and despair. *U.S. News and World Report*.

Krugman, P. (2009). *The conscience of a liberal: Reclaiming America from the right*. New York: Penguin Books.

Krugman, P. (2010a, April 22). Don't cry for Wall Street. *The New York Times*.

Krugman, P. (2010b, August 15). Attacking Social Security. *The New York Times*.

Kuhn, T. S. (1997). *The Copernican revolution: Planetary astronomy in the development of Western thought*. Cambridge: Harvard University Press. (Original work published 1957)

Let the deed show. (2011, March). *China Economic Review, 22*, 4–5.

Levitt, S. D., & Dubner, S. J. (2006). Freakonomics: *A rogue economist explores the hidden side of everything*. London: Penguin Books.

Levitt, S. D., & Dubner, S. J. (2010). *Superfreakonomics: Global cooling, patriotic prostitutes and why suicide bombers should buy life insurance*. London: Allen Lane.

Lewis, H. (2009, April 9). "American Idol" makes $15 million an hour. *Business Insider*. Retrieved from http://articles.businessinsider.com/2009-04-09/entertainment/30061440_1_ad-revenue-ad-buyers-viewers

Li, D. K. (2009, October 28). Schwarzenegger drops F-bomb in message to lawmaker. *New York Post*.

Liptak, A. (2009, June 29). Supreme Court finds bias against white firefighters. *The New York Times*.

Lister, S. (2006, June 27). Here's $31bn. Spend it how you will. *The Times*.

List of lawyers in the 111th Congress. (2009, May 28). *Daily Paul*. Retrieved from www.dailypaul.com/94514/list-of-lawyers-in-the-111th-congress

Llewellyn, R. (Presenter). (2008, February 25). Thames Barrier [Television series episode]. In R. Stansfield (Producer), *How do they do that?* (Series 1, Ep. 8). London: Discovery Channel.

Loomis, C. J. (2006, June 25). A conversation with Warren Buffett. *Fortune*. Retrieved from http://money.cnn.com/2006/06/25/magazines/fortune/charity2.fortune/index.htm

Lyons, T. (2006, July 6). Psychics Live seeks monthly fee. *Irish Independent*.

Maass, P. (2009). *Crude world: The violent twilight of oil*. London: Penguin Books.

Maddox, B. (2010, April 3). Merkel has saved Greece—but not the euro. *The Times*.

Malthus, T. (1798). *An essay on the principle of population*. London: J. Johnson.

Markopolos, H. (2010). *No one would listen: A true financial thriller*. New York: John Wiley. (Excerpted in *Bloomberg Businessweek*, March 22, 29.)

Marlowe, B., & Rushe, D. (2010, January 24). Man Utd blows €63m on bond. *The Sunday Times*.

Marsh, S. (2009a, November 24). Imagine no possessions. *The Times*.

Marsh, S. (2009b, December 9). Soak the rich: The millionaires' guide to a fairer society. *The Times*.

Martin, J. (2000). Greenspan: *The man behind the money*. Cambridge, MA: Perseus.

Reich, R. (2009). *Supercapitalism: The battle for democracy in an age of big business*. London: Icon Books.

Martin, T., & Associates. (2011, May). A profile of physics graduates. *Physics in Ireland: The brightest minds go further*. Dublin: Institute of Physics in Ireland.

Marx, K., & Engels, F. (1992). *The communist manifesto*. Oxford: Oxford University Press. (Original work published 1848)

Mason, P. (2009). *Meltdown: The end of the age of greed*. London: Verso.

Matson, J. (2010, January 8). Record 232-digit number from cryptography challenge factored. *Observations: Scientific American*. Retrieved from http://blogs.scientificamerican.com/observations/2010/01/08/record-232-digit-number-from-cryptography-challenge-factored/

May, B., Moore, P., & Lintott, C. (2007). *Bang! The complete history of the universe*. London: Carlton Books.

McAuley, J. W. (2004, Spring–Summer). Fantasy politics? Restructuring unionism after the Good Friday Agreement. *Éire-Ireland: Journal of Irish Studies, 39*, 189–214.

McDermott, L. C., Rosenquist, M. L., & van Zee, E. H. (1987). Student difficulties in connecting graphs and physics: Examples from kinematics. *American Journal of Physics, 55*, 503–513.

Miller, C. (Director). (2008, June 20). How a geek changed the world [Television series episode]. In C. Miller & D. Trelford (Producers), *The money programme*. London: BBC.

Miller, M., & Greenberg, D. (Eds.). (2009, September 30). The richest people in America. *Forbes.com*. Retrieved from http://www.forbes.com/2009/09/30/forbes-400-gates-buffett-wealth-rich-list-09_land.html

Mlodinow, L. (2008). *The drunkard's walk: How randomness rules our lives*. London: Penguin Books.

Monmonier, M. (1996). *How to lie with maps* (2nd ed.). Chicago: University of Chicago Press.

Moore, G. E. (1965). Cramming more components onto integrated circuits. *Electronics, 38*(8), 114–117.

Morello, C. (2010, September 16). About 44 million in U.S. live below poverty line in 2009, census data show. *The Washington Post.*

Mostrous, A. (2011, January 10). Plug pulled on stars who don't say when they're paid to tweet. *The Times.*

Moyo, D. (2011). *How the West was lost: Fifty years of economic folly—and the stark choices ahead.* London: Allen Lane.

Naisbitt, J. (1984). *Megatrends.* London: Macdonald.

Nave, R. (2011). Jupiter effect. Retrieved from http://hyperphysics.phy-astr.gsu.edu/hbase/tide.html

Ng, S., & Catan, T. (2010, July 1). We were 'prudent': AIG man at center of crisis. *The Washington Post.*

Office of Management and Budget. (2011). *Historical tables: Budget of the United States government, fiscal year 2010.* Washington, DC: U.S. Government Printing Office. Retrieved from http://www.gpoaccess.gov/usbudget/fy10/pdf/hist.pdf

Ohnsman, A., Green, J., & Inoue, K. (2010, March 22, 29). The humbling of Toyota. *Bloomberg Businessweek.*

Oil: Where does a barrel go? (2010, June 19). *The Times.*

Olding, P. (Director). (2010, March 28). Dead or alive [Television series episode]. In D. Peck (Producer), *Wonders of the solar system* (Ep. 4). London: BBC.

O'Mahony, J. (2000, August 31). Catching a mangy tiger by the tail. *Guardian Weekly.*

Ontario Hydro. (2009). Hourly demands [data file]. Retrieved from http://www.ieso.ca/imoweb/marketData/marketData.asp

Parry, R. L. (2009, February 6). Flamboyant businessman seized over £1.75bn 'fraud.' *The Times.*

Patterson, S. (2010). *The quants: How a small band of maths wizards took over Wall Street and nearly destroyed it.* London: Random House Business Books.

Paulos, J. A. (1996). *A mathematician reads the newspaper.* New York: Anchor Books.

Paulos, J. A. (2001). *Innumeracy: Mathematical illiteracy and its consequences.* New York: Hill & Wang.

Pear, R., & Herszenhorn, D. M. (2010, March 21). Obama hails vote on health care as answering 'the call of history.' *The New York Times.*

Per capita resource consumption. (2011). American Museum of Natural History: The Center for Biodiversity and Conservation. Retrieved from http://cbc.amnh.org/crisis/resconpercap.html

Perryman, M. (2006, November 6). *Our galaxy in three dimensions.* Dublin: Royal Dublin Society.

Philp, C. (2011, April 14). We won't dismantle welfare system to cut deficit, Obama tells Republicans. *The Times.*

Preis, T., & Stanley, H. E. (2011, May). Bubble trouble. *Physics World, 24,* 29–32.

President Obama's remarks on executive pay. (2009, February 4). *The New York Times.*

Preston, R. (2008, December 9). A crash as historic as the end of communism. *The Times.*

Pryce-Jones, J. (2011, July 20). The power of nice. *The Times.*

Purves, L. (2010, March 22). Take a 13% pay cut. You know it makes sense. *The Times.*

Reid, T. (2009, November 21). Unburied bodies tell tale of city in despair. *The Times.*

Reinhart, C. M., & Rogoff, K. S. (2010, January). *Growth in a time of debt* (Working Paper No. 15639). Cambridge, MA: National Bureau of Economic Research.

Riddell, P. (2009, April 17). Alarm in Westminster as voters turn against MPs over expense abuses. *The Times.*

Robinson, N. (2005, January 29). Four brainy white men who know that African poverty is stupid. *The Times.*

Rose, D. (2010, January 5). Smokers who quit twice as likely to develop diabetes. *The Times.*

Rosling, H. (2009, November 23). *Asia's rise: How and when.* TED2009. Retrieved from http://www.ted.com/talks/hans_rosling_asia_s_rise_how_and_when.html

Rowan, D. (2002, October 18). Hard sell, soft targets. *The Times.*

Rusche, D. (2009, February 1). 'Mini-Madoff' pyramid swindlers unmasked. *The Sunday Times.*

Rusche, D. (2010, August 8). Sub-prime crime leaves victims but no villains in its wake. *The Sunday Times.*

Sabbagh, D. (2009, November 11). They can't sing but they can make a record: The most expensive ad slots on television. *The Times.*

Schama, S. (Presenter/Writer). (2010, January 14). The end of the dream [Television series episode]. In A. Gething (Producer/Director), *Simon Schama on Obama's America* (Ep. 2). London: BBC2.

Schofield, K. (2006, May 12). Rise in higher education has widened the social divide. *The Scotsman.*

Schumacher, E. F. (1993). *Small is beautiful: A study of economics as if people mattered.* London: Vintage.

Selb, C. (2010, June 19). Don't cut back on the stimulus spending, warns Obama. *The Times.*

Shakir, F., Terkel, A., Khanna, S., Corley, M., Frick, A., & Armbruster, B. (2008, February 29). *Veterans: A G. I. bill for the 21st century.* The Progress Report. Retrieved from http://www.americanprogress.org/pr/2008/02/pr20080229

Sherwin, A. (2002, February 11). Winner fails to take all in £100m 'Pop Idol' bonanza. *The Times.*

Sigmund, K. (1995). *Games of life: Explorations in ecology, evolution and behaviour.* London: Penguin Books.

Singh, S. (2000). *The code book: The secret history of codes and code-breaking.* London: Fourth Estate.

Singh, S. (2008, November 12). *Cracking the cipher challenge.* Dublin: University College Dublin.

Smith, G. (Director). (2009a, September 10). The bank that bust the world [Television series episode]. In G. Smith & M. Tuft (Producers), *The love of money* (Ep. 1). London: BBC.

Smith, G. (Director). (2009b, September 17). The age of risk [Television series episode]. In G. Smith & M. Tuft (Producers), *The love of money* (Ep. 2). London: BBC.

Snow, J. (Presenter). (2009, March 12). *Channel 4 news* [Television broadcast]. London: Channel 4.

Snow, T. (2003, December 28). Bush leads country on spending spree: President urges restraint but approves new expenses. *The Detroit News*.

Soros, G. (2008). *The crash of 2008 and what it means: The new paradigm for financial markets*. New York: PublicAffairs.

Sparrow, A. (2003, July 23). 'They want a song. You should do it, darling.' *The Daily Telegraph*.

Spillius, A. (2010, September 14). Gulf arms race fear as U.S. agrees £40bn deal with Saudis. *The Daily Telegraph*.

Stanley, A. (2006, May 25). Surprise (well, not exactly)! 'American Idol' finale unfolds and unfolds. *The New York Times*.

State Elections Offices. (2001, December). 2000 official presidential general election results. *Federal Election Commission*. Retrieved from http://www.fec.gov/pubrec/2000presgeresults.htm

Stewart, I. (Presenter). (2010, February 23). Human planet [Television series episode]. In M. Dyas (Director), *How earth made us* (Ep. 5). London: BBC2.

Stiglitz, J. (2009, June 12). America's socialism for the rich. *The Guardian*.

Stix, G. (1997, April). Small (lending) is beautiful. *Scientific American, 276*, 16–20.

Story, L., Thomas, L., Jr., & Schwartz, N. D. (2010, February 13). Wall St. helped to mask debt fueling Europe's crisis. *The New York Times*.

Summers, L. (2010, December 12). Larry Summers remarks on the Great Recession. *C-SPAN.org*. Retrieved from http://www.cspan.org/Events/Lawrence-Summers-Remarks-on-the-Great-Recession/20380-1/

Sussman, D. (2010, February 11). New poll shows support for repeal of 'Don't Ask, Don't Tell.' *The New York Times*.

Swain, J. (2009, November 15). A disaster waiting in the wings. *The Sunday Times*.

Sylvester, R., & Griffiths, K. (2010, April 3). Lord Mandelson gets personal over banker's pay. *The Times*.

Taleb, N. N. (2004). *Fooled by randomness: The hidden role of chance in life and the markets*. London: Penguin Books.

Tennessee Valley Authority. (2011). TVA fact book. Retrieved from http://www.tva.com/abouttva/factbook.htm

Till, M. T. (2011). *Conversation with power: What great presidents and prime ministers can teach us about leadership*. New York: Palgrave Macmillan.

Turner, J. (2007, March 17). And today's game, kids, is spot the shyster. *The Times*.

Ungoed-Thomas, J. (2009, October 18). Blair cashes in on his Gulf War contacts. *The Sunday Times*.

United Nations. (2011). *The world at six billion*. Retrieved from http://www.un.org/esa/population/publications/sixbillion/sixbilpart1.pdf

University of California. (2003). *Atlas of global inequality: GNP per capita* [data file]. Santa Cruz: Center for Global, International and Regional Studies, University of California. Retrieved from http://ucatlas.ucsc.edu/gnp/gnp_capita.TXT

U.S. Census Bureau. (1890). *Eleventh census of the United States: Schedule No. 1; Population and social statistics.* Retrieved from http://www.census.gov/history/pdf/1890_questionnaire.pdf

U.S. Census Bureau. (2009, December 23). *Statistical abstract of the United States, 2010: The national data book* (129th ed.). Washington, DC: U.S. Department of Commerce, Author.

U.S. Census Bureau. (2010). International data base. Retrieved from http://www.census.gov/population/international/data/idb/informationGateway.php

U.S. Census Bureau. (2011a). Banking, finance, and insurance. In *Statistical abstract of the United States: 2011* (p. 740). Retrieved from http://www.census.gov/compendia/statab/2011/tables/11s1187.pdf

U.S. Census Bureau. (2011b). Population estimates. Retrieved from http://www.census.gov/popest/eval-estimates/eval-est2010.html

U.S. Energy Information Administration. (2011). *International energy outlook.* Retrieved from http://www.eia.doe.gov/oiaf/ieo/world.html

Veseth, M. (2005). *Globaloney: Unraveling the myths of globalization.* Lanham, MD: Rowman & Littlefield.

Wade, M. (2010, December 16). Christmas truce was 'tray bon'—and not just in 1914. *The Times.*

Watts, R. (2009, April 19). Google uses Irish base to avoid tax. *The Sunday Times.*

Webster, B. (2009a, March 14). Safety measures on Boeing jets 'won't stop another crash.' *The Times.*

Webster, B. (2009b, November 4). It's a dirty business—the new gold rush that is blackening Canada's name. *The Times.*

Wheen, F. (2004). *How mumbo-jumbo conquered the world: A short history of modern delusions.* London: Harper Perennial.

White, D. (2010, March 25). Empty estates could now be a lifeline. *The Herald.*

Wighton, D. (2011, April 19). Queen's historic visit will help to heal both our nations, says new Taoiseach. *The Times.*

Wilkinson, R., & Pickett, K. (2009). *The spirit level: Why more equal societies almost always do better.* London: Allen Lane.

Williams, D. (2008a). The Apollo 15 hammer-feather drop. *NASA.* Retrieved from http://nssdc.gsfc.nasa.gov/planetary/lunar/apollo_15_feather_drop.html

Williams, H. (2008b). *Days that changed the world: The defining moments of world history.* London: Quercus.

World Bank. (2011). GNI per capita, Atlas method (current US$). Retrieved from http://data.worldbank.org/indicator/NY.GNP.PCAP.CD/countries?display=default

Zakaria, F. (2009a, June 22). The capitalist manifesto: Greed is good (to a point). *Newsweek.*

Zakaria, F. (2009b). *The post-American world.* New York: W. W. Norton.

Zielenziger, D. (2003, December 24). US companies moving more jobs overseas. *Reuters.* Retrieved from http://www.commondreams.org/headlines03/1224-07.htm

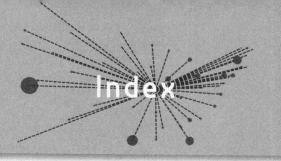

Index

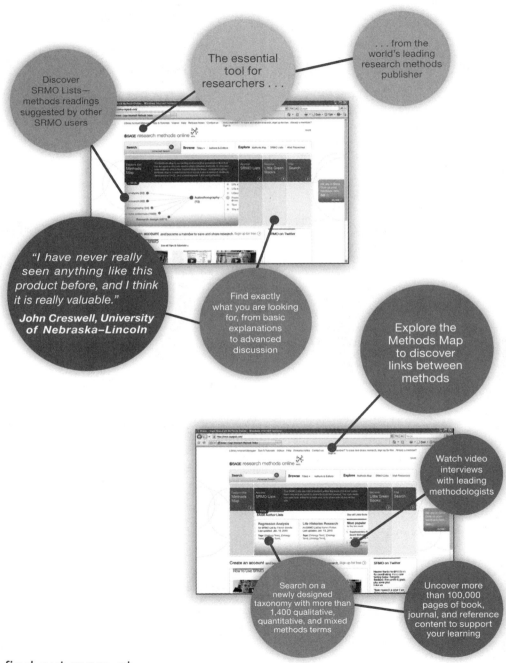

ⓈSAGE researchmethods

The Essential Online Tool for Researchers

The essential tool for researchers . . .

. . . from the world's leading research methods publisher

Discover SRMO Lists— methods readings suggested by other SRMO users

"*I have never really seen anything like this product before, and I think it is really valuable.*"

John Creswell, University of Nebraska–Lincoln

Find exactly what you are looking for, from basic explanations to advanced discussion

Explore the Methods Map to discover links between methods

Watch video interviews with leading methodologists

Search on a newly designed taxonomy with more than 1,400 qualitative, quantitative, and mixed methods terms

Uncover more than 100,000 pages of book, journal, and reference content to support your learning

find out more at
srmo.sagepub.com